D0402816

Among
the
Great Apes

Among
the
Great Apes

ADVENTURES *on the* TRAIL
of OUR CLOSEST RELATIVES

Paul Raffaele

 Smithsonian Books

An Imprint of HarperCollins*Publishers*
HARPER www.harpercollins.com

HarperCollins books may be purchased for educational, business, or sales promotional use. For information, please write: Special Markets Department, HarperCollins Publishers, 10 East 53rd Street, New York, NY 10022.

FIRST EDITION

Library of Congress Cataloging-in-Publication Data
  Raffaele, Paul.
    Among the great apes : adventures on the trail of our closest relatives / Paul Raffaele.
  —1st ed.
      p.   cm.
    Includes bibliographical references and index.
    ISBN 978-0-06-167183-8 (alk. paper)
    1. Apes. I. Title.
QL737.P96R35 2011
599.88—dc22
2009025858

10 11 12 13 14 OV/RRD 10 9 8 7 6 5 4 3 2 1

For Cecilia, Catherine, Andrew, and Elisabeth

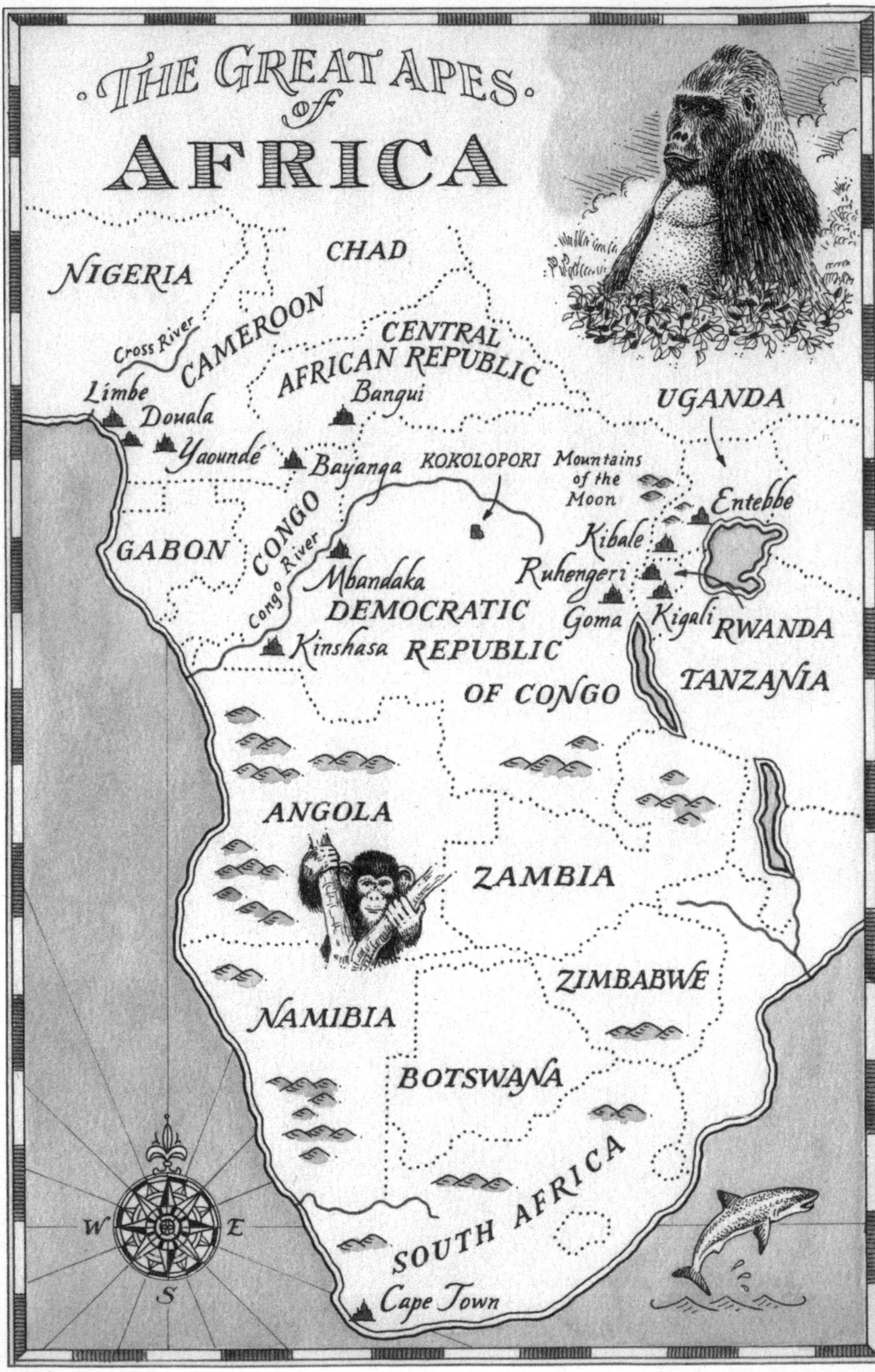

THE GREAT APES
of
AFRICA

NIGERIA   CHAD

Cross River

CAMEROON

CENTRAL
AFRICAN REPUBLIC

Limbe   Bangui

Douala   UGANDA

Yaounde   Bayanga   KOKOLOPORI   Mountains
of the
Moon   Entebbe

CONGO   Kibale

GABON   Congo River   Ruhengeri

Mbandaka   Goma   Kigali   RWANDA

DEMOCRATIC

Kinshasa   REPUBLIC   TANZANIA

OF CONGO

ANGOLA

ZAMBIA

ZIMBABWE

NAMIBIA

BOTSWANA

SOUTH AFRICA

Cape Town

W   E   S

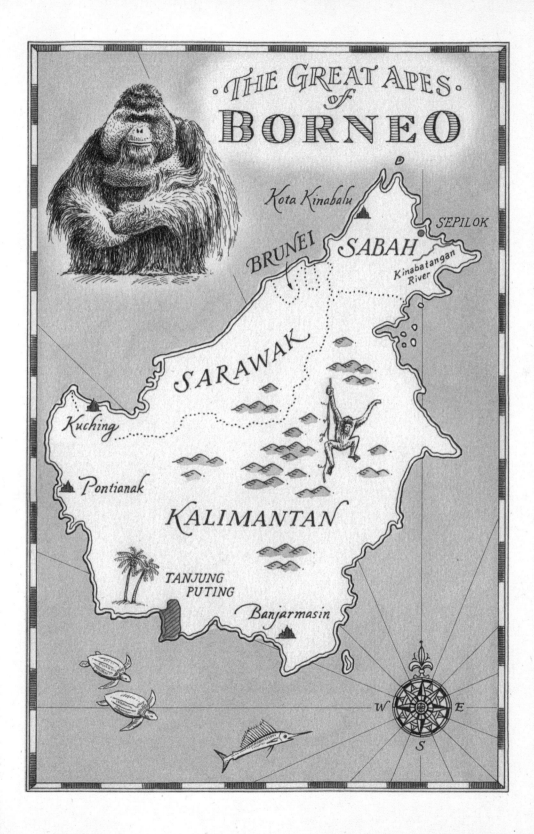

Contents

Prologue

Great apes are our wildlife cousins, the animals most like us. This common bloodline—we share up to 98 percent of our DNA—is why they are among the most popular exhibits at zoos around the world. It also explains the enduring popularity of Jane Goodall and Dian Fossey, the two most admired, even romanticized, figures in the world of primate natural science in the past 50 years.

Largely because of their studies, there has been a remarkable change in attitude toward the great apes, especially in the Western world. Until the 1960s, the best-known great apes tended to be depicted in highly contrasting threatening or friendly images. The gorilla was shown as a fierce man-eating monster, the cinematic King Kong, while the chimpanzee was cheeky and always amusing, as with Cheeta in the Tarzan movies. But once primatologists went into the central African jungles and devoted years to studying the great apes' behavior in their native habitats, a very different reality emerged.

Over the following decades, through the observations in Africa and Southeast Asia of scores of primatologists spawned by Fossey and Goodall, we have discovered that great ape species each have their separate character. The orangutans are introspective loners; gorillas laid back and largely undemonstrative; the bonobos gleeful hedonists; and chimpanzees the thugs, by far the most destructive and murderous, with males forever fighting tribal battles over territory and personal ranking within the group.

But just as we are beginning to discover the fascinating complexities of our wildlife kins' lifestyles, all the great ape species are now threatened with extinction in the wild. Because of widespread poaching, habitat loss, logging, and civil wars in their native lands, all are already on the International Union for Conservation of Nature's endangered list. The time when great apes can live unharmed in their native habitats is largely coming to an end—and fast. Primatologists estimate that within five decades the only great apes left will be in zoos, sanctuaries, and pockets of wildlife habitat in their native lands, most of them protected by armed rangers.

So, while there was still time, I set out to meet our charismatic cousins in their native habitats in Africa and Asia, in isolated jungles and misty mountain forests in some of the world's most remote spots. I also visited their captive kin in zoos and sanctuaries across the globe. The result, set down in the following pages, is the most comprehensive eyewitness portrait ever of the great apes in their native habitats.

During my journeys I witnessed not only their dramatic lifestyles but also the very different cultures of the countries where they live—including Rwanda, the Central African Republic, Borneo, the Democratic Republic of Congo, Cameroon, and Uganda. The tragedy for most of the world's great apes is that they live in lands of never-ending turmoil, countries in the grip of corrupt and violent governments.

The struggle for survival of the gorillas, chimpanzees, bonobos, and orangutans in the wild has to be seen within the context of the political, social, and environmental circumstances that surround them, and so those issues are essential to the narrative. Close contact with these great apes also allows us the chance to understand how different they are from other primates, and to appreciate their personalities and intelligence.

Ahead, you will meet unforgettable individuals such as Rugendo, the tragic patriarch of a doomed mountain gorilla family in the Congo, and his feisty young son Noel; Imoso, the dominant male in a clan of violent but fascinating chimpanzees in Uganda; Mlima, a western lowland gorilla, the subject of a four-year ongoing study in a central African rainforest, who was killed by a rival silverback; Jenny, a model orangutan mother, who roams a Borneo jungle as she diligently raises her infants; and Kanzi, the bonobo, or pygmy chimpanzee, whose extraordinary language abilities are being studied at a multimillion-dollar lab in snowy Iowa. Kanzi is possibly the most intelligent nonhuman that we know of on the planet.

You will also meet devoted human primates, such as Kanzi's "mentor," Sue Savage-Rumbaugh, who has spent three decades researching the remarkable extent of the super-smart bonobo's language abilities; American David Greer, a selfless gorilla researcher turned poacher-hunter, who daily risks his life to protect the great apes; Wasse, the greatest of the Bayaka pygmy hunters, who now shuns his tribe's traditional practice of killing gorillas; Jean-Marie Serundori, the head ranger who protects the Congo mountain gorillas, defying rebel groups who have killed more than 150 rangers; and the Congolese primatologist Mola Ihomi, who spends up to 10 months away from his family each year studying the behavior of the amazing bonobos in a remote and dangerous jungle.

Valiant efforts are underway to save the great apes before they fall victim to the destruction of their habitats. This is one of the world's most pressing conservation problems, and yet it is not given the publicity it deserves. The fate of the whales generates 100 times more column inches in newspapers and magazines and far more TV coverage. Yet most of the whale species are in nowhere near as much danger as are most of the great ape species.

There are up to 2 million minke whales and about 75,000 humpbacks in the waters of the Northern and Southern hemispheres. Contrast that with just 300 Cross River gorillas, about 700 mountain gorillas, and 25,000 bonobos. The most numerous great apes are the chimpanzees, but even they face extinction in the wild because of widespread poaching and habitat destruction.

The question must be asked: Why study and protect the lifestyles of the great apes at great cost and risk of life to the researchers and rangers? The famed paleo-anthropologist Louis Leakey summed it up, stating that "details about the behavior of one of the most manlike creatures living today, in its natural state, may give us useful pointers as to the habits of prehistoric man himself."

Jane Goodall, in contemplating the chimpanzee, said that one of the most important characteristics of the human species has been our desire to acquire knowledge through the ages, and that this goal has led to remarkable advances in the natural sciences and in technology. She wrote, "We do not need to justify, in terms of relevance to ourselves the study of a creature who is surely next to *Homo sapiens*, the most fascinating and complex in the world today."

Despite the hundreds of studies over the past half-century of the great apes in their habitats, we still have much to learn about these self-aware and intelligent relatives of ours. They have been on Earth for millions of years and have developed complex brains and complex societies. It is inconceivable that one day in the not too distant future they will have become largely extinct in the wild. Inconceivable, but probable.

The Mountain Gorillas

Gorillas and Guerillas in the Mist

1

For much of the past century, the artful chimpanzee and the gorilla were the creatures that first came to mind when most people thought about the great apes. Tarzan had the chimpanzee Cheeta as his bosom buddy. Circuses from Baltimore to Berlin had troops of chimps gallivanting around the ring, tumbling and riding ponies. Some circuses and zoos even trained chimps to amuse patrons with imitations of humans, clad in snappy clothes, puffing on cigarettes, and downing tumblers of beer. They were presented as a comical version of ourselves.

The hulking gorilla was seen as a stark contrast. Its gigantic size, huge canines, ear-splitting roar, and terrifying chest thumping prompted the myth of the man-eating monster among African tribal people who shared its habitat, and also among European explorers. This famously culminated in the 1933 movie *King Kong*, where an 18-foot-high gorilla ogre with a taste for blondes lived on the mysterious Skull Island deep in the Indian Ocean. King Kong was venerated as a god by a tribe of spear-waving natives who regularly fed him live humans.

Taken to New York in chains, the lovesick King Kong broke free and was shot down by fighter planes after he scaled the Empire State Building while tenderly carrying his blond damsel. This dramatic retelling of the age-old story of the beauty and the beast has been so popular, with its tweaking of humankind's ancient fear of the primeval monster, that it has been remade twice, in 1976 and 2005.

This myth was demolished in the middle of the twentieth century, when primatologists began living in the gorillas' habitats and discovered their generally peaceful nature and largely vegetarian diet. Around this time, a new star arose on the great ape landscape—the mountain gorilla. Until then, most people thought of gorillas as living exclusively in lowland jungles, but the mountain gorillas inhabit the alpine forests of high, misty mountains in East Africa, and they are bigger and hairier than their lowland relatives. They mostly live in family groups where a patriarch silverback, weighing more than 500 pounds and standing up to six feet tall, cohabits with several adult females and their offspring.

The mountain gorilla was largely brought to the attention of the outside world by Dian Fossey in her 1983 book *Gorillas in the Mist*. Since then it has become one of the world's most popular animals, because in many ways it behaves like we humans do, and because it is highly endangered. There are just over 700 mountain gorillas left in the wild, and none in zoos. They live high up in grassy woodland slopes along a 40-mile stretch of volcanoes that provide the borders between eastern Congo, Rwanda, and Uganda.

It is a place of incessant and bloody political instability, with government forces constantly battling more than a dozen murderous rebel armies. Mostly they keep their fighting among themselves, not interested in the local animals unless they can provide bushmeat that sells in the markets, and in this region gorilla meat is not regarded as tasty or desirable.

But occasionally, for whatever reason, the violence is directed at the rare mountain gorillas. Rebel soldiers and others slaughtered 10 percent of the mountain gorillas on the Congo side of their habitat in 2007. As if that were not bad enough, the mountain gorillas' habitat is under constant threat from poachers and charcoal traders who are chopping down the gorillas' forests to provide the massive quantities of charcoal lumps that villagers and city folk demand daily for their ovens.

I have decided to begin my journey to be among the great apes in the wild by traveling to the mountain gorillas' habitat. For me, the mountain gorilla is the symbol of all that is awe-inspiring and intriguing about the great apes and all that is threatening their existence in the wild.

My wife, daughter, and dog stand on our porch in Sydney soon after dawn, their smiles tight and eyes concerned as we await the taxi taking me to the airport. "I'll be careful, I won't do anything silly," I promise, a ritual that began four decades ago when my wife and I were courting. Understandably, they are never entirely happy about me going to some of the most isolated and perilous places on Earth.

This time my family has more to fear than usual. I am on the first step of an expedition to the Virunga volcano chain that straddles the eastern border of the Democratic Republic of Congo, one of the world's poorest and most turbulent countries. It is a sprawling hell on earth dominated by a corrupt and brutal president, Joseph Kabila, who won power in the 2006 nationwide elections. Its border with Rwanda, where I am headed, is infested with brigands and rebels who have come close to defeating the government troops fighting them. I am going to this tumultuous place because it is the home of most of the last remaining mountain gorillas on Earth.

After thriving for hundreds of thousands of years in their cloudy highlands, mountain gorillas are now facing extinction. A surging tide of humans has pushed these enormous apes into just two areas of mountain forest in eastern Africa. Because of political stability, the approximately 600 mountain gorillas in Rwanda and Uganda are relatively safe for now, but assailants in 2007 murdered 10 mountain gorillas living on the Congo side and have threatened to wipe out the remainder.

The former director of Washington's National Zoo, Lucy Spelman, now based in Rwanda at the Virungas and dedicated to protecting the mountain gorillas' health, told me by phone, "Their gene pool is so limited that even the loss of a few puts the species at peril. We're fighting desperately to save them."

I fly to Hong Kong to pick up the Kenya Airways overnight service to Nairobi, but on the 14-hour flight to the Kenyan capital I have trouble sleeping and spend much of the time thinking about the mountain gorillas. We humans evolved from a species common to all the great apes that existed a very long time ago. The orangutans went their own way about 12 million years back, while the gorillas split off to form a separate lineage about 7 million years ago. Around four million years later, the chimpanzees and bonobos departed to establish their own lineages.

Our ancestors evolved through several genetic modifications such as *Homo habilis* and *Homo erectus* until humans who looked much like us today, *Homo sapiens*, started to appear on Earth up to 200,000 years ago.

The mountain gorillas were among the last of Africa's great apes to be discovered by outsiders. In 1861, British explorer John Speke, who found the outlet of the White Nile in nearby Uganda, was warned by locals to stay away from the Virunga mountain slopes because they were inhabited by manlike monsters, in reality the mountain gorillas.

Two decades later, the mountain gorillas' habitat in the Virungas became the object of a geopolitical tug-of-war with Belgium grabbing the Congo side, Germany seizing the Rwandan chunk, and Britain taking the tail end in what is now southwestern Uganda. In 1902, Captain Oscar von Beringe, a German officer on an expedition, became the first European to see mountain gorillas when he spied a family on the Rwandan side of the Virunga volcano slopes above Ruhengeri, northwest of the capital, Kigali. At 9,300 feet above sea level, he spotted "black large apes" making their way down a ravine. He shot two of them, and to honor his discovery, and not his deadly aim, the subspecies was named *Gorilla beringei beringei.*

Over the following two decades, Western expeditions killed or captured 43 mountain gorillas. A pair shot by American naturalist Carl Akeley in 1921 can still be seen in a popular diorama at New York's American Museum of Natural History. Akeley ached with remorse when he saw one of the silverbacks he had just killed up close. He wrote: "As he lay at the base of a tree, it took all one's scientific ardor to keep from feeling like a murderer. He was a magnificent creature with the face of an amiable giant who would do no harm except perhaps in self-defense or in defense of his friends."

Because he feared that hunters would wipe out the rare mountain gorillas, Akeley persuaded the king of Belgium, then the colonial power, to create Africa's first national park, Albert National Park, in 1925 on the Congo side of the Virungas. Akeley returned the same year on a visit with his wife, but died from malaria and was buried in the meadow where he first spotted mountain gorillas. In a bizarre robbery in 1976, Congo poachers dug up Akeley's skeleton and took it away.

I am following the path of Akeley, and also that of Dian Fossey, who began her study of the mountain gorillas in the Congo Virungas four decades earlier. She had been there only a few months when she had to flee across the border from a bloody civil war waged by a warlord not far from where I am headed. The same region is now being devastated by yet another ruthless warlord who commands thousands of rebels.

In September 1967, Fossey pitched a tent on the Virungas' slopes on the Rwandan side. Mostly spurning the outside world, she spent the next 18 years in the jungle with her beloved mountain gorillas until she was murdered in 1985 when someone split her skull open with a machete, a crime that remains unsolved to this day. Fossey's best-selling book,

Gorillas in the Mist, a stirring account of her often obsessive and always possessive life with the mountain gorillas, and the follow-up movie, demolished the persistent belief that gorillas were man-killing beasts. This has sparked a multimillion-dollar boom in mountain gorilla tourism, confined at present to the Rwandan and Ugandan gorillas, because the Congo side of the Virungas is too dangerous to visit.

On the long plane ride I reread Fossey's book. She had her first ever encounter with mountain gorillas in an alpine meadow at Kabara on the Congo side of the Virungas, close to where I am going. She smelled the gorillas before she spied them, "an overwhelming musky-barnyard, humanlike scent." The gorillas, hidden by dense vegetation, warned off the intruder with high-pitched screams and thumping chest-beats. Creeping closer, she saw several black furry primates peeking through the vegetation. She wrote: "Their bright eyes darted nervously from under heavy brows as though trying to identify us as familiar friends or possible foes."

The dominant silverback rose to his full height to intimidate Fossey with his size and chest-beating, but other family members—his adult females, young males and females, and infants—were drawn from cover by curiosity to take a closer look at this strange creature. Fossey noted: "As if competing for our attention, some animals went through a series of actions that included yawning, symbolic-feeding, branch-breaking or chest-beating."

After each display, the gorillas peered at Fossey to see what effect it had on her. She wrote, "It was their individuality combined with the shyness of their behavior that remained the most captivating impression of this first encounter with the greatest of the great apes. I left Kabara with reluctance but with never a doubt that I would somehow return to learn more about the gorillas of the misted mountains."

2

We touch down at Nairobi airport, bleary-eyed, at 4:00 a.m. local time. The sky is still dark, and the taxi hustles along the road to the city at 90 miles an hour, well above the speed limit. "It's dangerous at this time," the driver explains. I know that too well. On the very day I flew in on an earlier visit, gunmen ambushed a taxi carrying newly arrived foreign tourists along this same road, shooting to death the taxi driver and all the passengers. The bandits coolly stole all their luggage, money, and passports, and got away.

"That's happened again," the taxi driver says with a shrug. "What else can you expect? Nairobi is one of the world's most dangerous cities. We call it Nairobbery. I keep as little petrol in the tank as possible so that if robbers hijack me and steal the taxi they won't get far."

"Have you ever been hijacked?"

"Twice. I've got a radio connection to the police in the taxi. The first time the robbers abandoned my taxi because it ran out of petrol, but the second time the police caught up with them and shot them dead. The police are merciless with robbers, and I support them. The more robbers they shoot, the happier I am."

"But Chairman Mao of China once said, in regard to thieves, that you only have to kill one monkey to scare off a thousand."

The taxi driver snorts scornfully. "He was wrong. You kill one of the thieves here, and the others smile at their good fortune. You've got to jail or kill all of them, and then we'll be able to lead peaceful lives. But the biggest robbers are the politicians, they control the police, and so what hope do we little people have?"

I always plan my stay in Nairobi to be as short as possible. Soon after noon I take a taxi to the city's northern suburbs, passing splendid mansions set amidst leafy streets and high schools boasting swimming pools, tennis courts, and cricket grounds. Once these plush residences were the homes of the colonial British and the academies educated their children, but the children I see spilling out of the classrooms are mostly Kenyan, the progeny of the new rich, the politicians and the businesspeople favored by the politicians. You see this all over the continent, and Africans with wry humor claim they belong to a new tribe they have named the Wa-Benzi. *Wa* is a prefix in Swahili, which can have the meaning "the tribe of," and *Benzi* means the Mercedes-Benz cars they favor.

My lunch companion is Kenyan-born 65-year-old Richard Leakey, chairman of WildlifeDirect, a nonprofit organization he set up in November 2006 to organize and fund protection for the Congo mountain gorillas. His paleo-anthropologist father, Louis, famously chose Dian Fossey, Jane Goodall, and Biruté Galdikas in the 1960s and '70s as his "three angels" to study our closest animal relatives, mountain gorillas, chimpanzees, and orangutans, to seek clues about human evolution. Louis Leakey believed that because we humans are so close in nature to the great apes, it was best to send females into their habitats to study them up close because adult male great apes might react aggressively toward male researchers.

It is not the first time the feisty younger Leakey has been a savior knight charging in to rescue a species in desperate trouble. He is credited with saving the elephants in Kenya, gaining world attention when in 1989, as head of the Kenya Wildlife Service, he publicly burned 12 tons of elephant tusks that had been poached for their ivory and were destined to be sold, mostly to China and Japan. He ordered his rangers to shoot back at poachers who were slaughtering elephants.

Leakey is still a bull of a man even though he lost both legs below the knee in a light-plane crash in 1993, and later needed transplants for both his kidneys. On his artificial legs, he stomps into the restaurant below his skyscraper office, a daunting sight with his massive head and bulky body. The accident seems to have carved a permanent frown onto his features.

With him is WildlifeDirect's executive director, Emmanuel de Merode, also Kenyan-born. In his 40s, de Merode is as gentle and mannerly as Leakey is blunt and devoid of pleasantries. His shy nature overshadowed by the gruff but ebullient Leakey, Emmanuel stays silent for most of the lunch.

Leakey has channeled his formidable charisma and energy into building the Web-based WildlifeDirect, which supports the Congo rangers who daily risk their lives safeguarding the Congo mountain gorillas.

Leakey lets his steak grow cold to sermonize, and he has every right because, although little known, his is one of the most inspiring and brave conservation efforts of our time. "Since the beginning of armed conflict in eastern Congo over 150 rangers have been killed on active service," Leakey tells me. "Many were not paid for years, and it's a credit to them that they remained there. The world has to help them so they can safeguard the mountain gorillas. We provide them with a steady income, proper equipment, and weapons to defend themselves in that dangerous place."

The following afternoon I fly southwest to the Rwandan capital, Kigali, and book into the Hotel Des Mille Collines, also known as the notorious Hotel Rwanda. It sits atop a hill, almost hidden by trees. Though the Mille Collines brands itself as a luxury hotel and charges high rates for the privilege of staying there, the cramped rooms seem eerily like prison cells, the narrow corridors dimly lit. I sleep uneasily, knowing that a terrified Tutsi family probably occupied the room in 1994 while Hutu *Interahamwe* militia waited outside the hotel to murder them during the three months of genocide. After breakfast the next morning I decide not to take my usual on-the-road dip in the hotel swimming pool, knowing that the trapped Tutsi families drank it dry, as their only source of water.

A decade after I was last here, Kigali is almost unrecognizable, its leafy hills studded with new skyscrapers, shopping malls, and modern offices. The president, Paul Kagame, is a largely benevolent dictator, unwilling to release the reins of power to the people because he believes the majority Hutu would then charge back into a revengeful rule over his people, the Tutsi.

On the way west to the gorillas, the car passes Kigali's prison, and I shudder in horror on spying through an open gate about 30 middle-aged and elderly men, clad in prison-issue pink short-pants pajamas. They are herded by a warden armed with a rifle. "They're *Interahamwe*, and they murdered Tutsis during the genocide," my driver says. Rwanda still holds more than 90,000 members of the killer militia in jails across the country, and the nation cannot shake off the trauma. He gives me a fierce look. "We Tutsi will never forgive and never forget the genocide."

Landlocked Rwanda, with 10,000 square miles, the size of Maryland, has 10 million people and is one of the world's most densely populated countries. Its people call it "land of a thousand hills" with good reason. As the car heads toward the Congo border, the road weaves in between lush, fertile mountains that look as if a giant had shoved their steep, verdant slopes close together. Ninety percent of the population lives in villages, and they have made the best use of their limited land by terracing the steep hillsides with crops. Rwanda's nickname is "the Switzerland of Africa."

Three hours after leaving Kigali we reach Ruhengeri, in the shadow of the Virungas' chain of eight major volcanoes that sweep about the town in a 40-mile arc of ragged peaks towering more than 14,000 feet above sea level. This was where Dian Fossey set up her headquarters when she fled the nearby Congo, though she preferred to spend most of her time close to the gorillas, living in a bare-bones bungalow high on the volcano slopes at a place she named Karisoke.

Soon after dawn the next day, I drive to the Parc National des Volcans headquarters, housed in a bungalow in the shadow of the Virungas on the town's fertile outskirts. Already gathered here are about 40 Americans and a scattering of other nationalities, waiting to trek up the volcanoes to visit the seven habituated mountain gorilla families on the Rwandan side. Above us loom the jagged peaks of several volcanoes, their summits marbled by drifting clouds, the mysterious fabled land of the mountain gorillas.

I have come to see an old friend, Justin Rurangirwa-Nyampeta, chief warden of the national park, a slim bespectacled man with a gentle manner and sparkling brown eyes. We had spent time together during my visit here 12 years earlier and he has hardly changed, just a sprinkling of white curly hair peeping through the black.

Justin had told me then that he and his rangers risked their lives in that troubled time because "the gorillas are so precious—they belong to the world." Appalled at his lack of equipment, on my last visit here I gave him my binoculars. Now, he apologizes for their loss, explaining, "I had them for years, they were very useful, but then the president came to visit the national park. With him was the minister for tourism. He asked to use the binoculars to see the mountain gorillas, and never gave them back."

The tourist groups are limited to eight trekkers for each of the seven habituated gorilla families, and each person pays $500 to spend one

hour with the great apes. Despite the steep price, Justin tells me worldwide demand is so great that there is a yearlong waiting list. The income
is vital to Rwanda's still feeble economy. "We earn about eight million
dollars yearly from the entrance fees, and more millions from our visitors' hotel, travel, and food costs," he says.

When I was last in Ruhengeri to report on the challenges facing the
mountain gorillas a few years after the genocide, Justin was struggling
to keep the Rwandan gorillas from harm. The Hutu rebels, the *Interahamwe,* used their territory, the forest-covered volcanoes, to move
between Rwanda and their Congo bases on raids. They also seeded the
mountain passes with land mines. He sighs at the memory and tells me,
"Despite the fighting, only one mountain gorilla was killed on our side
of the border because the army and the rebels vowed not to harm them.
It's never been our custom to eat gorillas. A silverback named Mrithi
was shot dead because a soldier stumbled into him during a night patrol
and thought he was a rebel."

On that first visit back in 1996, the Hutu rebels were still terrorizing
Ruhengeri and villages around the town. Each night of the two weeks
I was here I went to bed uneasily, knowing that my sleep could be in-
terrupted by a rebel attack on the town. A pair of German shepherds
patrolled the bungalow's grounds along with an armed guard, but we
all knew they would be useless in a rebel attack. A few months after
my departure, Hutu rebels stormed Ruhengeri looking for foreigners
and executed three Spanish doctors and badly injured an American aid
worker. Weeks later they killed a Canadian priest. But Justin says the
Hutu militia rarely cross the border now, and so the mountain gorillas on the Rwandan side are out of harm's way at present. The town is
safe to visit. What about the Congo side? Justin shrugs in a Gallic way.
"It's very dangerous; the rebels have killed gorillas and threaten to kill
more."

Poaching in the national park on the Rwandan side has also been reduced from a serious threat a decade ago to a low-level nuisance. Poachers set wire snares to capture small antelope by their legs, but these
snares can cause serious damage to snared gorillas' hands and feet.
"Our rangers patrol vigorously in the park, and that's a major reason
they rarely come across snares nowadays," Justin tells me.

Talking with Justin reminds me of when I first visited the mountain gorillas on this side of the border. Back then I headed for the volcanic slopes of Mount Sabinyo, passing villages whose people illegally went up the volcano each day to forage for firewood or find food for their cattle. This helped to despoil the pristine habitat.

Sabinyo is a dormant volcano, 11,919 feet high, and is the oldest in the Virunga chain. High above us, its five ragged peaks resembled the broken teeth of an old man. They were the remnants of the ancient crater mouth. With me was a tracker, Benjamin Mugabukomeye, a short, stocky Hutu, and Dr. Jonathan Sleeman, a veterinarian and field director of the Mountain Gorilla Veterinary Project.

As now, the MGVP was based in Ruhengeri and tasked with protecting the mountain gorillas' health. Dogging our footsteps were two Rwandan Army soldiers carrying submachine guns slung over their shoulders and wearing camouflage uniforms and gum boots to wade through the volcano's marshes and streams. Jonathan, a slim, boyish Englishman, told me the soldiers would protect us against ambush by Hutu rebels, the genocide killers, who used the volcanoes to move back and forth between Rwanda and the Congo.

Two hours into the lofty trek, our tracker raised a hand and halted. Bending, he plucked from the grass and destroyed a poacher's trap set to capture small antelope—a circle of rigid bamboo with a dozen sharp slivers about 12 inches in diameter. Other snares are sharp wire that sometimes catches gorillas by the hands or feet. They often do not know how to release the snare and keep pulling on it, cutting off their blood supply. The hand or foot falls off after about a week, leaving an infected bloody and potentially gangrenous stump.

Fossey saw several gorillas with missing hands and feet, as I did in that earlier visit. She wrote of gorillas who had lost a hand: "They learned to use their feet for the preparation and stabilization of food items; however each became noticeably weaker before disappearing from their groups. None were ever seen again." Figures collected in the Virungas over the three decades since give a rosier view. Of 67 injuries to mountain gorillas caused by poachers, 75 percent were from snares, with four of the gorillas perishing while 46 survived because of the MGVP's devoted care, though losing a hand or a foot.

We climbed higher and steeper for another two hours, plodding through marshy fields, traversing meadows carpeted with wildflowers;

moving in and out of banks of mist settled near ground level; brushing past towering bamboo groves; racing through giant nests of stinging driver ants and then picking them off our bodies for a long time after; and crawling across nets of vines suspended 10 feet or more above the ground to get across ravines. Torn by thorn-studded vines and stung by nettles, we continually gasped for air in the rarefied atmosphere of the upper slopes.

At about 11,000 feet above sea level, Benjamin led us to several oval indentations in the knee-high grass scattered around a clearing. They looked like shallow bathtubs fashioned from vegetation. "That's where the gorillas slept last night," Jonathan said. The tracker knew where to find them, alerted by walkie-talkie by other trackers who had followed the gorilla family to their night nests the evening before.

Gorillas are nomadic creatures, forever foraging, traveling up to half a mile each day. They sleep in new nests they make themselves every night. The least sophisticated of the great ape nests, they consist of a rudimentary weaving of surrounding vegetation to provide a springy platform. Dung was scattered in and around the nests, its size, khaki color, and segmented lobes reminding me of horses' dung. Jonathan scooped up several samples from the nests and put them into marked plastic bags. The gorillas suffer from a host of parasites such as jiggers, ticks, lice, tapeworms, and even a bacterium that results in yaws and is passed on by body contact. The dung was tested back at the MGVP bungalow.

Jonathan also collected samples of fur from the nests to use in DNA identification tests to build up a comprehensive gallery of the mountain gorillas here. Some of the nests contained fur from a mother and an infant. Gorilla youngsters generally sleep in their mothers' nests until they are about three or four years old, when the next baby usually arrives. Gestation takes about eight and a half months, and after giving birth, the females do not ovulate again for at least three years. The infant suckles for about six months and then begins eating solid food, imitating its mother. Its first taste is often scraps from its mother's meal caught in the dense hair on her body.

Most gorilla births occur in the mother's nest at night when the gorillas are at rest. Because the nest can be soaked in blood, expectant mothers sometimes build up to five nests adjacent to each other when they know they are about to give birth, moving from one to the next as the amount of blood builds in each. They seem to prefer the privacy of the nest so that the other members of the group cannot interfere during the birth. The mothers eat the placenta, but not the afterbirth. Dian Fossey believed the placenta has dietary and antibiotic benefits.

As we headed farther up the slope a sprinkling of rain fell, but ended in minutes as a jumble of dark clouds passed by. The air this high was very cold and damp, the primary reason for mountain gorillas catching pneumonia and respiratory tract diseases such as colds and inflammations of the laryngeal sac. These are frequent causes of death. Jonathan told me that the mountain gorillas like to sunbathe, nestled in the crook of a tree to soak up the warmth on sunny days or dry themselves after heavy rain. Lowland gorillas, however, typically seek shelter in the jungle shade to escape the burning sun.

About half an hour later, as we approached a stand of low trees hung with liana, long strands of jungle vine, and mottled with glistening moss, Jonathan whispered, "Gorillas," and pointed at a tree. It shook violently as an enormous furry arm reached for a stout branch and tore it off with the ease of a child snapping a twig. Moments later a potbellied adult female swathed in dark black hair climbed down the tree. Suddenly, we heard a warning grunt from behind a wall of vine streamers. "A silverback," Benjamin said softly, and pulled back the vines to reveal a family tableau. As he did so we fell to our knees in homage. Mountains gorillas do not like to be confronted by humans standing up, and Fossey usually went down on her stomach or on her side in their presence.

Mountain gorillas are the behemoths of the great ape world—the biggest, heaviest, and tallest. A massive silverback was seated with one hand cupped under his chin, the other resting on an elbow. He looked like a sumo wrestler imitating Rodin's famous statue *The Thinker*. Nestled around him were five adult females, one grooming his furry back, another nursing an infant. Two toddlers played by their father's side. Intrigued by the whir of my camera, one infant, about two years old, loped toward us. He was about to touch the camera when the silverback uttered a warning grunt. Like a naughty child, the young ape scurried back into the safety of his father's burly arms.

"Umugome's the dominant male now, but his father, who was dominant for many years before, still lives with the family," said Jonathan. This peaceful transfer of power is not unusual when a son takes over control of the family from his aging father. He pointed up the slope and murmured, "The grand chief." There, shadowed by foliage, a big male gripped a sturdy branch with salami-sized fingers, tore off the bark with his huge canines, and stuffed it into his mouth.

"He's about thirty-seven, and he'll die of old age in about ten years," Jonathan told me. If it avoids injury and disease, a mountain gorilla can live for between 40 and 50 years. The big ape's eyes were now cloudy and seemed world-weary, like those of a leader who had spent too many years battling to protect his followers and his own lofty position.

That the family had two silverbacks was not unusual in the Virungas. Among mountain gorillas, between 26 and 40 percent of the groups studied consisted of several males, dominated by the patriarch silverback. Fossey observed that such groups can have up to four adult males. In lowland gorilla families, only 8 percent of the groups observed had several adult males. The reason for this difference is not yet clear. However, adult mountain gorilla females who choose to abandon their group to move to another group are more likely to join one with more than one silverback.

Gorilla researchers believe a female abandons her family for a new group to maximize her chances of successful breeding. The lush vegetation in their habitats makes it possible for a female to make herself scarce when two groups stumble on one another. While the rival silverbacks are ferociously displaying their power at one another, it is sometimes easy for a female with a wandering eye to slip across the line. However, silverbacks try to keep a watch on their females during intergroup encounters. If a silverback notices one of his females moving toward the other group, he will charge her and sometimes bite her to keep her in line, and then bully all the adult females away from the rival group. Primatologists call this herding.

The mountain gorillas live in a feisty paradise and if it were not for humans, their only predators, the Virungas volcanoes would be their Eden. On the lush slopes, they pick among more than 70 types of leaves, bark, fruit, roots, fungi, flowers, and bamboo, with leaves, stems, and shoots making up about 68 percent of their normal diet. It is a pristine forest smorgasbord. They do not eat meat and rarely drink water; the moisture they require comes from juicy forest plants such as wild celery, a favorite. The adult males need to consume at least 70 pounds of food each day to maintain their majestic bulk, which also gives them their bulging bellies.

Jonathan explained that nettles and thistles were an important food source for the gorillas—as they were in certain parts of Europe where cooked nettles were for a long time a traditional food. On the trek up to the gorillas I had accidentally brushed against nettles several times, and their stinging hairs caused a long-lasting burning where they touched my skin.

Did the gorillas accept the thistles' pain as a swap for the pleasure of eating them? It seems not. The gorillas carefully fold the stinging hairs within the leaf, facing each other, and then place a bundle of the folded leaves in the mouth and crush them before swallowing.

Giant bamboo, up to eight inches in diameter, is also a favorite food, and the gorillas use their huge teeth to split it open and get at the inner pith. For about one-quarter of the year, from mid-October to mid-January, when the bamboo shoots are in season, they make up much of the mountain gorilla diet, a concentrated form of protein. The gorillas occasionally eat snails and even ants, using their fists to pound open nests and shoving handfuls of ants into their mouths.

Unhabituated gorillas usually flee at the first sign of humans, but over the years, Umugome and his family had become used to their presence. Still, irritated by our being so close, the dominant silverback abruptly rose and clambered down the hill to get away. The family followed him. A female rolled along the ground, dogging the silverback's footsteps and tearing at plants to eat as she went. Her left foot was missing. "A snare," Jonathan explained, shaking his head.

Some are killed by the snares. A four-year-old female from another group under observation by researchers was cheerful and boisterous until she was caught by a snare. The wire cut deep into her wrist. The other gorillas screamed, snapped branches, and beat their chests in panic. As the little one grew weaker by the day, the silverbacks slowed the pace of their foraging so she could keep up, but two months after she was snared, she was released from her pain and torment. She perished from gangrene infecting her wounded wrist and pneumonia.

One of the infants leaped onto the back of his mother, clinging to her dense black hair, as she followed Umugome down the hill. Until three years of age, gorilla infants feed beside their mother and are carried around like this as they learn the rules and niceties of gorilla society and how to survive and prosper. As the family disappeared into the undergrowth, the old silverback rose and followed in the rear to sound the alarm if danger threatened. By their mid–30s silverbacks can suffer from arthritis, and the old chief moved stiffly and slowly away. Respecting the gorillas' wishes, we went up the slope in the opposite direction.

At 12,000 feet my lungs felt ready to burst, but we relentlessly climbed on. Benjamin swung a *panga*, a machete, carving a narrow path through the vines and bamboo. We turned a corner to find our path blocked by a giant silverback, six feet tall and more than 500 pounds, perched on a slope above us. He was swathed in jet-black hair, except for his back, which glistened silvery white. His brawny arms were several times thicker than a weightlifter's, and his crested head was as big as a bull's. Jonathan pulled me to the ground and went down on his knees with his head resting on his chest, imitating a submissive pose used by lower-ranking gorillas. "His name is Kwakane," he whispered.

Suddenly, Kwakane roared a ferocious battle cry. We could not retreat, because to do so would probably provoke him to charge and attack us. Jonathan plucked a vine leaf and gripped it with his teeth. I joined him in this gesture that signaled we came in peace. Kwakane was not persuaded. He stood and pounded his great chest with a terrifying *thock thock thock* that turned my heart to water. Then, without warning, he charged straight at us on all fours, his massive feet and hands slamming into the ground.

Just before he reached us, he raised a huge, hairy arm to strike. He was so close it blocked out the sky. The moment froze. He had the strength to bash out our brains. Without breaking stride, Kwakane veered to the right and slapped Jonathan on the shoulder as he passed, a glancing blow, then halted and sat a few yards away, staring at us with smoldering eyes. I had no idea whether this was a silverback's roughhouse greeting or a warning to go away. Jonathan seemed to know what Kwakane meant, and he edged down the slope with me and Benjamin at his heels. Getting down the slopes was a lot easier than climbing them.

"Kwakane's favorite little brother got his hand caught in a snare and it had fallen off by the time we saw it," Jonathan told me when we stopped for a few minutes to drink water from our flasks. The tight snare had cut off the blood supply to the hand. "I darted the young gorilla not far from where we saw Kwakane just now so that I could inject him with long-lasting antibiotics. Kwakane saw this. So, I think he was giving me a warning just now, saying not to do that to his little brother again or he'd really beat me up next time."

Had I lost my nerve and fled, there was a strong chance Kwakane would have chased me down and taken a chunk out of my neck. It had happened to novice rangers. Their flight was compelled by fear, but one attack on the slopes came from stupidity.

For weeks, the cinematographer of *Gorillas in the Mist* diligently filmed Sigourney Weaver and a collection of real mountain gorillas and actors in gorilla costumes playing out the tragic life of Fossey. The one shot that eluded him was a silverback charging head-on at the camera. One of the Fossey researchers at Ruhengeri told me that near the end of filming, in desperation the cinematographer began throwing sticks at a silverback to provoke him to charge.

That was about as sensible as throwing punches at Mike Tyson in his prime. He got his wish, and the shot is in the film, a ferocious display by the silverback, but the big male's fury so spooked him that he abandoned his camera and fled. The researcher said that the silverback ran the cameraman down and sunk his big canines into the back of his neck, putting him in the hospital.

Three days after my encounter with Kwakane, I met David Greer, a young red-headed American gorilla researcher with the Fossey group, who was to take me to a gorilla family he was trying to habituate. As he drove along the dirt track leading to the foot of one of the volcanoes, Mount Visoke, Dave told me how he got here. Raised on the wrong side of the tracks in ethnically diverse Kansas City, he grew up in a tough neighborhood where disagreements among boys were settled by fistfights. To get Dave away for a time from this crime-ridden place, his father, an electricity linesman, constantly took him fishing, hunting, and camping, and he got his first gun license at age 13. "We shot mostly birds, but also rabbits and squirrels," he said.

A natural athlete, Dave played football, baseball, and basketball as a schoolboy and won a baseball scholarship to Baker University. After graduating with a psychology degree, he worked at Providence Medical Center as the psychiatric unit's director of admissions. A growing fascination with the great apes, and their closeness to humans, led him to volunteer each weekend to work with the chimpanzees and gorillas at the local zoo. In 1995, he abandoned his career and traveled to Tanzania to work with Jane Goodall at her world-famous center studying chimpanzees in the Gombe Stream National Park.

Dave's introduction to the wild chimps was brutal. He had been at Gombe only two weeks when he was confronted by the dominant male, Frodo, the forest's most notorious chimpanzee. Frodo glared at him, "a look that said it's time for your initiation into the community." Dave rolled himself into a ball, protecting his head with his hands, as the chimpanzee began pounding him with his powerful fists and stamping on him.

The Tanzanian government refused Dave a resident's visa and he moved to Karisoke, the mountain gorilla research center founded by Dian Fossey in neighboring Rwanda. He arrived not long after the genocide.

The first time he saw mountain gorillas he was instantly drawn to them. A silverback was sitting in the forest with his six adult females and their young, chomping on nettles and plants. "I felt the luckiest human being on Earth, I felt I was meant to be here, this was my calling," he told me. "Every time I saw the mountain gorillas after that my stomach would tighten with emotion. They're so big and furry and beautiful, and yet so peaceful."

The comparison between chimpanzees and the mountain gorillas was stark. "Chimps have this extremely violent side that is not unlike humans, and gorillas have this peaceful laid-back, take-care-of-the-basic-necessities-of-life mentality."

It was about this time, faced with the extreme brutality of what had happened in Rwanda and being with the mostly serene gorillas each day, that his own peaceful side emerged and he turned vegetarian. Yet he dared not ignore the *Interahamwe* gunmen, coming across their trails along the mountain slopes and even encountering bands of armed militants crossing into Rwanda from their hideouts in the Congo. "There were bodies all over the forest," he said. Once, while accompanying foraging gorillas, he came across a dead Hutu riddled with bullets. "The gorillas glanced at the body, then stepped around it."

Despite the danger, Dave stayed with the gorillas, monitoring their daily lives and herbaceous diet, and even logging the number of gunshots he heard echoing across the mountains each day. "The gorillas were so special, there were so few of them, and I felt we had to be with them every day to make sure they weren't harmed."

But he never warmed to Amahoro, an aggressive silverback Dave was now trying to habituate. He said, "Amahoro and his family are what we call a sacrificial group. The Rwandan government has pressured the Fossey organization to habituate Amahoro and his family so they can send foreign tourists up the slopes each day to visit them. They'd charge hundreds of dollars for each person to spend an hour with the gorillas and so earn plenty of foreign currency."

Amahoro refused to cooperate and regularly attacked and bit trackers accompanying Dave, putting some in the hospital. One day the silverback charged Dave in a rage, forcing him to dive over a ravine edge. "I envisioned rolling down the hill with this five-hundred-pound gorilla mauling my face, and so I thought I had a better chance with the side of the ravine," he said.

Dave broke his fall by grabbing a bunch of long, tough grass and dangled over the ravine as Amahoro leaned down and screamed into his face from just a few yards away. The massive gorilla threatened him for what seemed an eternity to Dave, who swung back to safety after Amahoro left.

He was lucky to get away unhurt. Decades before, Carl Akeley described a silverback as essentially "an amiable and decent fellow," but warned that anyone "who will allow a gorilla to get within ten feet of him is a darn fool."

That description certainly applies when a silverback feels threatened, especially one like Amahoro that is unused to or avoids the presence of humans. The chief tracker of a mountain gorilla habitat in nearby Uganda once crawled after a silverback who had entered a low, narrow tunnel formed from wet vegetation. Feeling trapped, the silverback lunged at the tracker and began to strangle him. Another tracker fired over the gorilla's head. Shocked by the loud bang, the silverback let go and fled.

An American gorilla researcher, Bill Weber, was not so lucky when he followed a gorilla trail into a tunnel made of vegetation and was confronted by an unhabituated silverback. The gorilla charged and slammed into him. Weber blacked out. The silverback dragged him into a ravine and left him there, unconscious, with broken ribs and deep bite marks puncturing his neck.

Even habituated gorillas can react with violent intent when surprised. Dian Fossey once stumbled without warning into a group who knew her well, shocking them. "Suddenly, like a pane of broken glass, the air around me was shattered by the screams of the five males of the group as they bulldozed their way up through the foliage towards me," she wrote. At close range the gorillas recognized a friend. The charging silverback pulled to a halt just a yard from Fossey, causing a comical pileup of the four charging males immediately behind him. Fossey dropped to the ground, but they continued to scream at her for half an hour.

Another time, a student of Fossey's casually approached a habituated group on a slope without observing the proper humbling protocol of bent-over body and leaf-eating. The silverback charged and grabbed the student to teach him a painful lesson. The gorilla rolled him down the hill for 15 yards and then bit him hard on the neck, just missing the jugular vein.

3

Dave and I left his vehicle at a mud-hut village nudging the foot of 12,172-foot Mount Visoke. Fossey named her headquarters Karisoke because the spot sat in the grassy saddle between the Karisimbi and Visoke volcanoes. At the village we met two veteran gorilla trackers, middle-aged Fidele Nshogoza and Jean Bosco Bizumuremyi ("JB"), who was in his late 20s. Fidele was one of Fossey's earliest trackers, and his eyes sparkled with delight as he told me of his first encounter with mountain gorillas when Fossey took him to see them. They came across more than a dozen sprawled in an alpine meadow, looking, he said, like a family on a picnic. Fidele watched as the brawny patriarch lay on his belly, resting his chin on blockbuster arms.

Gathered about the leader were his six adult females, typically half his size. An imp-faced youngster clambered on the silverback's shoulders, pulling his hair in fun. The father let the infant yank away. Fidele marveled at the sight. The silverback has the power of several men, he thought then, but was so gentle with his family.

Following the genocide in 1994, Fidele and JB had to flee with their families into exile in a vast field filled with Hutu refugees escaping the revenging Tutsi army not far from Goma, on the Congo side of the border. They had not been involved in the killing of Tutsis, and had even visited the gorillas each day during the homicidal madness gripping hundreds of thousands of their people to ensure that the great apes were safe. But to many Tutsi soldiers a Hutu was a Hutu, to be killed without mercy. From the refugee camp, Fidele and JB would gaze up at the cloud-streaked blue Virunga volcanic peaks and worry about the gorillas up there, unprotected amid the carnage.

Fidele and JB crossed back about a year later with their families, along with hundreds of thousands of other Hutu, and returned to work as trackers with the gorillas monitored by the Fossey researchers. "They defied Hutu *Interahamwe* rebels who warned both that they'd kill them if they continued working with us," Dave said. "The rebels knew the Rwandan government, their enemy, got a lot of foreign currency from foreign tourists visiting the gorillas, and wanted to disrupt this."

JB opened his mouth and pointed to a gap where he was missing his front teeth. A Hutu rebel leader, in warning him to stop working with the gorillas, had thrust a bayonet into his mouth.

Just after 8:00 a.m. we began our steep climb up to the spot where Fidele and JB had seen Amahoro and his clan make their nests the previous evening. The sky was a moody gray, and heavy dark clouds swirled around the volcano peak like restless ghosts. Rain was predicted, not unusual because the Virungas enjoy, if that is the word, an average of 72 inches of precipitation per year.

Spread around the base of the Visoke volcano were fields of daisy-like flowers used to make a natural pesticide, pyrethrin. "To provide the farmers with a cash crop, in 1969, the European Development Fund funded the pyrethrin projects and gave five thousand families plots totaling two thousand acres that were cut away from the national park and the land cleared," Dave told me. "The villagers now lead their cattle up the volcano slopes to graze, and I fear for the welfare of the gorillas because they disrupt and destroy their habitat."

As we trudged through lava fields bursting with fertility, with crops laid out in neat rows for as far as I could see, we passed many villages, their conical mud huts topped by straw roofs that had the capped shape of an acorn. Along the muddy paths, farmers were taking their cattle to graze. They were sturdy, plodding beasts with very long, heavy horns that curved like lyres, unlike any cattle I had seen anywhere else in the world.

Once we climbed the foothills, the muddy path soared almost vertically upward and I was soon panting with the effort. We splashed through icy streams and tramped across spongy marshes fringed with reeds and clusters of orchids. Every time I dared to look up—and that was not often—Visoke's peak soared high above. The lake that nestled in its crater remained invisible.

At about 9,000 feet we reached the bamboo belt that girdles the Virunga volcanoes with enormous and dense groves. At about 11,000 feet, a chilly wind sliced through my sweater and cut me to the bone. The trees here were stunted and twisted, kept in check by fierce winds. Mist drifted in and out of the trees, giving the ancient meadows an eerie primeval look, as if it were a place haunted by witches, banshees, and hobgoblins.

Soon after, we found the gorillas' nests from the night before. Dave took samples of their dung and fur and placed them in small marked plastic bottles. The dung was speckled with the tiny white eggs of flies. Amahoro's nest, much bigger than the others, was about 10 yards farther up the slope. It gave the silverback a good vantage point from which to spot intruders. From here on, the tracking was easy as the gorilla clan had foraged in earnest, enjoying breakfast, brunch, snacks, and lunch, tearing, yanking, and pulling vegetation from the ground and leaving behind half-eaten plant stems, leafy vines, and bark. In the soft earth, their prints were clearly visible as they knuckle-walked up the slopes. A trail of dung also marked their path.

"They're happy and sure of themselves," Dave remarked. "When they flee danger the dung turns diarrhetic. Mostly it's humans they fear, especially those gorillas not conditioned to our presence."

At noon, we crossed the rim of a ravine by scrambling across a wide skein of sturdy vine stretched about 10 feet above the ground. A few yards before I reached solid ground on the other side, I slipped and grabbed at the vine, dangling high above the grass. I risked breaking my legs and even my pelvis if I fell. Then, getting me down the volcano slopes would be a nightmare. Fidele edged back across the vine on his stomach to grip my wrists with both hands and haul me to safety. I nodded my thanks.

As we labored up a steep grassy slope, with Fidele and JB hacking a path through the tangled vine and tough head-high grass with their machetes, a terrifying roar suddenly came from behind a wall of vegetation about five yards away. We immediately dropped to the ground. Dave whispered, "It's Amahoro. We can't see him, but he can see us. If we make a wrong move he'll charge."

We remained perfectly still, hugging the ground. We heard grunts, grumbles, growls, sneezes, farts, and belches from just a few yards away as the family fed and rested. An hour passed, then another, but Amahoro refused to show himself. But a young male, about six years old, popped up above the foliage, glared at us, beat his sturdy little chest to cow us, and then disappeared. The resonant drumming sound is produced from inflatable air pouches under the ape's chest that are connected to his larynx. We smiled at the youngster's courage.

From far below came the sounds of villagers, children playing games and women calling across the fields to each other. Amahoro heard these strange sounds every day and knew instinctively that the creatures who made them meant him and his family no good.

After more than two hours hugging the jungle floor we quietly stood up, since Amahoro and his clan seemed to have gone. We struggled through the dense undergrowth to the other side of the ravine. Moments later, from nearby the great ape let out a mighty roar that spiked the back of my neck with fear, and yet made me feel for the poignancy of his last stand.

In the tropics night comes suddenly, and just after 4:00 p.m. Fidele began to lead us down the volcano. In a mud-hut bar in a village we sat on low benches and shared bottles of *pombe,* banana beer, as we talked about mountain gorillas. Fidele recalled a young tourist who foolishly went to pick up a gorilla infant for a cuddle. The mother screamed for help and the silverback came charging to the rescue. The tourist turned and ran, prompting the silverback and the mother to chase him. They bowled him over and ripped apart his clothes as they repeatedly bit him. He was lucky to survive, but bore deep scars on his legs and arms.

That night, I slept in a small bedroom in the house Dian Fossey used when she came down the slopes from her base at Karisoke to Ruhengeri to relax and shop. She always returned as soon as possible to the gorillas. Dave and the other Fossey researchers lived in the house. It was on the outskirts of town, a modest bungalow fronted by a lawn and a high trimmed hedge. "It's rumored Dian put away more than her share of whiskey when she stayed here," he told me as we chatted before bed.

The following day was a repeat of the first. The gorillas had moved about a mile across the slopes of Mount Visoke to another ravine, and once again Amahoro refused to show himself. We spotted a female with an infant riding on her back from across the ravine, but when we reached the place, the gorillas had gone. "It's going to take months if not years to habituate Amahoro," Dave told me as we shared more *pombe* at the village bar that evening.

"Do you really want that to happen?"

He looked wistful. "No, I'd rather he and his group remain free, having no contact with humans. But, if the Rwandan government can't earn tourists' hard currency from the gorillas, then they might open the mountains to agriculture and we'll be swamped by people. That would be the end of the gorillas."

After we each drank a few long-necked bottles of beer, I asked Fidele if he knew who had killed Fossey. He stared at the mud wall with sad eyes and then turned back to me. "It wasn't me, but I have heard talk in the village. That's all I can say. We called her the old woman who lived on the mountain and had no man. She was hard to be with, especially if you made her angry, but she showed me how to appreciate the gorillas."

4

More than a decade has passed since Jonathan, Dave, JB, and Fidele took me to the mountain gorillas at the Virunga volcanoes in Rwanda, yet those memories remain fresh in my mind on my return. After visiting Justin, the Rwandan gorilla-sector chief warden, I have dinner at the Ruhengeri hotel restaurant. The waiter seats me next to a tall, bulky woman clad Congo style in a colorful red blouse and matching long skirt imprinted with strutting red roosters, with a headdress cut from the same flamboyant fabric. Her name is Rosemary Rudasingwa. A Tutsi from Kigali, she works as a high-placed public servant in the capital. She tells me she is here for a workshop on the development of women's rights.

Rosemary was in Kigali in 1994, when Hutu hatred of the Tutsi exploded into one of the twentieth century's worst genocidal killings. A tenth of the population, mostly Tutsi but also sympathetic Hutu, were murdered by the mobs. "I got warning the Hutu were coming to kill us, and so I grabbed my children and hid in the tall grass near my home," she tells me. "My heart broke when I saw my cook of twenty years, a Hutu, leading the mob to my house to slaughter me and my kids."

She fled with her children to the Milles Collines hotel, where they were given sanctuary, and then driven to the airport through mobs of screaming Hutu to be airlifted to safety in Belgium. "My brothers and sisters were killed, and so I'm raising their surviving children, as well as my own," she says.

I was in Rwanda not long after the massacres and remember that the trauma still hung like an evil shadow over the country. "I'm sure you've found it very different today; there's a real sense of going places, helped by the outside world," she says.

The president, Paul Kagame, and several government ministers are Tutsi, even though they make up only 14 percent of Rwanda's population, and I ask whether there is power sharing with the Hutu. "No!" Rosemary growls, screwing her face in disgust, as if the question were obscene. "Maybe in the next generation, or the one after that, but for now we don't trust the Hutu and I doubt we ever will."

"But won't that cause the Hutu resentment to build to a dangerous level, and because they're the vast majority they'll inevitably rise up against you again?"

"Let them. We control the army, we've got the guns, and that's why we'll never let them have the chance to do it again."

Despite Rosemary's resolve, there is every chance the Hutu will rise up once more against their Tutsi rulers, though it may take decades. On this trip many Hutu have told me they have good reason to hold a centuries-long grudge against the Tutsi, as they were the land's original inhabitants. The Tutsi arrived just a few centuries ago, coming from the north with their cattle, and because they were better warriors enforced a feudal system upon the unwilling Hutu, headed by a Tutsi king.

The Hutu were largely farmers, and were forced to become serfs of the aristocratic Tutsi, paying rent with a share of their crops for what were once their ancestral lands. The Tutsi king and aristocrats were at the top of the totem pole, while the Hutu villagers had to grovel at the bottom. This twist of history can never exonerate the genocide, but it goes a long way in explaining the Hutu hatred caused by centuries of repression they suffered at the hands of the Tutsi.

What does this mean for the gorillas?

As Rwanda has almost half of the world's mountain gorillas, I ask Rosemary whether her people support their protection. "People are much more important than gorillas, and so most of the government ministers at first opposed protecting them just because Westerners told us to do that," she replies with a grim smile. "The Westerners didn't rush to protect we Tutsi when we were being annihilated. The gorillas take up valuable farmland. But when we saw that Westerners were silly enough to pay hundreds of dollars each for just an hour with a gorilla family, money that could support a Rwandan village family for a year, we realized we could make millions of dollars a year because of the gorillas. So, we've made their habitat a sanctuary, and even sent soldiers there to protect them."

Rwanda's yearly per capita income is U.S. $400, but much of that is trapped within the capital. Ninety percent of the population live in rural areas and practice subsistence agriculture, and so many Rwandan families do indeed earn less each year than the fee charged foreign tourists for one hour with the mountain gorillas.

Rosemary's opinion mirrored that of many Africans I have spoken to during a dozen trips to the continent over the past two decades. Most show little concern about protecting their precious threatened wildlife, even the great apes. They are understandably focused more on their own survival. However, big, charismatic wild animals such as those in the Masai Mara, the Serengeti, and the Virungas earn a country significant foreign exchange income, and there is a general consensus among educated Africans that they should therefore be protected.

The next morning I visit Lucy Spelman at the Mountain Gorilla Veterinary Project, which began in Ruhengeri in 1985 at Fossey's request. There were only 248 known mountain gorillas then, their numbers decimated by habitat loss, poaching, and disease, and she was worried about their rapid decline toward extinction.

On my visit to Ruhengeri a decade earlier, the project employed just Jonathan Sleeman and another veterinarian, who both worked out of an ill-equipped bungalow. Now, there is a modern base with a laboratory and seven veterinarians. The MGVP is headed by Lucy, a thin, intense woman with graying hair, who tells me its 21-person staff's primary aim is to save the mountain gorillas from extinction. Its major funding comes from Baltimore's Maryland Zoo.

The MGVP maintains the health of the mountain gorillas in Rwanda and Uganda, but for the time being cannot monitor the well-being of the gorillas in the Congo, just a few miles over the border. "It's too dangerous for our vets to go there now," she says. "There are thousands of rebel troops in the mountain gorilla habitat there. We fear for the gorillas' safety."

Lucy is a veterinarian, and climbs the Virungas' slopes on the Rwandan side on average every second day to see how the habituated families are faring and check whether any need treatment. "I look especially for limping, coughing, hair loss, and diarrhea," she tells me. A few weeks earlier, she had tried to save Umurava, a silverback studied over two decades by Fossey and her successors. The gorilla was ailing with a neck wound, probably from a fight with another silverback, and a cough. "I tranquilized Umurava with a dart, anesthetized him, examined and treated him on the spot for presumptive pneumonia, shock, and sepsis."

He was too far gone and died the following night. In the autopsy, Lucy found Umurava's lungs riddled with abscesses. Because mountain gorillas share 97.7 percent of our DNA, they are susceptible to many of the same diseases that harm us. "They can pick up diseases such as polio, malaria, measles, strep throat, tuberculosis, and herpes from humans, as well as salmonella, rabies, and other diseases from animals in the forest," she says.

Dian Fossey witnessed the death of a gorilla from malaria as the animal slowly wasted away, leaving blood spots each night in her nest. Toward the end the other gorillas began to kick and bite her, perhaps prompted by fear. Weakened, she was unable to defend herself and died. And yet Fossey observed the exact opposite when a high-ranking female was badly bitten on the neck, head, shoulders, and back in a fight. Her five-year-old daughter took on the role of medic, constantly licking the wounds until they were completely healed five weeks later. She refused help from the other gorillas in the group and even pushed away her weakened mother's hands as she went about her task.

Contact with humans can be dangerous. In 1985, MGVP veterinarians discovered many Rwandan mountain gorillas suffering from measles. They had all been in contact with visiting tourists. The veterinarians vaccinated 65 gorillas with a dart gun and stopped the infection from spreading. All the gorillas survived.

I ask Lucy about Kwakane, the young silverback bristling with testosterone a decade previously, expecting him to have become the patriarch of a flourishing family. "Kwakane never made it," she tells me. "He was a loner for many years, and then he disappeared not long ago."

Life is tough for silverbacks fighting for dominance, and though Kwakane scared the hell out of me with his fierce display, sadly for him it seems he never had the muscle and charisma to attract females. On my earlier visit, while with Rwandan trackers in the mountain forests, I had brief, poignant encounters with these tragic solitary gorillas. As we pushed through the undergrowth on a narrow path created over the decades by duiker, knee-high antelopes, the tracker would sniff at the air and go down on his knees. I would do the same, and then a few moments later smell the unique musky odor of a silverback. A minute or two after that a solitary silverback would shoulder through the bushes and pass us, sometimes inches away, giving us barely a glance.

With a dominant silverback keeping up to a dozen adult females in his family, there have to be many loner silverbacks roaming the mountains. It is one of the most forlorn lives I can imagine. For 13 or 14 years the males are raised within a loving, tightly bonded family, foraging for food, sleeping, playing, and grooming together. The young males learn by example from their father how to be a successful silverback, while the young females do the same with their mothers. But the time comes when they must leave the family.

The males are sometimes forced out, especially if related to the dominant silverback, but it is often a voluntary move conditioned by hundreds of thousands of years of evolution. The blackbacks, adolescent males between the ages of 10 and 12, begin to drift away from the family, spending days, then weeks, then months at a distance before departing forever. The fertile young females leave the natal group much earlier, at age eight or nine, instinctively knowing that they must avoid inbreeding by joining another family. The females do not have to roam the forests for long. They are usually swiftly taken into other silverbacks' families as new breeding females, or attracted by a loner into becoming his mate. It is the female's choice whether she joins a silverback and his family or keeps looking. But the young males must wander the forests, sleeping and foraging alone, with no other gorillas to groom them or share life's burdens and joys.

Year after year they are peripatetic loners. But when they become strong enough, they challenge other males in ferocious battles that can sometimes last for two or three days. Some young males grow into behemoths, bristling with charisma and muscle, and win the fights and attract the attention of the females. But many, like Kwakane, are destined to roam the forests alone until they die.

Lucy is raising a four-year-old orphan seized from poachers on the Congo side, the only mountain gorilla in captivity at the time, hoping eventually to return her to the wild, a world-first if she succeeds. She offers to show me the orphan on my return from the Congo. "If I manage to get back," I joke. I might not have been so flippant had I known how close I would come to life-threatening peril over in the Congo.

5

The border crossing at Goma in eastern Congo is an hour's drive west from Ruhengeri, and it is like descending from an earthly paradise to the outer gates of Hell on Earth. Mount Nyiragongo, a volcano soaring 11,384 feet over Goma, erupted in January 2002. Half a million people fled as the volcano spewed flaming lava down onto the town. The massive eruption destroyed 80 percent of downtown Goma and smothered it with a blanket of lava more than a yard thick, turning it into a modern-day Pompeii.

"Goma should not be rebuilt where it is now," warned world-famous vulcanologist Professor Dario Tedesco of Naples University after inspecting the devastation a few days following the disaster. He said Mount Nyiragongo was still active, and "the next eruption could be much closer to the town, or even inside it."

Despite his apocalyptic alert, Goma's residents rushed back, only to flee again for a short time in December 2006 along with the Congolese army when the biggest rebel group, led by ruthless Tutsi warlord Laurent Nkunda, threatened to overrun the town. Nkunda carries a scepter topped with a silver-plated eagle and a pin that reads "Rebels for Christ." A savage counterattack by United Nations peacekeeping forces based in Goma forced the rebels back into the jungles.

On the way to the hotel, my eyes sting and my nose clogs from the volcanic dust thrown up by a brisk wind and the many international aid workers' sport utility vehicles charging around Goma. Piles of volcanic rock run like guide rails along both sides of the streets, dug out from the innards of shops demolished by the fiery lava flow. Goma is a Wild West–style frontier town, booming and dangerous, and you get the feeling that if you take a wrong step or say a wrong word here you could get your head shot off.

It thrives because of the presence of 4,000 UN troops and more than 100 Western aid workers. The flood of money generated by this international tribe has attracted swarms of settlers from all over the Congo and neighboring Rwanda and Uganda. They throng the crumbling streets and ramshackle shops, and the tension of people living on the edge often explodes into violence.

As we head down the main street, lined on either side by shacks, grim-faced paramilitary police patrol in single file, clad in body armor and wearing visored helmets. They brandish assault rifles and look keen to use them. My driver tells me that in this town murder is an everyday hazard, and the previous night three men killed a priest and a carpenter. The police had to rescue the murderers from a lynch mob of relatives. The authorities have put on this show of force, fearing that the smoldering tension will flare into more bloodshed.

In contrast, the hotel's whitewashed bungalows perch by Lake Kivu like villas overlooking Lake Geneva. I am eager to swim in the lake, but the hotel manager warns me away from its flat, dark waters: "Swimmers die in there, because the poisonous fumes from volcanic mud sometimes bubble to the surface without warning and overwhelm the swimmer, who falls unconscious and drowns."

Samantha Newport, a tall Englishwoman in her late 20s with a cool manner, arrives for dinner; she is based in the Congo and works with WildlifeDirect. She tells me, "I'll take you to Bukima patrol post, about three hours' drive from here, where the rangers monitor the gorilla families who live on the slopes of Mount Mikeno. It's a dangerous place. Nkunda's rebels occupied it until a few weeks ago, and while there they killed two silverbacks for food. They stripped them of all their flesh, and threw the remains of one into an outdoor latrine. We've only just returned, and found they'd looted the patrol post and badly damaged the building."

At mid-morning the next day, I visit the headquarters of the UN peacekeeping force based here to ward off the many rebel forces occupying the nearby hills. Barbed wire tops the high concrete walls, and wary troops armed with assault rifles and shielded by sand bag emplacements guard the offices and barracks of the 4,000 Indian peacekeeping troops. Their commander is Brigadier General Pramod Behl, a tough-eyed, battle-hardened warrior from the Punjab, a veteran of the India–Pakistan wars. He cautions me that the region around Goma is still unstable and dangerous, especially in the mountain gorillas' habitat.

"There are about six thousand *Interahamwe*," he says. They haunt the Congolese jungles in and around the mountain gorillas' habitat, unwilling to return to Rwanda for fear of imprisonment or worse. They escaped across the Virunga volcanoes, fleeing from the victorious Tutsi army in 1994. "They call themselves the Democratic Forces for the Liberation of Rwanda, and they've killed and raped tens of thousands of villagers around here. They've abducted countless women to be their wives and concubines."

What hope, then, do the mountain gorillas have, living amid these hordes of cutthroats? The Brigadier General shrugs. "You've heard what happened to the two silverbacks in January? The gorillas up there clearly need all the help they can get."

Fighting the rebel Hutu and the Congolese army soldiers are the 10,000 rebel Tutsi troops of renegade general Laurent Nkunda. Behl says, "He parades as the savior of his people here in eastern Congo, the Tutsi, but he's just another bloody warlord. He formed his own rebel group here, calling it the National Congress for the People's Defense. Nkunda claims he's protecting the local Tutsi from being massacred by the Hutu, but he's in it also for the money. We know he's already a multimillionaire."

The UN Brigadier General cannot say so, but it is common knowledge here in Goma that Nkunda is a proxy for Rwanda, financially backed by the Rwandan government who want to keep the Congo destabilized so as to control the border area and root out the Hutu rebels. It matters little to the Rwandan government if Nkunda pockets the occasional million dollars.

A few months after my visit, an investigation by a team of independent experts for the United Nations Security Council found strong evidence of an ongoing collusion between high-level Rwandan government officials and Nkunda. The primary aim is to destroy Hutu rebels operating on the Congo side of the Virungas. The team also discovered that the Rwandan government sent recruits, including child soldiers, to fight with Nkunda's forces. At the same time, senior Congolese government officials were giving the Hutu rebels weapons and ammunition to join attacks by government soldiers on Nkunda's troops.

The Brigadier General tells me that Nkunda's rebels launched an attack on Goma in December 2006 and reached the city's outskirts. "The Congolese army, the bloody cowards, fled along with most of the people, but we took them on and pushed them back into the hills," he says. I ask whether there were any civilian casualties. He looks at the ceiling. "We don't keep numbers. You'll have to ask the Congolese Army."

I tell him that I read that his troops went at the rebels with guns blazing, killing more than 300 men, women, and children in the way.

"In war there are always civilian casualties," he answers.

Did he lose any of his troops in the battle?

He turns his hard-eyed gaze back on me. "Not a single one."

As if that is not enough turmoil, there is another group in the hills above us, the ruthless Mai Mai, about 4,000 rebels who appear to have no ideology except rape, pillage, and murder. When they moved into the Virunga National Park a decade ago, they targeted the 25,000 hippos living in the lakes and rivers for bushmeat. The latest aerial survey two years ago could find only 940 hippos.

As I leave his office, the Brigadier General gives me a farewell wave. "Good luck," he says. "You'll need it."

On Goma's outskirts, the shantytowns morph into green fields that border a potholed road patrolled by hundreds of soldiers in single file with assault rifles at the ready. Not one offers a smile as we pass, but then there is not much to smile about in this volatile and ugly place. About halfway to the gorilla slopes, we drive by the vast muddy fields where a million Hutu refugees settled in tents after fleeing the avenging Tutsi army in 1994 in one of history's biggest movements of people. Among them were Fidele and JB. Almost all returned home, assured of their safety by international aid organizations and the Rwandan government, and just a few scattered tents remain, inhabited by listless refugees who claim they have nowhere else to go.

At a turnoff from the Goma road, the four-wheel-drive van heads up a track studded with chunks of lava rock that bounce us around like pinballs. Despite the presence of so many rebel groups, the hills are dotted with mud-hut villages. Their sloping patchwork fields of fertile volcanic earth flourish with crops of corn, potatoes, peas, and tobacco. Looming over this deceptively peaceful landscape is the dormant Mikeno, the 14,557-foot-high volcano on whose cloudy jungle slopes many of the Congo mountain gorillas live. Its sharp, jagged granite peak looks like a giant wolf's tooth. No one waves to us on the way up the steep track, but children clad in ragged clothing keep pace with the slow-moving vehicle and with smiling faces beg us for money.

Two hours later, within sight of the Virunga National Park, we reach our destination, the Bukima patrol post, a dilapidated weatherboard hut, home to the rangers who accompany and guard the gorilla trackers each day. Samantha introduces me to Jean-Marie Serundori. Clad in a jungle-green uniform and beret, the head ranger of what is termed the Gorilla Sector of the national park is middle-aged and slim, with kind, fatherly features.

For 20 years he has led patrols protecting the mountain gorillas here, and has had to live with the everyday risks. "So many of our rangers have been killed by rebels and poachers in the park," he tells me in a soft voice. "Two months ago, hundreds of Nkunda's troops occupied this very spot and looted it, remaining until just two weeks ago. We fled at the time, and have only just returned, after the rebels left. They're still in the hills above just a few miles from here. They can see us right now."

"Then why did you risk your life by returning?"

"The gorillas are our brothers and sisters; I know them as well as my own family, and love them." He smiles. "If we don't check that they're safe every day, soldiers and poachers might harm them."

I stow my luggage in a tent in a clearing by the patrol post, hemmed in by high jungle trees, and eat a lunch of beans and bread washed down with bottled water. In mid-afternoon, Jean-Marie leads me through sloping terraced fields of potato and corn lined with rows of villagers upending the rich soil with hoes. Across the fields I see other villagers herding cattle to pasture. Up the hill is the jungle, the perimeter of the Virunga National Park, 3,012 square miles of pristine, moist montane forest and savannah.

"Rugendo, the silverback, and his family are nearby," Jean-Marie says, amazing me because I expected to find the mountain gorillas living deep in the highland forest, putting plenty of distance between themselves and the villagers. "The gorillas like to raid the fields, especially to eat corn," he explains. The locals who live next to the park are surprisingly tolerant of the gorillas' pilfering. Jean-Marie pauses to talk with a village chief, a tall man with craggy features. He tells me that his people have never killed gorillas because they regard them as sacred totemic creatures. He wears a dark, dirt-soaked two-button suit that has been frayed and turned grubby from years of daily wear. I had seen many other village men clad in such suits, even when they were tilling the field.

"How do you know that it was rebels who killed the two silverbacks earlier this year, and not villagers seeking revenge for the crop raiding?" I ask Jean-Marie.

"We know from informants, but there's nothing we can do because the rebels are too strong, and our army is too afraid to fight them. Every time the rebels challenge our soldiers they run away."

As we enter the national park, the dense canopy casts the jungle into a verdant gloom. At 10,000 feet above sea level, I struggle to breathe in the rarefied air, gasping as we climb a steep, rocky slope. Then, I catch a whiff of a distinctive musky odor. A silverback's behind has seven layers of endocrine glands that give out this odor. Males and females also have glands in their soles and palms that lubricate them.

I know the silverback's odor from my time with the Rwandan mountain gorillas and meaningfully nod at Jean-Marie. He nods back. Following the gorillas' trail, Jean-Marie utters their rasping call of peace. He points at the thickets ahead. "*Le grand chef,* the big chief, is in there."

Moments later we hear the *thock, thock, thock* of a silverback beating his barrel chest, a thrilling sound that echoes around the forest. Could it be a challenge? I tense as the silverback, six-foot-tall Rugendo, bustles through the thicket, but relax as he ignores us and heads deeper into the forest. He is followed by two of his children, an adolescent male and Noel, an impish infant male. Noel's liquid dark eyes and leathery mouth are all but lost in his fluffy black fur, giving him the look of a cuddly soft toy. "We called him Noel because he was born on Christmas day two years ago," Jean-Marie explains in a whisper.

Noel has begun some of the most enjoyable years of his life. Mountain gorilla youngsters spend the first six months virtually attached by an invisible string to their mothers, never straying from their sides, but then they begin to move away a few yards, still keeping their mother in sight. At about 18 months their world begins to open up, much like those of children at a kindergarten, as they spend much more time with the family's other youngsters. Life for young gorillas, aside from eating and sleeping, is a constant round of games such as mock fighting and "king of the mountain," gaining the high ground and defending it against all comers, often played on a fallen tree trunk. They tumble, wrestle, and sometimes scramble onto the broad shoulders of their father and pull playfully at his dense hair.

Through the gloom I see several more gorillas, adult females, peering at me. Their bodies are swathed in the shaggy black fur that allows them to survive nighttime temperatures that can fall below zero way up here. Mountain gorilla territory lies between 7,300 and 14,000 feet above sea level. Another young male crashes through the branches, performs a perfect "10" gymnast's roll, and scurries after his gargantuan father. A potbellied mature female, half Rugendo's size, breaks cover and waddles past, barely glancing at us. All four gorilla subspecies are sexually dimorphic, the adult males weighing about twice as much as the females. This size disparity probably evolved through the females preferring to live and mate with the biggest, brawniest males who can protect them best.

Jean-Marie leads me to Rugendo, who sits by a cluster of low trees, munching on fistfuls of leaves. Mountain gorillas are herbivores, apart from an occasional feast of ants and grubs. Everywhere they look in their habitat they find food such as bamboo shoots in season, stems of vines, and pith, but not much fruit. This results in the smallest range of all the gorillas, between 1.5 and 6.5 square miles for each family, which effectively becomes even less because they overlap extensively.

Despite the abundance, a mountain gorilla's life is largely a relentless daily search for food, the diversions being sleep, sex, play, grumbling, and mock fighting. The silverback heading a family group must always show vigorous leadership or the females could desert him for a stronger silverback, Jean-Marie tells me in a soft voice. If that happens he becomes a loner, wandering the forests, like Kwakane, hoping to attract young females who have left their natal group in search of a mate.

The silverbacks can explode into murderous rages when a loner challenges a patriarch to a fight. When you pit two 500-pound giants, each about 10 times stronger than an adult human male, and with huge canines and massive arms, it is no wonder the fight sometimes ends in death.

A tracker once spied a pair of screaming, chest-beating silverbacks charging each other and tumbling on the ground, biting and bashing each other with terrifying ferocity. The next day he followed two blood trails through the trampled undergrowth but could not find the gorillas. Eleven days later the battle resumed between the same silverbacks and the tracker, scared off by the screaming, fled.

He returned the following day and found a field of devastation with branches broken, grass ripped out, and vine streamers torn down. Hidden in the bushes where he retreated after the fight was the losing silverback—dead. He lay on his back with one eye shut and the other badly swollen. His throat and neck were badly bruised, suggesting that the victorious silverback might have strangled him.

There is no mercy for any gorilla who gets in the way of such a fight, even youngsters. After one battle between rival silverbacks, trackers found a 10-month-old infant killed at the scene, bearing many bite marks. One bite broke the infant's femur while another punctured the stomach, causing peritonitis.

Habituation of the mountain gorillas for tourism and research may have had some effect on their behavior. Ymke Warren, a gorilla researcher, spent a decade studying the mountain gorillas, and she told me that unhabituated lone silverbacks may have developed a fear of habituated silverbacks when they are accompanied by humans. She told me about a screaming silverback that charged into a group, but the encounter was very much influenced by the humans present. "The challenging silverback soon left the vicinity. His departure was cheered by the trackers, who were terrified he would hurt 'their silverback.'"

And yet, despite their fearsome appearance and fighting might, silverbacks are softies at heart with their family, and dedicated fathers. "The females' primary role is to look after the young, but if one dies or abandons her infant to run off with another male, then the silverback will take care of his child and raise it," Jean-Marie says as we watch the roughhouse antics of Rugendo's children, the youngsters playfully banging each other on the head. "I've seen it many times."

One of Dian Fossey's favorite gorillas, Pablo, was abandoned by his mother, Liza, who left the family suddenly one day, attracted by a big silverback named Brutus from another group. She never came back. Pablo was five years old at the time and his father, Beethoven, let him sleep in his nest at night and sit by his side during the day while he was feeding. A year later, the families of Beethoven and Brutus met by accident and there was a face-off on a hillside. Pablo was in full view of his mother, but Liza made no move to join him, not even for a brief reunion with her son.

Such behavior does not mean mountain gorilla females lack a motherly instinct, even though care of their offspring can range from the casual grooming of the little one to a mother who spends every waking minute with one eye on her baby. A researcher with the Karisoke Research Center in Rwanda's Parc National des Volcans once saw an infant dangling helplessly from a tree, her neck caught in a fork. Her family began screaming in alarm because she could have been strangled in a few moments, and her mother swiftly jumped to her aid, plucking her to safety.

Researchers at Karisoke also observed mountain gorilla mothers carrying around the corpses of their dead infants for many days after they had perished. The researchers noted that even the babies' "aunties," as they called them, adult females who had been "babysitting" the infants, joined in carrying around the tiny bodies.

Thirty-six days after an infant was born to a female named Umuyaga, it died bearing unexplained wounds under the arms and legs and on the lower abdomen. The mother recovered the body from an unrelated female, Mudakama, six days after its death, and carried it for a further six days. Yet another female, Guntangara, toted the body around for another three days until she and the others left it behind in a night nest. A clue to their behavior might be that Guntangara and Mudakama were pregnant when they carried around Umuyaga's dead infant. The researchers saw the behavior as natural "mourning," expressing a strong attachment to the infant.

But there is a darker side to the mountain gorillas' nature that is rarely mentioned in stories about them. Silverbacks acquiring a family after the death of its patriarch have been observed killing all the females' unweaned youngsters. Male lions do the same thing. There is method in the murder. Female mountain gorillas suckle and care for their young for up to three years, and during this time the urge to mate is suppressed. By killing the infants, the new silverback prompts the females back into estrus so that he can sire his own brood.

Alexander H. Harcourt and Kelly J. Stewart, authors of *Gorilla Society*, write that with all gorillas, "Infanticide almost always occurs when the protective male dies."

When the silverback Nunki died of disease, his group consisted of four adult females and their youngsters. Pandora, the mother of an infant named Kazi, transferred to a new group headed by another silverback, and two weeks later Kazi's body was found at the spot where they had met. He had many bite wounds, the most serious being in the groin, and his injuries indicated that he had been tossed about.

In another incident, poachers killed a dominant silverback, and his family then merged with another group headed by a silverback named Susa. Observers soon after found the body of an infant with injuries that pointed to infanticide. Females have been observed attempting to save their infants, but the silverbacks, at twice their size, are too powerful. Because the mother is unrelated to the other female adults in the group, she can expect no help from them.

Dian Fossey found that three out of 10 mountain gorillas die during the first three years of life, and that 34 percent of these deaths are caused by infanticide. She discovered most of the infanticides "took place during violent inter-group interactions that occurred when one silverback challenged another for possession of the females."

Blackbacks, adolescent males, can also be baby killers. In one case a blackback had migrated into a group, was not related to any of its females, and had not yet mated with the baby's mother when he killed it. In contrast, infanticide is rare among the other great apes—chimpanzees, bonobos, and orangutans—because their mating systems are quite different. Chimps and bonobos live in extended families where the many adults are promiscuous, and so any one of the males could be the father, while the orangutans live wandering, solitary existences in the distant jungles of Borneo and Sumatra. The fathers never remain with their offspring, who are raised solely by the mothers.

Rugendo looks almost indestructible as he sits by a profusion of vine tangled around a tree and with single-minded purpose, yanks down fistfuls of leaves to stuff into his massive mouth. We could be moths fluttering around him for all he cares. The shape of his broad, flat, circular nostrils, as shiny as black leather, is unique to him and is a feature that allows researchers to identify each individual. His dark brown eyes, set below a heavy, hairy brow, sparkle with primate intelligence. His massive pectoral muscles are as defined as a weightlifter's, and the silvery saddle of hair that extends from his bull neck to his massive buttocks gives Rugendo a mature, dignified appearance, the sense of an adult male who has lived through and been tempered by good times and bad.

Though he seems concentrated on feeding his massive appetite, Rugendo, like all dominant silverbacks, keeps his family under his watchful care day and night. "Around five a.m. the silverback wakes his family by beating his chest and charging them," Jean-Marie tells me. "He then leads them to their first feeding spot for the day. The family forages over about five hundred yards a day, eating plants, leaves, wild celery, fruits, and even thistles. They play a lot, and take mid-morning and mid-afternoon naps. Around six p.m. the silverback chooses a place for them to sleep for the night."

As if on cue, sated by his bulky snack, Rugendo rolls onto his side for a mid-morning nap. His easy acceptance of our presence bodes both good and bad for him. It allows the rangers to stay with his family for up to four hours each day, but it also lets poachers and rebels who want to kill them get near the gorillas without scaring them.

I edge closer to the giant, awed by his brawny arms and enormous fingers. His massive, furry crested head locks his huge jaw muscles into place. Like all the great apes, including humans, Rugendo has a pair of incisors, a single canine, two premolars, and three molars in each half of the lower and upper jaws. His feet are a larger version of our own, more so than our other great ape relatives. The mountain gorillas are largely terrestrial, in contrast to all the other great apes, who spend much of their time in trees, and so evolution has given them a similar-shaped foot to ours.

While the big chief dozes, two of his older sons and little Noel tussle in mock combat nearby, tumbling, growling, slapping, punching, and tugging. Kongomani and Mukunda, eight- and ten-year-old males, are still swathed in black fur, but when they are about 12 will begin to grow a distinctive saddle of silvery hair like their father's. It has been suggested that this makes the adult males look bigger and more imposing, an optical illusion. Dividing a black rectangle with a central area of white makes the rectangle look longer. It also signals maturity to the females and potential rivals.

Noel is especially aggressive, baring his teeth as he repeatedly bangs his fists on the ground and charges his brothers. He leaps on them, pulls at their fur, bites their legs and arms, and whacks them on the head. They humor him for a while but, like human big brothers, soon tire of junior's aggressive antics. Then each time he attacks, the closest brother grabs him with a brawny arm and tosses him back into the bushes.

After a few rejections he turns to peer at the pale-skinned stranger. Up close his dark brown eyes, amazingly similar to our own, buzz with intelligence. I sense that this little male's quiet, intense focus on me signals that he seems to feel a bond between us, hard to comprehend for him but difficult to deny.

The hour to which outsiders visiting the gorillas are restricted, to guard them from contracting human disease, passes much too fast, and Jean-Marie leads me away. About 20 yards from Rugendo I spy a female, almost obscured by vine leaves, holding an infant close to her breast. Their shaggy, dark fur is so mingled that I can pick out the infant only by its gleaming eyes and tiny, flat, leathery nose.

"That's Safari and her baby. When they're nursing the females stay out of sight of humans if they can, especially one like you they've never seen before," Jean-Marie explains.

Mountain gorilla females give birth to their first baby at about 10 years of age. Unlike chimpanzee females, whose spectacular pink rear swelling signals estrus, gorilla females have only a slight swelling, and the males detect their readiness to mate by their seductive body language. The female approaches the male backwards in a crouch, her rear end raised, and sometimes nudges him. Mountain gorilla males have been observed copulating between the ages of nine and 10, but begin in earnest only when they have become fully grown silverbacks at about the age of 15 and have willing females. Until then, the young males are kept in check by the family patriarch.

Jean-Marie leads me out into the fields and back to the Bukima patrol post. In the small clearing in the foothills of Mikeno, conservationist Rob Muir of the Frankfurt Zoological Society has just arrived from his home at Goma. He is project manager for the conservation of the Virunga National Park and has been here for four years. Rob had to collect the bodies of the two murdered silverbacks a few weeks earlier, and has just set up Camp Karema in the clearing here, named after one of the silverbacks. His organization has joined with WildlifeDirect in funding the rangers here to protect the mountain gorillas.

Tragically, it was not the only massacre of mountain gorillas in the Congo. In February 1999, in the forests west of Lake Kivu, rangers found gorilla bones scattered across the range of one of the groups, and they suspected poachers killed the gorillas for bushmeat. Two months later gunshots rang out in the range of another group, and poachers were then seen carrying the bodies of dead gorillas to their nearby villages.

As Rob and I share the standard camp dinner of beans, rice, and bottled water, he tells me about Nkunda's rebels murdering the two silverbacks a short time earlier. "The body of Karema was discovered near here, and another just a few days later. They were stripped of their flesh," he says. But rebels are not the only danger faced by the mountain gorillas here. "There's a threat from poachers as well as a threat from the exotic pet trade, with young juveniles smuggled out of the country and sold on the international black market."

But the biggest threat to the mountain gorillas here, he says, is the charcoal trade: "The charcoal traders have already destroyed about a quarter of the hardwood old-growth forests in the southern section of the Virunga National Park, and it is threatening a biological holocaust of the Congo mountain gorillas' habitat."

Traders daily sneak into the national park, chop down trees in huge numbers, and slow burn them in dirt-mound kilns. One kiln can produce up to 100 sacks of charcoal. The resulting charcoal is bagged into sacks and lugged out of the park by porters to sell as fuel for the villagers' and city folk's cooking fires. Rob tells me, "I've heard there are nineteen thousand sacks of charcoal in the park at the moment waiting to be carried out. It's a trade worth about thirty million dollars a year. The traders are a big threat to the gorillas because they don't want the rangers sniffing around where they burn the trees and then have porters carry the sacks out of the park. It's an illegal trade that's destroying the forests, and so the rangers go searching for them. I'm afraid they'll begin killing the gorillas as a warning to keep the rangers away."

"That's blackmail."

"Precisely. The Congolese army is not only unhelpful in this, even though the trade is illegal, but they actually protect the trucks taking the charcoal to town, and even use army trucks to transport the charcoal. At a cost, of course. The soldiers earn loose change from it, but the army big guns in Goma are paid off by the charcoal traders with large amounts of money."

"What happens to all the charcoal the rangers confiscate?"

"There's no point wasting it, and so the rangers give it to orphanages and refugee camps around Goma."

I ask the inevitable question. "If thousands of charcoal sacks are getting through the road checks, is it possible that high-level rangers in the park are involved?"

Rob looks troubled. "Yes, we suspect that. The head ranger in the southern sector of the Virunga National Park, the gorilla sector, Paulin Ngobobo, is incorruptible, but I can't answer for one or two others. Paulin is in Kinshasa now, but he should be back by the time you return to Goma."

However, I am forced to go back to Goma much sooner than planned. After breakfast the next morning as I get ready for another trek to the gorillas, lacing up my hiking boots and putting a couple of flasks of water in my backpack, a worried-looking Samantha comes to my tent. She says, "There's an emergency, Paul, we've got to evacuate here immediately and go back to Goma. Rangers at another patrol post have just told us by radio that a large force of Rwandan soldiers has crossed the border illegally and are heading this way to occupy Bukima. They'll reach here this afternoon or tonight, and I don't want to be here when they arrive."

I tell Samantha that I have been in more crises than I can remember, and do not get the sense now that we are in danger. I am prepared to risk trouble by staying here and going to the gorillas with the rangers over the following days. I explain that even if the Rwandan troops arrive, I believe they will not cause us harm. Any attack on the patrol post would earn international media coverage, and would prompt the United Nations to demand that the Rwandan government justify their actions.

Samantha shakes her head. "I'm responsible for your safety, and you'll have to come back to Goma."

Jean-Marie and the other rangers are staying at Bukima whatever the threat because, he tells me, it is their duty. I have grown fond of him; he is a gentle man, but I sense his smile masks a steely resolve. His dedication to the mountain gorillas under extreme duress has won my admiration, and I wave to him as our van bounces down the hill heading for Goma.

Two days pass with no word from Bukima, and at dinner overlooking Lake Kivu, Samantha tells me that we must wait for at least another week before we dare return. I suggest that I go back the following day to see what is happening and she agrees on the condition that I do not stay the night, but return to Goma before dusk.

Jean-Marie is at the patrol post. He tells me that about 30 Rwandan troops did come here the same day we left, but moved back toward the border soon after. He is happy for me to stay the night and go with him to the gorillas the following morning, but I keep my promise to Samantha to return to Goma.

6

Emmanuel de Merode, WildlifeDirect's executive director, has just flown in from Nairobi, and we meet at the lake hotel restaurant for dinner. Born in Kenya, Emmanuel's first job was as a park ranger, and on one of his first patrols poachers shot at him. "Welcome to the trade." He smiles. He has since worked in conservation, become a biological anthropologist, and then got his doctorate with a thesis on the destructive bushmeat trade in this part of the Congo.

Emmanuel at first insists that I cannot go back to Bukima for at least a week because it is too dangerous, but after we discuss it some more he smiles. "Okay, I'll take you back to Bukima tomorrow."

The following morning we take the same road, past hundreds of Congolese troops strung out in single file on patrols, or in sandbag emplacements and brandishing machine guns. After the turnoff we rattle and roll up the mountain track. Mikeno looms over us as Jean-Marie greets us at the patrol post with a big smile. Before WildlifeDirect and the Frankfurt Zoological Society began sponsoring the rangers in November 2006, he and his colleagues often went for months without pay and wore uniforms that were falling apart. They are employed by the Institut Congolais pour la Conservation de la Nature (ICCN), the Democratic Republic of Congo government's conservation body, and like most government workers suffered from the economic disruptions caused by a decade of civil war and extreme corruption. Now, because of WildlifeDirect's innovative aid program, the rangers are paid without fail every month and are gradually acquiring a full set of equipment, including weapons to defend themselves.

Emmanuel points to the satellite dish that has just been installed in the clearing at Camp Karema. He tells me, "Each ranger has his own blog, telling readers worldwide about his daily life, and we encourage individual donors to help sponsor a ranger. It costs thirty-three dollars a month for a ranger's food rations, eighty-seven dollars for his salary, while tents and uniforms add another thirty-seven dollars to the total."

Richard Leakey had told me in Nairobi that he made sure that whatever money is donated to keep the rangers effective through Wildlife-Direct reaches its target within a few days, without a cent deducted for administration costs. "We have a cash reserve in Goma, and so we buy food and equipment for the rangers within days of a donation arriving," explains Emmanuel.

He says the children at Stratton Elementary School in Colorado Springs heard about the rangers' bravery and now pool their pocket money each month to pay for a ranger's costs. They also bake cakes and cookies, and make dill pickles to sell, to fund the ranger. They frequently read WildlifeDirect's website, and e-mail messages to the ranger they sponsor, Joseph Aloma.

Over dinner, Emmanuel tells me that the Virunga National Park is, in his opinion, the greatest national park on Earth. He has spent years here, and is co-author of an illustrated book on the park. He says, "The Belgian colonial power in the Congo set it up in 1925. It's about the same size as Yellowstone Park, and it's unrivaled in the world for the diversity of landscape and animal species. It has active volcanoes, savannah, rainforest, alpine forests, and moors. Among the animals is the endangered okapi, a striped antelope the size of a pony. We even have hippos, though the Mai Mai have wiped out most of them. But the jewel in the crown of the park is the mountain gorilla."

At mid-morning the next day, Jean-Marie speaks with a clearly worried Emmanuel and they stride out of the camp, returning half an hour later. "About two hundred of the warlord Nkunda's rebels have turned up at the patrol post," Emmanuel tells me. They're just a hundred yards away, through the undergrowth. "I don't know why they're here or what they're planning. Best we keep calm and hope they'll go away."

I feel ashamed that I have brought him to such a dangerous place. Emmanuel shrugs. "I planned to come to Bukima sometime this week, and it might as well be now. Let's sit tight and see what happens."

Within an hour the rebels leave. Then, just before lunch, 10 heavily armed Congolese government troops stride into the camp without warning carrying assault rifles and long-barreled machine guns, with bandoliers of glinting bullets hung from their shoulders. They spread out and face inward, pointing their weapons at us, as a major wearing a broad-brimmed straw hat walks into the clearing. His base is by the road at the bottom of the hill, and he has come to find out why the rebels were just now at Bukima. Satisfied that they have gone, and that we had nothing to do with their visit, he leaves. Emmanuel decides that afternoon to visit another family of gorillas, led by Humba, Rugendo's brother, while we still have the chance.

This time we enter the Virunga National Park at a different place, where the perimeter is marked by a knee-high straggly wall of lava rock. Villagers can go into the national park but are forbidden from destroying vegetation or harming animals. The World Wildlife Fund's office in Goma is marking the perimeter with lava rocks, but because funds are limited, it has completed only 20 miles of the 300-mile boundary. The fields that nudge the edge of the national park are full of villagers industriously plowing their crops.

As we head up Mikeno's slopes, Rwanda is just two miles away on the other side of the volcano. The path to the Congo mountain gorillas is tortuous and not for the faint-hearted. For two hours we climb almost vertically up a narrow, rocky trail through dense jungle, at the risk of stumbling into one of the murderous rebel militia who swarm through these mountains more or less unchecked.

A couple of times I slip, and my knee hits one of the small, rough-edged lava rocks studding the path. Blood trickles down my leg and into my trekking boots, but I ignore the pain. Clambering over the rocks, we plunge into the virgin forest, thrusting aside lush bush and bamboo stands overhanging the trail. The trackers slash and chop a pathway with machetes.

At about 11,000 feet above sea level, the trackers pick up signs of Humba's nine-member clan. They hack a path through the walls of creepers, bamboo, branches, and prickly vines, following the messy trail of the gorillas who are feeding on plants. Two hours into our march, Jean-Marie points to several circular patches of flattened and bent grass spread around a grove of tall bamboo and observes, "The gorillas slept here last night."

The trackers say the gorillas are very close. "Keep seven meters back from the gorillas, and if any approaches move away," Emmanuel tells me. This is to guard the gorillas from human disease.

We sit on the track for a breather and for Emmanuel to instruct me in gorilla etiquette. I know much of it from my previous visits, but listen politely. He repeats what the veterinarian Jonathan Sleeman told me a decade earlier. Never stare at a silverback, because he regards that as a challenge and might charge. If he does, go down on your knees and bow your head. Put a leaf into your mouth, because it is a sign that you come in peace. "It's probably connected with the sense that you're eating and are therefore no danger to the silverback and his family," Emmanuel adds. "But if the silverback starts screaming and beats his chest, you know you're in trouble."

A few minutes later I smell the silverback's musky odor. Jean-Marie holds up a hand, halting us. He signals that we are very near by uttering a few hoarse coughs, gorilla talk. "Humba means 'placid' in our language," he whispers to me. "He's Rugendo's brother. Their father was killed in the crossfire in a fight between the army and rebels a few years ago."

Humba, the 500-pound dominant male, yanks aside the creepers about 10 yards ahead and stares at us with imperious grace. The silverback's muscled body ripples with power, and his massive head has all the gravitas of a Mount Rushmore president. Humba bares his fearsome canine teeth at us. "Don't be afraid, he's used to people, and has never hurt anyone," Jean-Marie reassures me.

Not far from here two months ago, either Nkunda's troops or Hutu rebels killed and ate the two silverbacks. Four more gorillas are missing, three silverbacks and a mature female. The rangers have been looking but cannot find them. Humba peers at us for a few moments, then goes back to his feeding, tearing down strings of vine leaves and pushing them into his mouth. Given the big chief's all-clear, the family's adult females and youngsters climb out of the undergrowth to stare at us for a few moments, then resume feeding. I smile as an infant, the very image of Noel, jumps onto the back of her much bigger brother and whacks him repeatedly on the head, growling in pleasure, until he scurries away from the little terror.

Each time the family moves as it forages through the forest, an eight-year-old blackback the rangers have named Nyakamye plops down between us and his family, keeping us under watch until they have all disappeared up the slope. He then scurries after them. "He's the sentry," Jean-Marie tells me. "He's there to see them safely away, and to raise the alarm if he thinks we pose a danger." Again and again Nyakamye dutifully takes up his post, blocking our path, peering at us until he sees that the others are out of sight.

Life is not all duty for the blackback. Nyakamye and the other sub-adult males do occasionally sneak the chance to mate with the adult females, usually while the dominant silverback is not there. They are not related to the adult females in the family group, with the exception of their mother, and so there is no risk of inbreeding. Emmanuel had earlier told me about watching a young male steal a tryst with a female while the silverback was foraging 100 yards away. When he finished mating, he glanced across at Emmanuel with what seemed a triumphant look.

For about 10 minutes Humba is nowhere to be seen, but then he suddenly bustles through the bushes about five yards ahead. He glares at us and shuffles forward, bristling with aggression. He growls, bares his teeth, and sways from side to side as if trying to decide whether to charge. We drop to our knees and freeze for several moments, and then relax as he ambles back into the thicket. Jean-Marie is puzzled by his sudden change of character, then points to the bright red floppy hat worn by an African observer with us and says that it is most likely the cause. The giant silverback has probably never seen such a strange sight before, and Jean-Marie believes it might have spooked him.

Although Humba is the undisputed boss, the adult females have their own strict ranking and will fight to maintain it. Bill Weber and Amy Vedder, pioneers in mountain gorilla studies, observed a family they were studying, noting that "Effie was the highest ranking of the clan's females. In conflicts over food, Effie's posture, cough grunts, and apparent willingness to fight if necessary always carried the day. She was one of the first to enter the clay cave and had a priority seat at the feast of the roots." Her allies bolstered her high ranking. "If reinforcements were needed, she had at least three offspring in the group; four if the young silverback, Icarus, was hers too."

I do not see much affection among the adult females in Rugendo's and Humba's families. They mostly sit apart from each other during the day, feeding and nursing their infants, and sleeping in separate nests at night. They each groom the silverback, but I rarely see them grooming each other. This contrasts with chimpanzee families I have been with in the wild where there is a dominant male, but there are many other adult males in the group and they all mate with the adult females. All the adults groom each other.

In their book *Gorilla Society,* Alexander H. Harcourt and Kelly J. Stewart explain: "Unrelated female gorillas, which include most adult gorillas in a social group, can be described as mostly tolerating one another. They rarely interact in any way and especially not in a friendly manner. In fact their most common overt social interaction is probably mild threat."

Blood bonds play an important part in relationships within a silverback's family. Mothers and daughters, or full sisters, frequently stay near each other, and they back each other up in fights with unrelated females and often groom each other. Researchers have observed constant tension between adult mountain gorilla females, but for most of the time I am with them they enjoy peace and harmony, with just the occasional grumble or vocal threat. The mock fighting and wrestling is confined to the youngsters.

Whenever Humba's clan settles down to feed or play, I am entranced by their "language," about 20 recognizably different sounds that include plenty of growling, grumbling, hooting, and grunting. Emmanuel studied wild gorillas for eight years and is an expert translator. He tells me that a single grunt or a resounding belch says the gorilla is feeling okay with the world. A rasping cough made in the back of the throat means "I come in peace." A deep-bellied chuckle signals the gorilla is happy and content with the clan. But watch out when the gorilla emits what Fossey labeled a pig grunt—a part growl and part rough grunt uttered with bared teeth. It means "get out of my hair because you're annoying me."

A study by Fossey researchers at nearby Karisoke found that adult mountain gorillas vocalized about once every eight minutes. The vocalizations seemed to be mostly statements and mood indicators intended for the other members of the group, a clear "I'm here." The most intriguing vocalizations are higher-pitched tonal calls, similar to humans humming and singing, which are most frequently performed by the younger gorillas.

Amy Vedder, the American mountain gorilla researcher, spent thousands of hours with the gorillas in the Virungas and sometimes observed such "performances," which occurred every few months, usually in a place where there was plenty of high-quality food. She and her co-author husband wrote, "One individual would start a low rumbling sound, breathing in and out in a modulated tone. This might remain a solo performance, and last no more than a minute. Often, however, others would join, adding gender- and age-specific basses, baritones, tenors and sopranos in a mix. The result was a chorus of entwined melodies, rising and falling in a natural rhythm that might continue for several minutes; a gorilla Gregorian chant in a Virunga cathedral."

The authors added that the performances probably had some purpose beyond the artistic—what sounded like gorillas' choral singing—but "we chose to enjoy them as unrestrained expressions of individual happiness and group harmony."

All the gorilla subspecies hum according to Daniela Ludwig, a primatologist with the Max Planck Institute for Evolutionary Anthropology in Leipzig. At a western lowland gorilla study site, Bai Hokou, in the Central African Republic, Ludwig has made tape recordings of gorillas humming. When I visited the remote rainforest in mid–2009, she played me a recording of a blackback clearly humming his own brief but identifiable tune. She told me, "He'd just found a good food patch and seemed to be communicating to the other gorillas that he was happy and content."

Other mountain gorilla vocalizations are familiar to us humans. I watch as one of Humba's infants whimpers to get attention, perhaps wanting to suckle or be carried on his mother's back. When that does not work he throws an attention-demanding tantrum, screaming just like a human baby. The silverback ignores the ruckus, but his mother gives him an open-mouthed threat. The infant ignores the hint. His mother then nips him on the arm. With peace restored, the mother soothes her infant by grooming him.

Humba focuses on his feeding, but now and then he stops and sits with a hand cupped under his chin like the silverback I saw on the Rwandan side. Humba, too, seems deep in thought, and the many moments of introverted contemplation I witness among the mountain gorillas are in striking contrast to the behavior of the hyperactive chimpanzees.

From the sidelines we watch their mostly peaceful everyday life, the silverback and his adult females and youngsters eating, playing, and dozing in their uncertain paradise. Every 10 or 15 minutes, Humba knuckle-walks farther up the slope in search of tasty plants, herbs, and thistles. The adult females dutifully follow him with their infants, fluffy bundles of black fur, riding on their backs like tiny jockeys. With Nyakamye watching us, the gorillas lope on all fours up the slope, and then he turns and follows.

Humba chooses the next place to feed, a wall of vine eaves dangling from a line of trees, which he rips down with gusto. He tears off the bark with his teeth and crams it into his mouth. A female decides tree leaves are more flavorsome and climbs up a tree with a brawny arm-over-arm method that has her rolling from side to side, her potbelly comically bumping against the trunk. Her infant clings to the fur on her back.

Once again the family grunts, grumbles, and chuckles their way through the snack.

Ten yards from the tree, Jean-Marie calls me over and points to the ground. "A snare," he says. At first I see nothing, but then I spy a length of wire running along a branch bent like a bow close to the ground before disappearing into the soft earth. He prods the ground with a stick and pulls back as the wire encircles the stick with a snap, and flings it into the air. He digs at the earth with his machete and shows me a shallow pit about the size of my palm, covered with slivers of bamboo.

None of Humba's family has been caught by snares, but Jean-Marie says that gorillas in other families on these slopes have been snared. Glancing back at the female who has been joined by two blackbacks in the tree, I recall Fossey's comment that some gorillas caught in snares manage for some time with their disability but eventually fail to cope and disappear forever. They did not have the care of the Mountain Gorilla Veterinary Project then, but because the situation is so dangerous at present, the veterinarians do not visit the mountain gorillas on the Congo side of the border.

Had the female nearby been caught in the snare, the infant would have perished with her. Had it happened to one of the blackbacks, then he might not have lived long enough to become a silverback. And had it been Humba who fell victim to the snare, the family might have panicked and fallen apart.

I leave reluctantly, turning my head back for a last glimpse of the gorillas. Coming down, thankfully, is much easier than going up the steep slope, though the strain on my aching knees is greater. As we near the perimeter of the national park, through the jungle I hear voices and glimpse dark faces and camouflage cloth through gaps in the foliage. We turn a corner and a few yards on, bump into a band of about 40 troops clad in camouflage uniforms and brandishing assault rifles, rocket-propelled grenades, and machine guns. Bandoliers of bullets are strung about their chests, Rambo style. They have no identifying patches on their uniforms, and all have the tall, slim, long-faced build of Tutsi.

They seem to be as surprised to see us as we are to see them, gripping their assault rifles at the ready with their fingers on the triggers. Emmanuel's forehead furrows with concern as Jean-Marie murmurs reassuring words to them in Kinyarwanda, the lingua franca on both sides of the border. "They're Rwandan army troops, Jean-Marie tells me," whispers Emmanuel. "They've illegally crossed into the Congo, so don't take any pictures or they'll probably shoot you and me too."

Emmanuel and I stand powerless as Jean-Marie talks but gets little response from the soldiers. To end the standoff I stride across to the most important-looking soldier, a lofty, lean fellow holding a machine gun, with my hand outstretched in friendship. He stares back at me with dark, edgy eyes. I offer a handshake. He seems unsure what to do and fingers the butt of his machine gun. My wide smile and "G'day mate, howya goin'" visibly softens his wariness. The Aussie greeting must surely be impossible to translate fully into Kinyarwanda, but he seems to sense the warmth in my voice. The soldier breaks into a cautious smile. As he shakes my hand, Emmanuel says, "We'd better leave before anything bad happens."

About 10 minutes later, as we climb over the lava rock barrier on the way back to Bukima, I notice that the fields are empty; the hard-working villagers who filled them on our way into the park are gone. Experience has taught them that when rebels or Rwandan soldiers are nearby, the only safe action is to swiftly make yourself scarce.

We visit Humba and his family one more time, the next day, and something remarkable happens. Just as we are about to leave, the silverback stuffs his mouth with leaves without eating them, a mountain gorilla sign of peace, and then without warning, ambles to within a yard of me. With the leaves still in his mouth, he pauses to stare meaningfully into my eyes for a few moments. It is too late for me to retreat. Perhaps my imagination is given free rein by my delight at being so close to this enormous and powerful cousin of ours, but I get the sense that Humba seems to be saying that he intends no harm to us, and wishes that we do not harm him and his family. Then he disappears into the undergrowth.

As we walk back to camp, I ponder the sublime feeling of peacefulness I experience when I am with the mountain gorillas. Their habitat is one of the most dangerous places I have been to in a career of journeying into remote and dangerous places, and yet I feel no fear when I am with them. Even after encountering the soldiers in the jungle, and the threat of running into Hutu or Mai Mai rebels, I am neither frightened nor deterred from going back. My usual caution seems somehow to have been swept aside by the privilege and thrill of being so close to these behemoths.

A few hours later at our camp by the Bukima patrol post, Emmanuel and I are still talking about what we saw during our visit to Humba and his family when we are visited again by the Congolese army major and his bodyguards. He heard somehow about our encounter with the soldiers in the jungle and wants to check what happened. After he leaves,

Emmanuel decides to return to Goma before we are confronted with even more serious safety problems.

The Congo mountain gorillas' future largely depends on a significant improvement in the security of their territory, especially the crushing of the charcoal trade and the defeat or disbandment of the rebel militias roaming their highland jungles. But the prospects do not look good. I return to the UN peacekeeping troops' headquarters in Goma, where Brigadier General Behl tells me he is not optimistic about a quick resolution. "It's a very difficult task for the DRC [Democratic Republic of Congo] government," he says with a frown. "It will take a long time before they can bring all these groups back into the mainstream."

I describe my encounter with the soldiers in the jungle, adding that they wore no identifying arm patches. "Even though you thought they were Rwandan army soldiers, it looks to me more likely they were Nkunda's rebels because they're based in that area, and they were wearing new combat uniforms that had no patches," he says. "You are probably the first foreigners they've ever met in the field, and so their response could have gone either way. You were very lucky to get away alive."

The senior warden of the national park's southern sector, Paulin Ngobobo, has just returned to Goma from talks with senior government officials in Kinshasa, and we meet in Rob Muir's garden. Paulin, a middle-aged, broad-faced, soft-spoken man, is clad in a ranger's green cotton uniform, web belt, and snappy beret. He tells me that the DRC president, Joseph Kabila, has vowed to protect the mountain gorillas. He then offers an eloquent shrug and says, "The problem is that after two civil wars our country is very poor, and we need outside support to save them."

I ask whether Kabila is truly interested in saving the mountain gorillas, or is it just talk? He surely has enough problems simply with remaining in power. Paulin shrugs again. Rob, who is with us, suggests that for all the evil ways of the notorious former President Mobutu, stealing billions of dollars from the national treasury and killing and jailing numerous political enemies, he was a savior of the Congo mountain gorillas: "He hardly knew they existed, and so conservationist groups in Kinshasa devised a plan to bring their plight to his attention. They put up billboards all over Kinshasa thanking Mobutu for saving the mountain gorillas in the far-off Virungas. Mobutu's flunkeys, seeing these billboards, began praising him to his face for saving the mountain gorillas, and the tame media lavished more praise on him. This appealed to Mobutu's giant ego. After that, he'd fly to Goma in his personal jet two or three times a year bringing suitcases full of cash to pay the rangers' wages, and bringing them new uniforms and equipment. That all ended when he was overthrown."

Paulin smiles wryly at the story. "Life's tougher for us now," he says.

This reliance on Western conservation groups to protect and conserve their great apes in the wild is common among African rulers. Billions of dollars flow into their treasuries from oil, gold, diamonds, forestry, and other natural resources, but a considerable amount of it exits through the back door into the politicians' bank accounts abroad. In its latest annual report on worldwide government corruption, Transparency International states that of the 10 most corrupt nations in the world, six are in Africa and one is the Democratic Republic of Congo.

This massive robbery of Africa's assets has held back the continent's advancement for decades. In his first trip to Africa as president in July 2009, Barack Obama deliberately visited just Ghana, one of the very few relatively uncorrupt and democratic countries on the continent. In a challenging speech reported worldwide, he told Africans across the continent, "No country is going to create wealth if its leaders exploit the economy to enrich themselves. . . . No one wants to live in a society where the rule of law gives way to brutality and bribery. That is not democracy, that is tyranny and now is the time for it to end."

When Paulin became warden of the southern sector a few years ago, he began to investigate the charcoal trade, aware that it was destroying the park and threatening the gorillas, and found that it was run by the Hutu rebels, arm in arm with the Congolese army. The Hutu were using the money to buy weapons and entrench themselves and their families within the park, knowing they could never return to Rwanda.

Back on the Rwandan side of the border at Ruhengeri, I meet again with veterinarian Lucy Spelman, head of the Mountain Gorilla Veterinary Project. She tells me the subspecies is on the road to recovery despite the threat to the Congo mountain gorillas. There are just over

100 on the Congo side of the Virungas, 270 on the Rwandan side, and 350 in nearby Uganda. "The most recent survey shows the numbers have increased by seventeen percent in Rwanda and Uganda since 1989," she says.

Lucy has combined with the Dian Fossey Fund's Karisoke Research Center and the Rwandan wildlife service to attempt a world-first—introducing a mountain gorilla orphan successfully back into the wild. On previous attempts, the orphans perished after they were rejected by their intended foster families. Mountain gorillas usually do not accept stranger infants into their fold.

Two decades ago, Dian Fossey attempted to introduce back into the wild an orphan baby gorilla named Bonne Annee (Happy New Year), a female seized from poachers who wanted to sell her. The gorilla family chosen had six adult females with their own infants and they rejected the orphan. A few weeks later, Fossey again took her up the mountains and introduced her to the same family. This time they accepted her, but six months later she died of malnutrition.

It helps if the little one is from the same family. In 2002, when poachers shot and killed a female, her 13-month-old baby clung to her, but rangers rescued the infant. After medical treatment the rangers returned her to the group. Her kin surrounded her in agitation until the silverback inspected the infant and then moved away. None of the females adopted her, but her 10-year-old brother recognized her and decided to take care of her. Observers have since seen her snuggled between her brother and the silverback on cold nights.

On Ruhengeri's outskirts, Maisha, a four-year-old mountain gorilla, lives with seven little lowland gorilla orphans in a quarantined, open grassy compound surrounded by high walls. Maisha's rescuers did not know which group she came from, and she had been kept hidden by the poachers for several months before she was confiscated. "This time we're trying something different," Lucy tells me.

Only the orphans' keepers and the vets are allowed inside so as to protect them against human disease. If Maisha catches something as simple as a cold from a human, she can never be taken back into the wild for fear of spreading an illness that could devastate the mountain gorilla population. "We undergo strict regular health checkups here, and when we travel overseas we must wait three weeks after our return until we can enter the compound," Lucy explains.

I watch from the roof of Lucy's SUV by the compound wall as Maisha, a furry bundle of energy, swings from the jungle-gym tangle of ropes and plays whack-your-head with another little orphan. Their keeper, Simon Childs, grabs Maisha and plays gorilla roughhouse with her, then swings her onto his back, piggyback style, just like her mother would in the wild. Her mother was probably killed by poachers when they snatched Maisha on the Congo side of the Virungas near Mikeno two years earlier.

"They kept her in a sack in a cave for several months while they tried to sell her," Simon tells me. A big, bluff Englishman, he learned gorilla behavior when he spent a decade as a keeper at a gorilla park in Yorkshire. "Maisha was deeply traumatized when rangers rescued her and brought her to us. It took us months of round-the-clock companionship with gorilla trackers who knew how to imitate gorilla vocalizations and behavior to bring her back to normal."

Lucy will keep Maisha with the other orphans so she does not forget gorilla behavior until she is about six years old and has begun her estrus cycle. "Then we'll take her back to the wild and hope a silverback will accept her into his family to mate with him," she says. That seems a lot of effort for just one little gorilla, but there are so few remaining. Lucy reminds me that "their gene pool is very limited, and so each mountain gorilla is precious."

As I depart Ruhengeri for the long journey home, I turn for a last lingering look at the Virungas, shimmering like translucent blue glass in the misty air. On the other side of those steep slopes, Humba, Rugendo, and their families are playing, mating, caring for their young, or sleeping off a hefty snack.

The relative calm does not last long. I receive shocking news just weeks later: four mountain gorillas in the Congo have been shot dead by unknown assailants. As details leak out, I am horrified to learn that the dead were the mountain gorillas I had visited: Rugendo and three adult females from his group—Neeza, Mburanume, and Safari. Two weeks later, the remains of the group's last adult female are found; her infant is presumed dead. It is the worst massacre of mountain gorillas in more than 25 years. Jean-Marie led rangers to track down six survivors, including feisty little Noel and his two brothers, Mukunda and Kongomani, who were caring for Safari's surviving infant.

Rugendo had a gentle nature, allowing me to approach close to him while he ate leaves as his children played nearby. He was so trusting of humans that he even fell asleep in front of me. The villagers and rangers who knew Rugendo clearly respected him. About 70 villagers carried the mountain gorillas' massive bodies from the forest to bury them near the rangers' headquarters. A photograph in *Newsweek* showed Rugendo roped onto a makeshift bamboo stretcher as the villagers carried him through a cornfield. In a moving tribute, the villagers had placed leaves in Rugendo's mouth, a sign in gorilla language that he wanted peace.

While habituation of the mountain gorillas on the other side of the Virunga volcanoes in Rwanda through gorilla tourism has saved them, it is a different story in the Congo. Habituation allowed the killers to approach Rugendo and his family without prompting them to flee, as mountain gorillas unused to human presence would have done. The bodies were not stripped of flesh, but left untouched or burned where they had been murdered. Clearly, the killers were not poachers. As I look at the tragic pictures of the bodies of Rugendo and his four females, I cannot help wondering whether they might still be alive if we had just left them alone.

7

Months after the deaths of Rugendo and his females, I return to Goma to see what happened to the surviving members of his family, including little Noel and his two big brothers. On an overnight stopover in Kigali, I stay once again at the Mille Collines hotel, and open up the local English newspaper at breakfast to see the ruthless president Paul Kagame still dominating the front page.

The driver taking me through the mountains to Ruhengeri wants only to be known as Peter, a false name to protect himself. This is not yet a country where you can openly criticize Kagame and expect life to go on as usual. A number of local journalists have tried and found themselves in jail on trumped-up charges. Peter tells me his mother is

Hutu and his father is Tutsi, and so he claims to understand the hidden but widespread ethnic tensions much better than his father's people.

"What do you expect?" he asks me. "Of course the Hutu are unhappy because they are more than eighty percent of the population, and yet the Tutsi have all the power. If there were a free vote, of course the Hutu would vote one of their own into power as president. Yet in the last presidential election, Kagame got ninety-four percent of the vote. So democracy in Rwanda is a fraud. The election of Kagame was staged, and yet the Americans hail him as a democratic hero."

An hour into the journey, I spot the creative way in which Rwandans are solving the problem of the huge number of prisoners who have yet to be tried for genocide 14 years after it happened. As we near a small roadside village, I see three men in pink shorts and pink tops, prisoners' garb, in a clearing in front of a huddle of mud huts. Two are seated on a bench while one stands addressing three elders, a woman and two men. The elders sit on another bench and are clad in formal clothes and wearing sashes. The other villagers are seated around them on benches, listening intently to the prisoner.

"It's a *gacaca*," Peter says.

I ask him to stop so we can see what is going on. The *gacaca* village courts are at the heart of Rwanda's traditional system of community justice, and are probably the only way of dealing efficiently with the clogged judicial aftermath of the genocide. According to the New York–based Human Rights Watch, in 2006 the *gacaca* courts tried 10,000 alleged killers. Another 50,000 had already been tried and released in 2004 and 2005.

The three sash-wearing village elders hearing the cases are called *Inyangamugayo* (the uncorrupted) and are given basic legal training before they judge cases. We are close enough to hear, and Peter tells me the suspect is admitting that he was in a mob that killed his neighbors, Tutsi men, women, and children, in the village. He now asks for forgiveness. "Because he has owned up and begged forgiveness, it's likely he'll be set free today after fourteen years in prison. Then, he'll live side by side with the families of the Tutsi he killed. This does cause tension in the village—imagine seeing every day the man who hacked to death your parents or your wife and children. But in a small country like ours with limited resources, it's the only way."

The *gacaca* will go on for some more hours, and so we leave. My heart beats faster when we reach Ruhengeri and I see the high, blue shadowy shapes of the Virunga volcanoes through the mountain mist. They hem in the city like the ramparts of a giant fort. On the other side of the slopes, in the Congo, are Rugendo's murderers.

I visit Lucy Spelman again and ask her about the fate of Noel, his two brothers, and the other juveniles in the Rugendo family. "It was clear they intended to make a statement, a warning, and also break up the family," Lucy says. "With the silverback and the adult females murdered, the others would usually disperse and join other groups, but the little ones like Noel might not survive. Happily, Mukunda, the ten-year-old blackback, seeing the need, promoted himself to be the family head four or five years before he would become a silverback. We think they're now okay, though Nkunda's troops control the area and so we can't yet visit them."

"What about Rugendo's infant daughter, found by the rangers at the murder site clinging to the body of Safari, her dead mother?"

"Without her mother she would have died, and so we brought her to Goma and she's under quarantine at present. I hope she can join Maisha, the young mountain gorilla female you saw here with the or-phaned lowland gorillas, but it's a sensitive question. The Congolese authorities want her to remain over there."

Rob Muir is in Kinshasa consulting with the government when I cross the border into Goma the next day. On the Rwandan side it looks like an African idyll, fine old mansions settled on hills overlooking a placid Lake Kivu. But just a few hundred yards beyond the immigration barri-ers, I am confronted once again by the chaos and confusion of Goma. I check in at a lakeside hotel rumored to be the haunt of Congo govern-ment spies. At lunch overlooking the lake, a trio of hard-eyed, bulky men in suits that seem at the point of bursting their seams hunch over their steak Diane and red wine.

The hotel's one suite is reached by grandiose white marble steps that sweep up to a balcony and a door with a fake-gold handle. The suite is occupied by a senior general from Kinshasa, and his dozen guards in flashy new combat uniforms strut around the hotel entrance brandish-ing their assault rifles and machine guns at passersby. This is the brave

Congolese army that harasses powerless citizens but scurries away like frightened rats time after time when confronted by Nkunda's jungle-toughened rebels.

When I go for a walk, the soldiers at the marble steps wave me on with their weapons. Suddenly, a stocky soldier draped in bandoliers of bullets strides toward me. He shakes a long-barreled machine gun inches from my face and screams something in French. Up close his eyes blaze with anger, and I back away. Moments later a new white SUV roars down the muddy road and swings into the hotel driveway, stopping right where I had been standing. An open-backed pickup carrying half a dozen soldiers armed with machine guns and rocket-propelled grenades pulls in behind it. The soldiers jump out and surround a senior army officer as he steps out of the SUV and walks up the steps into the suite. "The general is staying here," a porter tells me.

A tall, heavy-set middle-aged man clad in a new tailored green uniform, he looks like a rich thug playing at being a soldier. I smile wryly at the grand display as he disappears into the suite because while the general may have come to pick up his share of the charcoal spoils, and may be here to indulge in the intrigue of Congo politics, he is surely not in Goma to lead a fight against Nkunda's rebels who occupy the hills just two hours' drive from here.

Rob arrives the next day, and at mid-morning I visit him at his compound guarded by high walls, a secure lock on the gate, a security man, and a big, fierce dog that roams the grounds. He tells me that Paulin Ngobobo, the chief warden of the Mikeno sector of the Virunga National Park, has moved to Kinshasa for his own protection, explaining, "He was too effective against the charcoal traders, and threatened to put them out of business. They in turn threatened to kill him, and it was no idle boast."

A few weeks earlier Paulin and seven rangers had been attacked by Hutu troops and Congolese soldiers, because they were attempting to stop the charcoal trade inside the national park. They escaped by hiding in the forest.

Paulin lodged a complaint with the Congolese army but was arrested. He told *National Geographic* what happened next: "It was raining, the colonel's bodyguard took me outside and stripped me. The colonel had spoken to Honore Mashagiro (director of the Virunga National Park), and Mashagiro had said I was undisciplined and needed correction, and I must be given 75 lashes."

He had come to believe that his own boss, Mashagiro, was the kingpin of the illegal charcoal trade when he uncovered fake books, faked records, protection schemes, payoffs, and charcoal taxes. Paulin went on: "The guard removed my jacket and my belt and boots and made me lie facedown in the mud. He counted one, two, three as he was whipping me."

The whipping did not deter Paulin's efforts to destroy the illegal charcoal market. He arrested six senior rangers on suspicion of being involved, but Mashagiro had them released and sent back to work. Days later, a female gorilla was executed, Mafia style, with a bullet to the back of her head. To let the killer get so close, she must have been familiar with the person who shot her. Suspicion fell on the rangers. Then, three months later, Rugendo and his adult females were shot dead.

"It was a clear warning from those involved in the charcoal trade to stop harassing them or more gorillas could be killed," Rob Muir now says. "Paulin was sent to jail, forced to remain in a darkened cell without talking or moving all day, but he was released and is safe now."

Nkunda's troops now occupy the mountain gorillas' habitat, including the Bukima patrol post. They forced out all the rangers and have threatened to execute any ranger that they find back with the gorillas. Nkunda must be aware of the river of gold that the mountain gorillas are for his Tutsi kin across the border in Rwanda, and he may hope that in any peace settlement he will retain control of the mountain gorilla habitat in the Congo.

The next morning Rob drives me along a familiar road, through the scrambled traffic of the town and out past the muddy paddocks within sight of the volcanoes. Here, the refugee camps with heart-weary certainty are filling up once more with tens of thousands of villagers fleeing from renewed fighting between all the warring factions who rape, pillage, and murder villagers when they can. Human-rights investigators found that at Kwanja, Nkunda's forces executed 150 people and torched refugee camps holding 30,000 people. They forcibly conscripted boys aged 13 to 15 and gave them training, Kalashnikov rifles, and front-line positions.

Looming over the fields is the imposing peak of the Mikeno volcano, and I feel a pang of sadness, knowing the mountain gorillas up there are now under the control of Nkunda's rebels. The Humba clan's brave blackback sentry, Nyakamye, may be able to protect his family from feeble humans like myself, but he and the other gorillas are no match for assault rifles.

We pass by the turnoff to the steep track that leads uphill to the Bukima patrol post. "We believe Nkunda has mined its entire length, and it's now too dangerous to go up there," Rob says.

Not long after, I spy a tall, slim soldier clad in camouflage uniform with an assault rifle slung over a shoulder and carrying a spear. His long, thin face marks him as a Tutsi. I return his wave. "That's one of Nkunda's rebels," Rob says. "As Tutsi they're proud of their long tradition as warriors. Many carry a spear into battle as a Tutsi symbol, as well as an assault rifle."

He points to the hills sloping down to the road on the right. "That's Nkunda's territory, and on the other side of the road is Hutu territory. You can see how close the battle lines are drawn."

About 30 minutes later, and still in the divided Nkunda and Hutu territories, our SUV breaks down. Within moments we are surrounded by children begging for money. They seem well fed, and so I resist the temptation to give them any. Laughing and smiling, they run and skip and walk back to their roadside village. Rob assures me that we are unlikely to be attacked here, either by Nkunda's men or the Hutu, because a ceasefire is in effect.

He radios for a backup vehicle, but SUVs are scarce around here and we wait for three hours, stuck in no man's land. Another SUV finally arrives from the rangers' base, about 30 minutes away. The base is set on a hillside overlooking a lush valley, and for the time being the rangers and their families are housed in tents. Jean-Marie is away on patrol, searching for illegal charcoal, but the handful of rangers I see are all spick-and-span in freshly laundered uniforms. "It's important they keep up their morale while we sort out the military problems, and that means coming to work each day in uniforms even though some haven't much to do," says Rob.

An old colonial mansion by the tents serves as the park rangers' headquarters. In a cavernous room, I speak with a ranger who was trekking daily to Rugendo and his family at the time of the killings. He does not want his name mentioned. I have read reports of the killings, but this is the first time I have spoken to a ranger who investigated the atrocity. He tells me, "At the patrol post, we heard shots in the night, and at daylight went to investigate. We found Rugendo's females shot dead, and the rest of the family scattered. The next day we found Rugendo's body. He'd been shot through the heart. Noel and his two brothers survived." He confirms Lucy's news that Mukunda, the 10-year-old blackback, had taken over as head of the family.

When I ask whether he knows who had killed the gorillas, his dark eyes turn wary. "We know, but we can't tell. Their leader is a high person." Unstated but understood is his fear that he and his family will suffer if he gives me their names.

On the way back, we pass an open-backed truck pulled to the side of the road. Three soldiers point their rifles at about 20 villagers unloading vegetables from the truck. Stacked nearby are about 40 sacks of charcoal, each worth $25. "The soldiers have stopped the truck and commandeered it to take the charcoal to town," Rob says. "You've seen the roadblocks manned by our rangers, who have the authority to seize any charcoal they find. But when soldiers are driving the trucks and guarding them, our rangers dare not stop them."

I leave Goma with a sense of hopelessness, but a few weeks later I learn that Honore Mashagiro, the park's director, has just been arrested and imprisoned in Goma for trafficking in charcoal. The police allege Mashagiro had stashed away hundreds of thousands of dollars from the trade. Under questioning, it emerged that he masterminded the murder of the mountain gorillas in 2007 as a warning to the park rangers, his underlings, not to hinder the charcoal trade. At about the same time, rangers seized 100 tons of charcoal being carried out of the park.

Soon after, the Congolese Wildlife Authority appointed Emmanuel de Merode, my companion on the trek to Humba's family, as the new director of the Virunga National Park. The park's 680 rangers are under his control. It was, the authority declared, "a move to strengthen the rule of law, step up anti-poaching, and prevent forest destruction for charcoal in Africa's oldest national park."

Emmanuel well understood what he was facing. He said, "The intensity of the conflict in and around the park makes this a daunting challenge, but it is a great privilege to be working with a dedicated and courageous team of rangers."

He moved swiftly to exert his control. Just a month after his appointment, he negotiated the withdrawal of 1,000 Congolese army troops from inside the park. They had been living there with 5,000 family members. "De-militarizing the Virunga National Park remains our greatest and most difficult challenge," he explained. "The Congolese National Army has taken the first step, which represents a major breakthrough at a time when the threats to the park have never been greater."

The mountain gorillas in Rwanda and Uganda are safe for now, their survival protected by political stability in those countries, and by the high price charged the thousands of foreign tourists to visit them in their lofty forests each year. But the Congo mountain gorillas' safety continues to be precarious. As I write this, General Nkunda has fled across the border to Rwanda, where he is said to be under benign house arrest. The rangers have returned to the gorilla sector, but ferocious battles continue between the rebels, the government troops, and their Hutu allies in the surrounding hills.

The first news from the rangers has been encouraging. Against all expectations, little Noel and his brothers, Kongomandi and Mukunda, have survived together and two silverbacks are competing to become their family patriarch.

It is time to leave the mountain gorillas and journey to neighboring Uganda to be among the chimpanzees in the wild. The contrast between the ways of mountain gorillas and chimpanzees is troubling, as I shall soon discover.

The Wild, Wild Chimpanzees

1

In the dark, dank silence of a remote East African jungle, a gang of male wild chimpanzees is on the prowl. Using narrow paths trampled into the turf by tiny antelopes, the five hulking chimps patrol their territory's boundaries. The chimps pay no heed to me and senior field assistant Francis Mugurusi as they move silently in single file on all fours through the jungle, hour after hour, their cold, hard eyes alert, muscles tensed, swarthy black hair on end. Above us soars the canopy, the place where chimpanzees in the wild spend most of their lives, feeding, sleeping, and socializing. But once or twice a month, the senior males head out on a boundary patrol of their territory, about five square miles of jungle.

Francis tells me that the chimpanzees carry out their patrols in silence, but at mid-afternoon Imoso, the biggest male, who is leading the party, suddenly breaks into frenzied screams, what primatologists call "pant hoots." The others follow his lead so that the jungle rings with their cries. He dashes forward and slams his foot repeatedly against the buttresses of a big fig tree, producing a loud thumping sound that carries a long distance across the jungle. Again, the others follow his lead. The chimpanzees then sit on the ground, listening for any return sound. None comes. After a few minutes they resume their patrol.

We keep a distance of a few yards because an adult chimpanzee is about six times stronger than an adult male human, and these creatures could tear us apart if they wished. As we follow, Francis tells me that Imoso is the dominant male in a clan of 46 chimpanzees here at Kibale Forest National Park in Uganda studied by Harvard professor Richard Wrangham for the past two decades.

Wrangham's observations of chimpanzee behavior in the clan called Kanyawara played a crucial role in a seminal and highly influential book he co-wrote called *Demonic Males: Apes and the Origins of Human Violence*. It explored for the first time in detail similarities in extreme violence among male humans and male chimpanzees, behavior that may have been buried deep within our respective genes for millions of years. The book opened an evolutionary window on men's compulsions through the ages. In most human cultures, however different, men typically kill other humans at a rate 20 times more often than women.

In *Demonic Males*, Wrangham and his co-author wrote: "Kibale is providing the latest evidence that lethal violence is characteristic of chimpanzee males across Africa." His study witnessed the first documented attack by a chimpanzee using a stick to bash another chimpanzee. It happened in 2002, and there have been many more such attacks witnessed since at Kibale. Stanley Kubrick foresaw this discovery, the use of weapons by great apes, in his epic film *2001: A Space Odyssey,* in which chimpanzee-like apes using the jawbones of zebras battered a rival clan to death.

"The chimpanzees make the noises now and then to listen for a return of the same sounds by other chimps," says Francis as we continue to follow the patrol now on its way home. "If they hear pant hoots and foot thumping they'll investigate. If it's their own group, they'll greet each other with embraces and kisses and then groom each other's fur. But, if it's strange chimps, they'll attack them and try to drive them out of their territory. If it's just one chimp they'll try to kill him, battering his body with their fists and feet. I've seen this happen."

This is far removed from the Hollywood/circus image of cute, lovable chimpanzees, such as Tarzan's Checta. These ape "actors" tend to be infants or adolescents who have yet to reach their threatening adult bulk.

In their homicidal ways, the chimpanzees clearly resemble us, their closest great ape relatives. We share 98.4 percent of DNA, which makes chimpanzees closer biologically to us than they are to the much less violent gorillas and orangutans. But whereas humans are swarming across Africa, chimpanzees, like all the great apes, are suffering a disastrous drop in numbers in their native habitats. They are the most adaptable of the great apes and more than 2 million chimpanzees inhabited sub-Saharan jungles, swamps, montane forests, and savannah at the beginning of the twentieth century. But their numbers have plunged to about 250,000, with up to 10,000 disappearing each year.

The International Union for Conservation of Nature, which monitors endangered species worldwide, has tagged chimpanzees as endangered. It said they are "facing a very high risk of extinction in the wild in the near future."

The UN-sponsored *World Atlas of Great Apes and Their Conservation* explains that "most chimpanzees live outside protected areas, where they are vulnerable to disturbance of their forest habitat by logging; to habitat destruction by settlement, fire and farming; and to hunting that supplies the increasingly entrenched and powerful bushmeat trade." It warns: "Meanwhile, their fragmented populations are becoming increasingly subject to disease outbreaks as they come more often into contact with people."

This is not an alarmist assessment. In the Ivory Coast, the western African chimpanzee numbers have plummeted because of bushmeat hunting and deforestation of their habitat. Since 1989, the human population of the Ivory Coast has increased by 50 percent to just under 18 million, to the enormous detriment of the chimpanzees there. A survey of the Ivory Coast chimpanzees in 1988/1989 found between 8,000 and 12,000, already a dramatic decline in numbers from the beginning of the twentieth century. Two decades later, a new survey found that their numbers had plummeted by 90 percent, to between 800 and 1,200 in the wild. That total is more than half the remaining world population of the subspecies, which looks doomed to extinction in the wild unless urgent action is taken. And urgent action seems unlikely.

According to Chicago Lincoln Park Zoo primatologist Elizabeth Lonsdorf, a major limiting factor to their survival is that chimpanzees are relatively slow to reproduce in the wild: "Chimpanzees have their first offspring at 13 years on average, with a three to five year interbirth interval." In chimpanzees, "the long life spans, slow reproductive rates and behaviorally complex societies result in populations that are particularly vulnerable to declines and less likely to recover from them."

Chimpanzees have been given the amusing scientific name *Pan troglodytes* (cave-dwelling gods of nature), and their forest, savannah, and grassland habitats range across Western, Eastern, and Central Africa, encompassing 21 countries, from Senegal in the west to Tanzania in the east. There are four subspecies—the eastern chimpanzee, the central African chimp, the western chimpanzee, and the very rare Nigeria-Cameroon subspecies.

Though Africans have shared their territory with these apes for thousands of years, the first recorded contact between chimpanzees and Europeans took place early in the sixteenth century. Portuguese explorer Duarte Pacheco Pereira observed chimpanzees in the wild in 1506. In 1640, the first known chimpanzee to reach European shores was presented to Frederick, Prince of Orange, in the Netherlands for his amusement.

In Charles Darwin's book *The Descent of Man,* published in 1871, the father of the theory of evolution suggested that we humans had evolved from a common apelike ancestor. In Dayton, Tennessee, in 1925, famed politician and lawyer William Jennings Bryan, a believer in divine creation, was counsel for the prosecution in the famous "Monkey Trial" of a young biology teacher, John Scopes, in an unsuccessful attempt to stop him from teaching the theory of evolution in the state's schools. Key to the prosecution's case was the "unacceptable" belief that man was descended from apes. Newspaper cartoonists at the time depicted our alleged common ancestor as a bestial kind of chimpanzee.

The more we get to know about chimpanzees, the harder it is to deny our similarities. According to the Jane Goodall Institute, humans and chimpanzees shared a common ancestor up until 8 million years ago. The chimps' life span in the wild is about 45 years, though in captivity in zoos, chimps have lived up to 60 years.

Many shared traits show that the biological link between humans and chimpanzees (as well as the bonobos or pygmy chimpanzees) is stronger than humans and chimpanzees' link with gorillas and orangutans. The chimpanzee brain is closer to our own than that of any other living creature. Their impressively documented intelligent use of tools; cooperative hunting; complex, volatile, and hierarchical social life; cultures that differ from place to place; a primitive kind of language, and the males' murderous violence aimed at enemy chimps point to a connection between the chimpanzees and humans and imply a common ancestor.

Chimpanzee society is aptly described as male-bonded and male-dominated. Chimpanzees live in communities of about 20 to 150 individuals, all sharing a home range that the adult males defend with extreme violence, sometimes to the death. A male chimpanzee can weigh up to 132 pounds and can stand five feet six inches tall. Males and females embrace and kiss, show affection toward each other, playfully tickle each other, develop lifelong relationships, hold lifelong grudges, and mourn their deceased. When one of Jane Goodall's best-known study chimpanzees, Flo, died in 1972, her devoted son, Flint, was so overcome with grief and depression that he fell sick and also died.

Chimpanzees make crude tools, and although they commonly knuckle-walk on all fours, they can walk upright. They are also self-aware, recognizing themselves in mirrors. Researchers in what is called The Mark Test have anesthetized chimpanzees in laboratories and fixed a red dot that could not be felt and was odorless to an upper eyebrow ridge and the ear opposite. On waking, when the chimpanzees looked into a mirror they attempted to remove the marks. They also used the mirror to explore themselves, making exaggerated facial expressions at it and using their fingers in association with the mirror to take a close look at their eyes, teeth, and anal-genital area.

This self-awareness is remarkable, because not all humans identify themselves when they see their own image for the very first time in a mirror. A violent Stone Age tribe in the Amazon called the Korubo ("the head-bashers") has virtually no contact with outsiders and lives in a rainforest the size of Portugal. The Brazilian government has quarantined the Korubo land, cutting off access by outsiders, so that the tribe remains undisturbed. Men, women, and children go about naked, live in a communal hut, and still make fire by rubbing sticks together.

I once spent a week with a Korubo clan, accompanying the then director of the Brazilian government's Remote Indians Department, Sydney Possuelo. I made friends with Shishu, husband of the tribe's chief, an Amazonian woman named Maya. Not long before I went there, on first contact with the tribe, Shishu was captured on film peering into a camera lens where he saw his own reflection. He pointed to it and asked Sydney, "Who is that little man in there?"

The chimpanzees are the most omnivorous of all the great apes and consume fruit, leaves, plants, flowers, eggs, and even the fresh meat of a large number of vertebrate species, most commonly red colobus monkeys, which make up about 80 percent of their mammal prey. They suffer many of the same diseases as we humans, such as malaria, cold sores, ringworm, influenza, tuberculosis, heart problems, hepatitis, dysentery, arthritis, and AIDS.

Females reach reproductive age at about 13, males at 16, and the females commonly give birth to a single infant at a time. It is dependent on its mother until about five years of age. They even benefit or suffer from hierarchical social structures headed by a dominant male who shows no reluctance to use extreme violence on his own, or helped by allies, when his position is threatened. So, humans share more traits with chimpanzees than with any other creature, though thankfully we are a lot less hairy.

As far back as the fifth century A.D., Saint Augustine was aware of this similarity. He wrote: "If we did not know that apes and long tail monkeys . . . are not human beings but beasts, those same natural historians who pride in curious lore might with unscathed vanity foist them upon us as diverse distinct tribes of men."

Our closeness to chimpanzees even allows a human with the same blood match as a chimpanzee to receive a blood transfusion from it and vice versa. The transfusion can be carried out only once, because then the human and chimpanzee develop antibodies that prevent a second transfusion. In contrast, orangutans and gorillas cannot give blood transfusions to humans or chimpanzees.

But this closeness has caused thousands of captive chimpanzees great suffering. They have been used extensively in medical research as human substitutes, often held individually for many years in small concrete cells with no windows and subjected daily to intrusive, painful medical procedures. In the United States, according to Steve Ross of Chicago's Lincoln Park Zoo and chair of the Chimpanzee Species Survival Plan for U.S. zoos, about 1,100 chimpanzees are held in medical laboratories. He told me that there are presently 269 chimpanzees in zoos, as well as about 500 in sanctuaries and 250 privately owned.

In many countries such as my own, Australia, keeping great apes as pets is banned. However, according to Ross, "there are no federal laws prohibiting owning chimps as pets in the U.S. There are some states that do not allow chimpanzees to be pets . . . but they are few. I estimated that there are only six states that it is completely illegal . . . others have loopholes."

So, in 44 states anyone can own a big, brawny male chimpanzee and keep it at home. But chimpanzees are not meant to be pets. In 2009, a 200-pound adult male pet chimpanzee in Stamford, Connecticut, attacked a friend of his owner and almost tore off her face as well as biting off her hands. Four teams of doctors spent seven hours in surgery to stabilize the victim, and she may need a face transplant.

The chimpanzees' owner, widow Sandra Herold, raised Travis since he was young and was like a surrogate mother to him, living alone with him in her house and sharing a bed. They kissed and cuddled each other. Travis brushed Herold's hair each night. She dressed him in baseball shirts and took him for rides in her car. He knew the victim, but on the day of the attack she had on a blond wig, which may have confused him. "If there is another person entering his space, he might consider it a threat to his territory, or even his mate," says Stephen Rene Tello, executive director of Primarily Primates, a Texas chimpanzee sanctuary.

Richard Wrangham of *Demonic Males* fame has studied the true nature of male chimpanzees for more than two decades. He told me, "The Travis tragedy is a wake-up call. Chimpanzees, especially males, are intelligent, caring, and cooperative, but they are also inherently dangerous. It is crazy that people are allowed to keep them as pets in the U.S.A. It is time to change the law, for the sake of both humans and chimpanzees."

2

Richard Lynch Garner was the first Westerner to attempt seriously to observe the chimpanzees' behavior up close in the wild. To study what he called "the language" of chimpanzees and gorillas, he left New York by ship in 1892 and a few months later began his research in Gabon, on Africa's west coast. A picture in his subsequent book, *Gorillas and Chimpanzees* (1896), shows Garner looking the archetypal Western explorer of his day, clad in a topee, gaiters, boots, and paramilitary outfit, and lugging a shotgun for protection against the jungle's dangers.

To further safeguard himself from "serpents and wild beasts of divers kinds" and from "being devoured by leopards or panthers," Garner perched in the jungle for 112 days and nights in a purpose-built six-foot-six-inch square steel wire cage. Inside, he had a canvas bed, a folding chair, a stove, and cooking utensils.

Garner observed one chimpanzee who "plucks a bud of some kind, tears it apart with his fingers, smells it and throws it aside." However, up close, Garner was less charmed by the chimpanzee: "What a brutal visage! It has a scowl upon it, as if he were at odds with all his race."

Detecting 20 "speech" sounds made by chimpanzees, Garner claimed that he learned to use 10, including the chimp words for food and for an infant calling for its mother. He used this limited vocabulary to speak to stranger chimps and said they understood him. In the book, he set down the sounds in a dispassionate technical manner that, oddly enough, reminded me of Professor Henry Higgins coldly analyzing the cockney speech of Eliza Doolittle in the movie *My Fair Lady*.

Garner was followed in 1930 by another American, Henry Nissen, who went to what was then French Guinea to study chimpanzees in the wild but lasted only four months. Accompanied by a chattering team of porters, he had little success in getting close to the skittish wild chimpanzees.

Now we know considerably more about chimpanzees in the wild, and as a consequence have more clues to our own evolutionary history, because five decades ago a petite and energetic young Englishwoman, Jane Goodall, was seized by the desire to journey to Africa to work with animals. When she was two years old, her father bought her a lifelike toy chimpanzee named Jubilee, after the first captive-born chimp at the London Zoo. Goodall fell in love with the chimp toy, and to this day it still sits on a chair in her England home.

As a little girl she saved her pennies to purchase secondhand books about animals, and wrote down her own observations about the habits of animals and birds near her home in Bournemouth. When she was four, she upended a hen to find out where an egg came from. She even took earthworms to bed.

Goodall delighted in *The Jungle Book* and the Doctor Doolittle series, hoping that one day she, too, could talk to the animals, and she yearned to study them in the African wild. She loved the Tarzan movies and was quite sure he would prefer her to that other Jane. In 1957, at the age of 23, she took the first step toward her wish when she traveled to

Kenya. In Nairobi she worked as secretary to Louis Leakey, the paleo-anthropologist who headed the Coryndon Museum of Natural History in the Kenyan capital.

Because there was such a dearth of information about chimpanzee behavior in the wild, Leakey, aware of Goodall's desire, suggested she go out to where they lived and undertake a long-term study of mankind's closest relative. He told her that "details about the behavior of one of the most manlike creatures living today, in its natural state, may give us useful pointers as to the habits of prehistoric man himself."

Leakey would make a similar suggestion to Dian Fossey, to study mountain gorillas in the wild in their habitats, and Biruté Galdikas, to study the orangutan in Borneo. Leakey knew that all the great ape species were in danger of extinction and wanted to gather as much information as possible about the way they lived, as soon as possible.

In mid–1960, Goodall journeyed from Nairobi to Tanzania's Gombe Stream National Park, which is about half the size of Manhattan. There, she revolutionized great ape research by sacrificing personal comfort and risking personal safety to live beside and study groups of chimpanzees in the jungle. The park was a narrow band of 13 steeply sloping forested hills overlooking the shore of Lake Tanganyika.

Its forests were home to several groups of chimpanzees. Getting to know the wild chimpanzees was easier said than done, because unhabituated chimps almost always flee from humans the moment they spot them in their habitats. This is understandable, because for millennia African villagers have hunted them with spears, bows and arrows, nets, and snares for food. Their shamans used the chimpanzee skulls and other body parts in magic rituals.

This flee response was common in all chimpanzee research sites except for one, in a dense, uninhabited swampland in a remote Congo rainforest, the Goualougo Triangle, where it was likely that few if any humans had been before. As if it were an African Garden of Eden, the wild chimpanzees on first contact in 1999 allowed researchers to get close to them. Far from fleeing, the chimpanzees were intrigued by these strange creatures. Their most common response, in 218 encounters, was to move closer to get a better view of the researchers, to stare at them, slap tree trunks to gain their attention, and throw branches at them to see if they could get a response.

"Such an overwhelmingly curious response to the arrival of researchers had never been reported from another site," said one of the researchers, Crickette Sanz, an American primatologist.

Goodall was not so lucky, and at first she could only observe chimpanzees from a distance through binoculars. "They'd never seen a white ape before and were horrified," she recalled. But her motto was drummed into her when she was young by her mother—"Never give up"—and so she persevered.

When she tried to get near, the chimpanzees filled the air with their terrifying fear vocalization, a loud, wailing *wraaa*, and fled through the trees. She persisted, but it took several months at Gombe for the chimpanzees to trust her and allow her to come near them. To tempt them closer she resorted to a technique known as "provisioning," or placing food dumps out in the field or in unlatched boxes to draw the chimpanzees near and accustom them to her presence. It worked.

In an early attempt to attract the chimpanzees at Gombe, Goodall held up a large stalk full of ripe bananas in view of a pair of big male chimps. She says they did not react for about ten seconds, as if they could not believe their own eyes, but then they started screaming, hooting, and embracing, pressing their mouths against each other's bodies—the chimpanzee's way of expressing high excitement.

Chimpanzees are nomadic, generally wandering up to four miles a day in search of food, and here was a feast for the taking with hardly any effort. The pair rushed toward the bananas. By now their piercing screams had caught the ear of every chimpanzee within hearing, and they came scampering in from all over the valley. Chimpanzee hooting can carry for more than a mile. A big male can eat up to 60 bananas at one time, and so the others hurried in to get their share before they were all eaten.

"We scattered bananas liberally on the ground and the chimps, although they gathered them up, seemed too emotional to eat. Instead, they continued to embrace, pat and kiss," Goodall wrote. They eventually calmed down and began to stuff the bananas into their mouths. By the time Goodall left the reserve, nearly all the chimpanzees had arrived at the feeding area. Eventually, the chimps received up to 600 bananas a day, or about 40 percent of the group's nutritional needs.

A couple of decades on, provisioning of feeding sites for great apes under observation in the wild ceased, because it stimulated unnatural behavior by a group competing for the introduced food. Even Goodall eventually admitted that it could have exaggerated the chimpanzees' natural behavior and made them more aggressive toward each other. Critics such as Italian primatologist Paco Bertolani claimed that it had a profound effect on the chimpanzees and their relationship with their ecology, changing their travel patterns and skewing their social connections.

But in those pioneering days of great ape study, it was the only way known for Goodall to get close to the chimps, to accustom them to her presence. Within months of arriving at Gombe, she made some of the most startling discoveries ever about free-roaming wild chimpanzees in their natural habitat. Her finds linked them more closely to us humans than even Charles Darwin imagined.

Primatologists assumed that chimpanzees were peaceful vegetarians, eaters of leaves, fruit, and shoots, like the gorillas. That changed when Goodall, during her first year at Gombe, saw a big male she named David Greybeard, the first chimpanzee there to come close to her, up a tree with the carcass of a young bush pig. He shared the meat with a female. She also witnessed another big male, Rodolf, grab a young baboon by the leg and bash its head against rocks, killing it. The feat seemed to embolden Rodolf, who was usually submissive to the dominant male, Mike, and when approached by Mike he refused to share the meat. Mike would normally have attacked him for his disobedience, but this one time he seemed cowed by Rodolf's behavior.

Rodolf began to bite at the fresh meat. He followed each bite by placing leaves into his mouth and chewing the mixture. He then climbed down from the tree with the carcass slung across his shoulder. Goodall wrote that he looked "like some prehistoric human after a successful hunt." When he spied Goodall observing him, Rodolf charged her, terrifying her with his screaming. Another high-level male whacked her on the leg.

Courageously, Goodall stood her ground and saw, late in the afternoon, Rodolf abandon what was left of the young baboon. The chimps stared at the carcass for a few moments and then mobbed it, screaming, pushing, and grabbing. Mike "leaped onto the backs of the scrabbling mob and attacked one and all alike, stamping on them, and pulling their hair." He grabbed what was left of the meat, the head and shoulders, and ran away, swinging his grisly trophy in the air.

That clearly finished off the primatologists' belief that chimpanzees were nonviolent plant eaters. Richard Wrangham says, "There is far more meat eating among chimpanzees than any other primates except humans."

But Goodall's most important discovery came just three months after setting up camp. She had noticed that David Greybeard was a close ally of the dominant male Goliath, a highly excitable chimp. Goliath fretted whenever Goodall came too close to him, and then David Greybeard, who had a calm nature, would touch him or groom him for a short time, and this reassurance always calmed Goliath.

In one of her most memorable experiences during her three decades at Gombe, David Greybeard allowed her to come right up to him one day. She held out a palm nut as a gift. It was useless to the chimpanzee, but he took the nut, held it and her hand for a few moments, putting a gentle pressure on her fingertips, and then dropped the nut onto the ground. To Goodall, this gesture sparked the first known bond between a human and a chimpanzee in the wild.

So it seemed destined that David Greybeard would lead Goodall to another major discovery. One day in her first year at Gombe, she saw him trimming the edges of a length of sword grass with his teeth. She could hardly believe her eyes when he pushed the stem down a hole into a nest of termites, withdrew it, and licked off the termites swarming on it and ate them. He threw away the grass, then picked up a piece of vine and stripped off its leaves. Using his index finger, he made a new hole in the termite nest and pushed the vine stem into it. Once again he withdrew it and ate the termites that had latched onto it.

It seemed such a simple act, and yet it had momentous importance for anthropologists and primatologists. This was a eureka moment, as Goodall realized that for possibly the first time a nonhuman creature had been observed fashioning a tool, however rudimentary. Until Goodall's observations at Gombe, scientists believed that toolmaking separated human beings from all other animals.

Goodall later observed a male she had named McGregor using a variation on the technique she termed "termite fishing," thrusting a stick into a nest of ants, pulling it out, and eating them, while ignoring their sting. When Louis Leakey heard about Gooodall's discovery, he commented, "Ah, now we must redefine tool, redefine man or accept chimpanzees as humans."

When David Greybeard disappeared during a pneumonia epidemic in 1968, Goodall's heart ached. She said, "I mourned for him as I have for no other chimpanzee before or since."

Goodall and the primatologists who came after her have since observed countless such tool-making incidents in chimpanzee study sites across Africa. Chimpanzees in zoos also learn the trick, and eagerly thrust thin sticks into concrete mounds to withdraw tasty jam and honey secreted inside by the keepers.

Elizabeth Lonsdorf, director of the Lester E. Fisher Center for the Study and Conservation of Apes at Chicago's Lincoln Park Zoo, spent four years at Gombe observing how culture was passed on from the older chimpanzees to the younger ones. She found that young female chimpanzees learned much earlier than young males how to fish for termites from their mounds: "The females sat with their mothers and watched carefully how they did it, while the males spent their time playing." Using imitation and trial and error, the young female chimps were on average 31 months old when they learned to fish out their own termites from a nest, while the males averaged 58 months on their first successful try. The females were usually successful in getting more termites at each attempt.

Lornsdorf told me, "The learning behavior difference between the young females and males was striking, and we see the same difference among young humans." Girls are better at precision tasks such as writing and drawing, while boys are better at gross motor skills such as running, throwing, and jumping. She believes this learning difference might date right back to the common ancestor of chimpanzees and humans. "You'd expect that because in the first human societies, men were the hunters, and they were expert at running, jumping, and throwing, while the women excelled at gathering edibles such as plants, and that needed fine precision skills."

Five decades on from Goodall's discovery of termite fishing, a pair of American primatologists recently observed chimpanzees in the Congo's Goualougo Triangle fashioning a set of tools more complex than those at Gombe. The apes collected plain sticks and made a particular type of probe by using their sharp incisors to fray a stick end into a rudimentary brush. They employed the plain stick to break open a termite mound and then poked the brush down into the nest, capturing 10 times more termites with the probe than when using a plain stick. "The chimps seem to show some sort of planning before they arrive at the nests," says one of the researchers, Crickette Sanz.

Chimpanzees in the wild also use many other forms of tools. Goodall later observed chimp mothers using handfuls of leaves to clean their infants when they dirtied themselves. They also use leaves to wipe off sticky substances from their bodies such as mud, blood, and food smears. One Gombe chimpanzee, Melissa, even used a leaf to mop up blood from her nose. Chimps have been seen using leaves to scoop up tasty algae from ponds and leaf sponges to soak up drinking water collected in tree hollows. Others use twigs to clear out the snot in their nasal passages, and some have been seen putting down large leaves as seats on the wet forest floor.

Chimpanzees at Tai in the Ivory Coast use rocks as hammers and anvils to smash open nuts. A chimp carefully selects a flat-surfaced rock on which to place the nut, but the hammer does not need to be flat. If the chimp cannot find a flat-surfaced anvil, it will use a third stone to support the anvil. A Tanzanian chimpanzee was even spotted wearing a "necklace" fashioned from a chunk of skin from a red colobus monkey. The chimp had tied the skin in a single overhand knot.

One of Goodall's most disturbing discoveries was the homicidal impulses of the adult male chimpanzees. In 1972, a new group emerged from a split among the Gombe chimpanzees. The group was named Kahama and became the rival of an existing clan, Kasakela. A year later, males from the Kasakela group attacked a female and her infant from the rival clan. They almost killed the mother, and killed and ate the infant.

The two clans had once been friendly, but in 1974 full-blown warfare broke out between them in what was called the Four Year War. The Kasakela had eaten bananas together with the Kahama at Goodall's feeding station, but in a series of murderous raids into Kahama territory, they went after the Kahama males. The stronger Kasakela annihilated their rivals, killing almost all of the males. No one knows what sparked off the systematic slaughter.

This is not the only account of all-out chimpanzee clan warfare. In 1985 at Mahale, not far from Gombe, Japanese researchers witnessed a group they labeled K subjected to violent attacks by a group named M, their much more powerful neighbors. The males of the K group suffered near extinction.

At Kibale in 1988, an American anthropologist, Martin Muller, witnessed the immediate aftermath of the murder of a chimpanzee from a neighboring community by a group of Kanyawara males on patrol. Muller heard the males screaming with excitement and the sound of fists pounding. When he reached them he found that they had torn out the stranger chimp's trachea, ripped off his testicles, torn out his toe- and fingernails, and inflicted up to 40 puncture wounds and lacerations on his body. The murdered chimp's ribs were sticking out of his rib cage. Muller said there was evidence that some of the killers had held the chimp down while the others tortured him. Richard Wrangham says there were echoes of human behavior in this attack: "The males who attack, they do seem to take a certain joy in the attack, the drinking of the blood sometimes, and the gripping of the skin on the arm with their teeth and then rearing the head back and taking the skin with it, tearing it all the way around. It looks like they were more in an intense state of excitement, and maybe joy."

At Gombe, Goodall received a further shock when a mother and daughter, Passion and Pom, stole, killed, and ate at least five newborn chimpanzees from their own group over a four-year period. One day in 1976, as dusk neared, Passion was seen chasing a new mother down from an oil-nut palm tree. Melissa was gripping her three-week-old infant daughter, Genie. Pom was waiting on the ground. Mother and daughter joined in the 10-minute fight to capture the infant. Passion grabbed Melissa, holding her down as she bit and clawed her, while Pom struggled to snatch the infant. Despite being badly wounded, Melissa fought fiercely, with all the chimpanzees screaming, but then Pom bit the baby on the forehead, probably killing it, and carried it up a tree. Passion joined her daughter and began to feed on the infant as its mother, bloodied and bruised, watched the gruesome feast from the ground. It took Melissa four weeks to recover from her wounds.

This was one of the first but not the last case of researchers witnessing chimpanzee cannibalism. A Japanese researcher saw an adult male up a tree in the Budongo forest eating an infant chimp that was still alive and crying weakly.

Francis tells me about another incident: "A young female came from the northern part of the forest, about three miles away, and settled with our group. A male tried to fight her and grabbed her baby. He tore it apart and took it into a tree to eat it."

This seems quite different from mountain gorilla infanticide, where a silverback who wins a dominance fight or steals and herds females kills the infants of a rival silverback to bring the nursing mothers back into estrus. Chimpanzee groups are larger, and there tend to be multiple chimpanzee females in estrus at any time.

In 1997, a hulking male named Frodo grabbed power as dominant male over a group of 50 chimpanzees at Gombe when he overthrew his ailing brother, Freud. Goodall remarked that Frodo ruled the group "with an iron fist." When he was an infant, Goodall called him a little frog, because he was always bouncing about. Frodo became a bully early in life, throwing rocks at the other chimps and Goodall when he was three years old. He grew into the handsomest, biggest, and heaviest chimpanzee ever seen at Gombe.

My primatologist friend Dave Greer told me that the first time he met Frodo the great ape immediately attacked him, punching him to the ground. He says, "He was showing me his dominance, so I curled up in a ball to minimize the damage." Frodo also attacked *The Far Side* cartoonist Gary Larson when he visited Gombe, scratching and bruising him. A year later, Frodo bashed Goodall and nearly broke her neck. From then on, she always was accompanied by a bodyguard when she went near him.

In 2002, Frodo earned worldwide notoriety when he snatched a 14-month-old human baby girl from her mother, Rukia Sadiki, as she walked through the national park with the infant strapped to her back African-style. Frodo took the tiny girl up a tree and killed and partially ate her before rangers chased him away from the mutilated body. The attack made world headlines, and people called for Frodo to be put down. Goodall defended him. He was the best hunter in the group; his major target was monkeys about the size of human babies, and to him a human baby was prey. Years earlier, when Goodall's son, nicknamed Grub, was an infant, she sometime kept him in a cage at the camp because she feared marauding male chimpanzees might seize him.

Frodo's downfall came at the end of 2002, when he suddenly became sick from intestinal parasites. He lost weight, his bones were visible, and the ailment turned him mild and mellow. He got back some of his old swagger when he was treated with antibiotics, but he offered submission to the new dominant male, Sheldon.

Goodall now lives in England, far from the African jungles. In her late 70s, she visits Gombe about once a year. "Had I stopped after only ten years [of study at Gombe], I should have continued to believe that chimpanzees were very like us in behavior," she said. Goodall added that we would have been left with the impression that chimpanzees were far more peaceable than humans. Now she claims, "We're rather nicer."

However, the Gombe chimpanzees also had their kind days. Goodall said, "We have learned of the extraordinarily enduring affectionate bonds among family members, which often persist throughout life, and we have witnessed the extent to which close kin will help and support one another." Goodall saw adult males protect their females and young, "and above all we have become more aware of the advanced cognitive capabilities of the chimpanzees which have led to sophisticated social interactions, cultural traditions and pronounced individual variation."

A singular example of family bonds occurred at Gombe when Freud injured his ankle. His mother and brother, the terrible Frodo, slowed their speed during foraging and made frequent stops so that Freud could keep up. Frodo groomed Freud more often than usual while staring at his ankle. In another instance, when an old female was stricken with polio and could barely walk, her two daughters climbed a fruit tree and brought some down for her to eat.

Those caring bonds even carry over between the species. Robert Yerkes, the pioneer primatologist, heard from one man how his pet chimpanzee, Toto, cared for him during a severe attack of fever in 1925: "All day he would sit beside me, watching with a care that seemed almost maternal and anything that I wanted he would bring to me. In the afternoon he would lie down on the bed beside me, put his arms out to protect me, and go fast asleep."

At Gombe Stream National Park, full-time provisioning of the chimpanzees with bananas went on for 15 years, and then sporadically for 22 more years until it ended in 2004. The site still hosts numerous chimpanzee research projects, and that makes it the longest-ever study of any group of wild animals. But Gombe is now a shadow of what it was when Goodall arrived five decades earlier. Ever-growing human settlements press against the chimpanzees' forests, disrupting their lives in the wild.

3

Far from the African jungle where Imoso rules his Kanyawara clan with bludgeon and bluster, I stroll down a peaceful street lined by brownstones in Cambridge, Massachusetts. I ascend the steps of Harvard University's venerable Peabody Museum and take the creaky elevator to the fifth floor to meet Professor Richard Wrangham, an Englishman with a courteous and courtly manner at odds with the subjects of his long-running study at Kibale. Richard has made major discoveries about the chimpanzees' lifestyle, notably the brutality and penchant for warfare of the Kanyawara males. In his office, he tells me, "I'm a biological anthropologist and began my study of the Kibale chimpanzees because they're our closest relatives, and so give us a revealing look at our own evolution."

Richard studied with Jane Goodall at Gombe, where he witnessed the frequent violent rampages of the male chimpanzees. In the late 1980s, he journeyed to the Kibale Forest National Park in western Uganda to set up his own study of chimpanzee behavior in the wild. If you want to be a primatologist, one of the most important requirements is extreme patience at the site of your study—not measured by days or months, but by years. "At first the chimpanzees fled as soon as they saw me coming, typical behavior of wild chimps that have not been habituated," he tells me. "It took me four years before I managed to see them for the first time up close on the ground."

That is when he witnessed the wide array of sounds and facial expressions chimpanzees use to communicate. Primatologists have found that chimpanzees also have 32 separate calls, conveying a broad range of emotions from fear, excitement, anger, and friendship to food calls and warnings to strange chimps. They pat each other to communicate love, and embrace and kiss. They laugh and giggle, and can feel grief when a friend dies.

At Kibale, Richard saw up close the males' brutality that gave him core material for *Demonic Males*. "I saw rapes, savage beatings, and fierce warfare between groups defending their territory," he reports. He wrote: "Male violence that surrounds and threatens chimpanzee communities is so extreme that to be in the wrong place at the wrong time from the wrong group means death."

He noted the clear similarities between humans and chimpanzees: "What makes this social world so extraordinary is comparison. Very few animals live in patrilineal, male-bonded communities wherein females routinely reduce the risks of inbreeding by moving to neighboring groups to mate. And only two animal species are known to do so with a system of intense, male-initiated aggression, including lethal raiding into neighboring communities in search of vulnerable enemies to attack and kill. Out of four thousand mammal species and ten million or more other animal species, this suite of behaviors is known only among chimpanzees and humans."

Goodall, as we have seen, documented this aggression at her research site at Gombe: "I had known aggression could flare up, sometimes for seemingly very trivial reasons; chimpanzees are volatile by nature, yet for the most part aggression within the community is more bluster and threat than fierce fighting—a whole lot of sound and fury signifying nothing. Then suddenly we saw that chimpanzees could be brutal—they, like us, had a dark side to their nature."

The males' primary weapons when beating other chimpanzees are their powerful hands and feet, but researchers at Kibale have documented many occasions when males used sticks to whack females. In 2002, Carole Hooven, Richard's colleague, witnessed the first known attack by a chimpanzee using a stick as a weapon. Imoso, the dominant male, had tried to get near Outamba's female baby, but the dominant female thrust him away. A furious Imoso picked up a stick. Hooven's field notes go on: "IMS (Imoso) first attacked OU (Outamba) with one stick for about 45 seconds, holding it in his right hand, near the middle. She was hit about five times. He beat her hard. The attacks ceased for a few seconds and then Imoso resumed the beating, this time with two sticks. The then two-year-old Tenkere, Outamba's daughter, leaped to her mother's rescue, jumping on Imoso's back and pounding the big male with her fists." Hooven witnessed more such attacks, including one by Johnny, a key ally of Imoso. He attacked Outamba with a stick for about three minutes.

Richard says the males clearly exercised restraint so that Outamba was not badly hurt. They were showing the dominant female who was boss rather than belting her to inflict serious damage. "Sexually active females have received most of the beatings," he tells me. "There's no point in a male killing or hurting seriously an adult female, especially Outamba, who is a favorite among the adult males."

Later, when I was in Kibale, I discovered a fascinating fact about the group. Adult males are not really interested in young females who have just become sexually active, preferring experienced females such as Outamba. She and other adult females have shown they are fertile by giving birth to plenty of infants over the years. "They have then shown they're capable mothers by raising them," Francis Mugurusi later told me. "The young females who've just become sexually active have yet to prove that, and so the adult males are not so interested in them."

Anthropologist Martin Muller led an eight-year study of this phenomenon with the Kanyawara clan at Kibale and says of the adult males: "We've seen them just ignoring the younger females who are all over them."

Muller's research team separated the females into three groups—pubescent, mothers under the age of 30, and mothers over the age of 30. The male chimpanzees gathered around the older females, and the dominant male made his choice plain by preferring to mate with the oldest females.

"Male chimpanzees do not merely disdain younger females, but actively prefer older mothers to younger mothers," Muller says. He believes that because chimpanzees do not experience menopause, the chimp males find the older females more seductive because they are likely to be far more experienced and better mothers.

One of Jane Goodall's best-known chimpanzees, old Flo, had a similar effect on the males. Goodall found her to be "incredibly ugly" with her bulbous nose, worn-down teeth, and torn ears, and yet when she was in estrus she had more suitors than any of the other females. One time, as the males swarmed around her, Flo's three-year-old daughter, Fifi, tried to thrust them away, but they were too strong for her and she gave up.

Mating often occurs when the chimpanzees find a new feeding spot, or when two groups from the same clan meet in the jungle. Goodall described the excitement among the males when they found a female among their group in season: "One by one [the six males] swung through the trees with exaggerated and stylized bounds and leaps before the final stage of the courtship when they swayed and swaggered towards [the female is season]. After they mated with her in turn, they settled down and started to feed."

4

On my way to Kibale from Nairobi, I fly over the lush forests of Uganda in eastern Africa and land at Entebbe airport near Kampala, the country's capital. This was the scene in 1976 of a famous raid by Israeli commandos who rescued a planeload of Israeli tourists being held hostage by Palestinian guerillas with the connivance of the bloodthirsty Ugandan president Idi Amin Dada.

Idi Amin died in exile in Saudi Arabia, but Uganda still suffers under another so-called Big Man, Yoweri Museveni, who stormed the capital a quarter-century ago with his guerillas. He took power as a Marxist rebel but quickly morphed into an enthusiastic capitalist. Museveni has ruled Uganda by hook or by crook ever since. During the elections in 2006, when the exiled opposition leader bravely flew into Entebbe, Museveni had his goons arrest him on the airstrip and haul him away to jail, charged with treason and rape. The allegations were nonsense and easily disproved, but it kept his rival in jail during the election campaign.

Nudging the airport is the vast Lake Victoria, source of the White Nile. Sixteen miles from its shores is one of the world's best chimpanzee

sanctuaries, funded by the Jane Goodall Institute, and I will be returning to it soon. But first, the wild chimpanzees at Kibale beckon.

I have been here many times and am surprised to see that the road
from the airport to Kampala, which was until recently falling apart and
potholed for the full length of the forty-minute car journey, is now
perfectly smooth. The taxi driver chuckles at my surprise: "We've just
had the queen of England here for the Commonwealth heads of state
conference. Museveni paved every road the queen was to travel on, including this one, but left all the others potholed."

The meeting was a three-day extravaganza, costing this impoverished nation more than $250 million, and allowed Museveni to strut
what he thought was the world stage. My driver, like so many Ugandans, has turned into a cynic under Museveni's rule. "Museveni told us
we'd get a billion dollars' worth of publicity worldwide, but few in the
international press turned up at the conference."

This wasting of the nation's wealth contrasts with its spending on
conserving its precious forests and wild animals. Foreign groups such as
the World Wildlife Fund largely pay for the country's conservation efforts. Museveni is infamous for ordering pristine forests to be chopped
down, whatever the threat to the species living there. When I was in
Kampala one time, I had to scamper out of the way of a demonstration
that turned ugly. Five people were shot dead by police when thousands
of Ugandans marched along the capital's streets protesting Museveni's
granting of 25,000 acres of virgin forest to a wealthy Indian/Ugandan
company.

The company planned to log and sell off the valuable hardwood timber
and then plant sugarcane. The forest was home to rare animal and bird
species. Museveni described his action as part of his self-proclaimed
"Uganda's New Vision," saying, "It is the short-sighted people who
don't understand that the future of all countries lies in processing."

With this attitude, what hope have the chimpanzees and other great
apes of Uganda?

We pass through the streets of the capital, which thankfully is quiet
today. Still, the U.S. State Department advisory website for Uganda
warns that visitors "should be aware of threats to their safety from insurgent groups and banditry," particularly along its western and southwestern borders with the Democratic Republic of Congo, where I am

headed. Insurgent groups there have engaged in murder, armed attacks, kidnapping, and the placement of land mines. Such incidents occur with little or no warning.

Kibale is to the west of Kampala and the bus ride takes a bone-shaking seven hours. Mud-hut villages hug the road on both sides along the entire route. Scattered along the way are open-air market stalls selling everything from soap powder to hand-me-down clothes to cheap footwear and stacked cans of corned beef. Butchers stand at stalls offering to slice chunks of meat from the bloodied carcasses of cattle and goats they have drawn and quartered earlier that morning and hung on hooks.

Stick-limbed boys play barefoot soccer in the dust, women in flamboyant cotton gowns return from the markets bearing parcels balanced on their heads, and grizzled old men huddle on stools doing what such veterans of life do all over the world—solve the world's problems day after day over puffs of tobacco and jugs of local beer.

At the bustling town of Fort Portal I hire a taxi, and it turns onto a dirt track that clearly has never been used by the queen of England because we bump and shake for 30 minutes until we reach the entrance to Kibale Forest Reserve, 215 square miles of tropical, deciduous, and montane forests. It is home to about 500 chimpanzees. Tall trees form an archway as we drive toward a huddle of bungalows spilling down a grassy slope. They are set aside for scientists who come here to study chimpanzees and other wild animals. The reserve is closed to the general public. Pressing against the slope's edges is a dense forest of high leafy trees.

In his small office piled with academic journals, the director of the park's field station, Professor John Kasenene, tells me that a French veterinarian studying the chimpanzees' use of plants as medicine and an Italian primatologist researching chimpanzee evolution are also staying here. "I'm sure you'll get to meet them," he says.

The following morning at 5:00, Francis Mugurusi, one of Richard Wrangham's most trusted field assistants, arrives at my bungalow. He is a slender man with thoughtful features. We trek through the forest to where Imoso, the dominant male of the Kanyawara group, and some of his clan have spent the night. The group has been under continuous study since 1987, and its home range extends for 13 square miles of moist deciduous forest grassland and swamp that was selectively logged in the 1960s.

It is still dark and Francis, with the aid of a headlamp, leads me along the muddy paths cutting through the forest. Even this early, the humidity is stifling as we clamber over fallen trees covered with moss. All around us, small flickers of phosphorescent light mark the path of fireflies as if they are tiny fairies darting about the jungle. We veer off onto a narrow dirt path surrounded by trees hung with vines, the jungle canopy blocking the sky. Mist drifts through the trees sodden from an overnight downpour.

Up ahead through the gloom I see human silhouettes. "They're my colleagues," Francis whispers. "We come every day to monitor the chimps' behavior and give the information to Professor Wrangham. We followed the chimps yesterday until they built their night nests, and so we knew where to come."

We sit on the damp soil under the trees waiting for the chimpanzees to wake. A big spider crawls across my leg but looks harmless, and so I let it pass. I spy a detachment of army ants scurrying over the ground on some mission and move a few feet to get out of their way. Just after 6:00 a.m. I hear rustling in a tree about eight yards above my head. An almost human silhouette perches at the edge of a commodious oval nest made from leafy branches woven together to form a comfortable, springy place to sleep. Another field assistant, Solomon Musana, has a plastic bag on a long stick, and when the chimpanzee urinates over the side of the nest he neatly catches some of the urine. "It's Outamba, and her urine is taken daily for the lab to check her hormone levels," Solomon tells me.

Though I am obviously no threat to the burly chimpanzees, Francis says they still may try and involve me in their power plays. "If one of the males charges you, don't run away because he won't hit you, he'll be showing you his dominance," he warns. "It will be a bluff." I remember Dave Greer's advice. If a chimpanzee attacks me, I should roll into a ball and cover my head with my hands to lessen the damage.

Outamba yawns and returns to her nest, but 10 minutes later as darkness seeps from the sky, she agilely climbs down and knuckle-walks toward a huge fig tree laden with fruit. The buttresses that surround its base are about two yards high while the tree itself soars more than 20 yards, and its profusion of thick, leafy branches spread across the forest for at least another 20 yards.

Seven more chimpanzees climb down from their night nests nearby and head for the same tree. The chimps have extra-long arms and stumpy legs. Their feet act like hands, with toes like fingers, as they grip the slippery tree trunks on their ascent. Unlike gorillas, there is no great difference in size between the male and female chimpanzees. An adult male of the eastern subspecies here at Kibale weighs on average 95 pounds, while a female averages 73 pounds. But while the males lack bulk when compared with gorilla silverbacks, their ferocity more than makes up for it.

The adult males are stocky, with burly shoulders and a thick neck. The eyes of both the males and females sit close together and shelter beneath heavy brow ridges. Their noses are flat and their pink protruding lips resemble highly mobile flaps.

Outamba, the clan's dominant female, has her infant female, Omusisa, riding on her back and by her side a young daughter, the brave Tenkere, who pummeled Imoso six years earlier. "Outamba's a really good mother, and that's why the males want to mate with her the most," Francis says. "The infants of some other females died because their mothers were careless. Chimpanzee males stay with their families, but once the females are sexually mature they move to new families. It's pretty tough at first for a newly arriving female, because she has to figure out the power structures and where she fits in. This causes a lot of stress."

The chimpanzees each settle on stout branches high above the ground, and I see through binoculars that they enthusiastically tuck into their breakfast of round green figs. Their diet also includes nuts, flowers, leaves, insects, small birds, eggs, bark, and meat as an occasional treat. The chimpanzees live in a society primatologists label fission-fusion, meaning that a big group will break up into smaller foraging parties. At this time of the year their favorite fruits are not in season, and so the Kanyawara clan has spread out in small groups over its territory to find food. There can be as many as 120 chimpanzees in a group, which comes together most often when fruit is bountiful.

The jungle is wakening, and the buzz of millions of insects sets up a noisy, sibilant backdrop as a troupe of black-and-white colobus monkeys, with spectacular bushy tails, leap from branch to branch in a nearby tree. Richard's three African field assistants note every movement of the chimpanzees in log books, but the apes' routine varies little from day to day. After two hours of feeding on the figs, at about 8:00 a.m. each chimpanzee makes a rudimentary day nest of branches and leaves in the fig tree, then settles in for a two-hour siesta.

Unlike their close relatives the Congo's bonobos, known until recently as pygmy chimpanzees, who researchers have reported make sex very much a part of their daily lives, the Kanyawara chimps do not seem interested in mating on this day. "They mate the most when there's plenty of fruit in season, when all the chimps in the family come together to eat, but now they've only got mostly figs," says Francis. That is why there are just 10 of the 46-strong group here this early morning. The others are spread out over their territory. But when a female comes into estrus for a few days each month, "the male chimps gather together and become very excited, and come down each morning from their night nests long before the females." That is probably because the peak period for chimpanzees mating in the wild is usually early in the morning.

Courting behavior varies from place to place and is learned by imitation. A male will stand his hair on end and bang on trees or wave branches to grab the female's attention in one place, while elsewhere the males have developed a seduction technique known as "leaf-clipping." These males put a leaf in their mouth and make a loud noise by biting it to attract a female. Seduction is mostly initiated by the male and is sometimes signaled by a male standing upright and swaggering toward a female. In one study at Gombe, of 213 observed matings or attempted matings, the male was the initiator 176 times while the female initiated sex 37 times. Females in estrus were intensely receptive to male advances. Once a male mates with a female, almost all the other adult males want to mate with her, and each male can couple with her several times a day. This is possible because the act lasts on average nine pelvic thrusts by the male.

Lower-ranking males dare not openly mate with the females without the approval of the dominant male, but they sometimes use deception to fool the group leader. A subordinate male will open his legs to show his erection to a female in estrus, but quickly cover the erection with his hands if the dominant male passes by. If the female is willing, then the pair will sneak away out of sight. A female chimp typically makes a loud noise during climax, but if the mating is surreptitious she remains quiet.

The black-and-white colobus that I see in a tree near the Kanyawara clan are often too big for the chimpanzees to tackle, but the red colobus, half their size, are a favorite prey. The chimpanzee males have devised cooperative strategies to successfully hunt them. "The chimpanzees surround the tree where they see the red colobus monkey," Francis tells me. "Some chimps stay in the lower branches and on the ground, while others chase the colobus down from the high branches using hunting barks to scare them. They ambush them as they try to escape, then kill and eat them."

The chimpanzees' success rate in observed hunts at Kibale is 84 percent. The males are in the thick of it, though females sometimes take part, more likely at the periphery of the hunting party. Their major responsibility is to care for their infants, and once a female reaches maturity she will usually have one infant or juvenile in tow and perhaps one on the way. A study has shown that at the climax of the hunt, 90 percent of the killers of the prey are adult males, with just 8 percent adolescent males.

The slaughter has purpose. The UN's *World Atlas of Great Apes and Their Conservation* claims, "Hunting [is] seen as a form of social display, in which a male chimp tries to show his prowess to other members of the community." At the Mahale mountains near Gombe in Tanzania, researchers found that the dominant male "used colobus meat for political gain, withholding it from rivals and doling it out to allies." The females also benefit: "In Gombe, female chimpanzees that consistently receive generous shares of meat after a kill have more surviving offspring, indicating a reproductive benefit tied to meat eating."

Tragically, as with Frodo, Passion, and Pom at Gombe, some Kibale chimps have a taste for human flesh. "On three occasions, chimpanzees have grabbed a baby from its mother in nearby villages," says Francis. "They escaped through the trees, still gripping the infants, and tore them apart and ate them."

Not far from the Kanyawara clan a male chimpanzee, nicknamed Saddam by the locals, did indeed kill and eat three babies and severely injured many more. Michael Gavin, who studied the killer chimps, told the BBC's *Wildlife* magazine that the way the chimps murdered human infants was exactly like the way they killed other prey, suggesting that they deliberately snatched the babies to eat them: "In most cases they bite off the limbs first before disemboweling them, just as they would the red colobus monkey which is among their favorite prey."

The Ugandan Wildlife Authority reported that all the attacks involved children under the age of five, and occurred when there were no men present.

"Does the killing and eating of human babies make the villagers determined to kill the chimpanzees in revenge?" I ask Francis.

"Maybe, but it's against the law. Though poachers do kill chimps and feed them to their hunting dogs. They say chimp meat makes them stronger. We once had about sixty chimps in the Kanyawara group, but it's now down to forty-six because of poaching and disease."

That night I met a slim youngish French veterinarian, Sabrina Krief. She has just performed an autopsy on a chimp from Imoso's Kanyawara clan who died in the forest. She told me, "She had tuberculosis, and could have picked it up from a villager because these chimps are used to humans. The chimps are susceptible to many of the same diseases as us." There have been frequent outbreaks of a coughing disease, a respiratory disease caused by a virus, among the chimps since 1998. It killed several of them, four in 2007 alone.

Sabrina has come to Kibale several times to study the chimpanzees' use of medicinal plants. She says, "We know the chimpanzees here eat about one hundred and sixty different types of plants, and we believe they use at least twenty of them as medicine to combat malaria, intestinal parasites, skin infections, and respiratory diseases. Local people use many of the same plants for the same reasons. The most common is *Aspilia*, and the chimps will often go a considerable way to get the plants, and ignore other good-tasting plants nearby."

Researchers have noted chimpanzees using plants possibly for medicinal purposes at several sites across their continental range. In Sierra Leone, on the other side of Africa from Kibale, a researcher found that chimps there who were suffering from internal parasites were restored to health when they ate bitter piths. At Gombe, female chimps have been observed ingesting the plant *A. pluriseta* at significantly higher rates than male chimps. The plant is used traditionally by villagers as an abortifacient, or an inducer of labor, and researchers have suggested that female chimpanzees eat the plant to regulate their fertility.

Richard Wrangham, a pioneer in this study of the chimpanzees' medicinal plants, told me that the local Africans also use the leaves of *Aspilia* as a medicine, brewing it into a tea to treat stomach ailments. He sent samples he collected at Kibale to a pair of laboratories in the United States, and they reported back that the leaf contained a powerful compound, thiarubrine-A, that was antiviral, antifungal, and antibiotic, and was more than 200 times more powerful than DDT. He believes the chimpanzees could be using the plant to expel parasites from their bodies.

Sabrina tells me the leaf is rough-edged. "Chimpanzees gulp down most other leaves, but with *Aspilia* they carefully fold each leaf in half and place it in the mouth," she says. Instead of chewing the leaf, the chimpanzees use their tongues to rub the leaves against the inside of the mouths for about five seconds and then swallow the leaves. Her examination in the small lab at Kibale of the chimpanzees' dung after they had eaten the plant suggests it sweeps intestinal parasites out of their bodies, and that, confirming the U.S. finding, it also acts as an antibiotic and an antifungal.

Local people use a woodland shrub, *Lippea picata*, as a medicine to combat malaria. "The chimpanzees also use it, perhaps for the same reason," says Sabrina. "But we have much more to learn, and I'll be coming back to Kibale many more times."

Chimpanzees across Africa in the wild commonly use plants for medicine, and this is one strand of evidence among many that they possess a common culture, though their behavior can differ in large and small ways. Researchers at the University of California carried out an extensive survey of chimpanzees in different localities and found that they shared 39 behavior patterns that could only have come about by imitation or learning.

One dramatic example is what researchers call "rain-dancing," which has been seen in several chimpanzee communities that are far away

from each other. It usually takes place when the chimpanzees experience a heavy rainfall. The Gombe chimps are expert practitioners. Goodall described one spectacular dance as the rain poured down, thunder
boomed, and lightning flashed. In the heavy rain, chimpanzees of all
ages climbed down from the trees by a ravine, and the adult females
and their offspring moved to a hillside where they sat and watched a big
male named Paleface do his steps.

Paleface stood upright and started to shift from foot to foot rhyth-
mically. Then, Goodall wrote: "[The] chimpanzee first crouches, then
stands erect and plunges down a slope, yelling all the while. Grabbing
a branch, he slaps the ground as he charges towards a tree, and climbs
the trunk. Hurling himself downwards, he snaps off a bough and drags
it behind him. To break his headlong rush, he swings around a tree.
Finally, he plods uphill to charge again."

One by one, the other adult males did their version, much the same
as Paleface, with the display lasting 20 minutes. Goodall never saw a
female perform a rain dance, but she did see males perform similar
rhythmic displays during other thunderstorms and also by the stream
bed immediately below a roaring waterfall. Hooting and screaming,
they stared up at the rush of water. Other researchers have seen chimpanzees react to heavy rainfall by going into a quasi-trance and doing
their "dance" alone, with no spectators.

The next day passes without incident as I join Francis and the other
field assistants in tracking the group from the moment they wake, at
about 6:00 a.m., to when they build their night nests, at about 5:00
p.m. The day begins with a leisurely breakfast in the fig tree, then a
snooze among the branches, more feasting, and a fierce, noisy display of
dominance by Imoso. He has broad shoulders, short, stubby legs, long,
brawny arms, and a face that seems at the same time eerily calm and yet
aggressively brutal. Imoso dashes at the other males as he screams and
bares his teeth. Johnny, an adult male, leads the other males in swiftly
climbing trees to get out of the way, descending only when Imoso has
calmed down. Then they approach the big male, bow their heads submissively, and place their lips against his neck.

I once witnessed a powerful display of dominance and submission
by a pair of male chimpanzees at the Taronga Park Zoo in Sydney,

Australia. Snowy, a young male from a zoo in New Zealand, had been brought into the group to expand its genetic diversity, something that never occurs in the wild. Because of the dominant male's hostility toward the stranger male, Snowy had been kept in an adjacent cage by himself for almost a year. The dominant male had almost torn off one of Snowy's fingers when he poked it through the bars separating them.

When the head keeper felt that the dominant male was ready to accept Snowy, the pair was led separately into a small walled cage, and as they faced each other their screams of excitement were so loud that I felt they would burst my eardrums. The tension in the air was very thick. The head keeper was visibly fearful that the dominant male would attack and kill Snowy. But Snowy knew his chimp protocol and crouched, presenting his behind to the dominant male in the same way a female gets ready when a male is about to mate with her.

The dominant male mounted the newcomer and immediately simulated vigorous sex. Snowy bent forward, hooting as the big male thrust at his back. This simulated sex lasted for a few minutes and when the dominant male climbed off Snowy's back, the younger chimp submissively bowed and groomed him. The keeper told me that the dominant male, by receiving this homage, signaled that he accepted Snowy into the group.

Each male's place in the hierarchy is fluid and Mike, a lower-ranking male at Gombe in Goodall's early days there, used a clever strategy of bluff to overthrow the dominant male, Goliath. Mike had discovered a pair of kerosene tins near Goodall's camp and when he banged them together they made, for a chimpanzee unused to the sound, a terrifying clatter. Seizing on this, in the presence of Goliath and other high-ranked males, Mike began hooting and rocking from side to side to gain their attention. He then stood upright and banged the cans together. The loud noise startled the other males. Mike tossed the cans in front of him at Goliath and the other chimps, chased the cans, and began banging them together again.

Cowed by this noisy display, Goliath and the other males approached Mike and offered him obeisance, panting, bowing, and crouching in submission before him. At that moment, Goliath and the others accepted Mike as the group's new dominant male.

Mike held the position for a long time, primarily Goodall says because he had "guts." He turned out to be a relatively benign dominant male, assiduously grooming his subordinates and sharing meat with them. When he grew too old, feeble, and world-weary to resist the challenge of a male named Humphrey, Mike slid rapidly down the rankings and paid homage with submissive behavior to even the lowest-ranking males.

A dominant chimpanzee male's rule brings plenty of advantages, particularly the admiration of the mature females, but it comes at a cost. By examining Imoso's urine, researchers at Kibale have found that he has significantly higher levels of testosterone and cortisol, subjecting him to increased levels of stress as he plots, bluffs, and fights to remain the group's dominant male. I suspect you would get the same result if you tested any of the CEOs in the Fortune 500.

After his dominance display by the big fig tree, Imoso settles his clan with pats, embraces, and kisses, producing low-pant grunts by the subordinate chimps, and mutual grooming. Each chimpanzee picks a partner and they use their fingers to search intently through each other's fur for ticks, grass seeds, burrs, pieces of dirt, and flakes of dry skin. Contrary to common belief, chimpanzees in the wild do not suffer from fleas.

All chimpanzees in the wild groom each other more for social bonding than hygiene, and they have evolved different ways of treating the bothersome ticks. Here in Kibale, the chimps squash them with their teeth and eat them. The Gombe chimpanzees remove the tick and squash it inside a leaf, while chimpanzees of the Tai forest in the Ivory Coast on the other side of Africa squash ticks with their forearms.

Imoso has signaled that at present, all is peaceful under heaven. But if he wishes to issue a warning to another male that deliberately falls short of a charge, he can employ a wide variety of facial expressions, postures, and gestures. He can jerk his head upward and backward, accompanied by a soft bark. A low-intensity threat gesture warns a lower-ranked chimp that he is coming too close to the dominant male while he is feeding. Imoso can also motion with the back of his hand as a warning toward a chimpanzee who is bothering him.

The chimpanzees use a wide range of gestures to communicate with each other. Adults and juveniles will present the back of their hand to the lips of an infant, a gesture to reassure them. To appease or reassure, a chimp will crouch and pat the head, rump, or back of another chimp. A higher-ranked chimp will pat a distressed inferior chimp, while a submissive chimp will bow as it presses its lips against the body of a higher-ranked chimpanzee.

The group, today led by Imoso, build rudimentary day nests in the trees for an hour-long snooze and then climb down to go for a stroll along the forest floor. More grooming follows, and then the group pant-hoot and thump trees to see if there are any other chimpanzees around. The chimps then lay on the ground for a quick nap, followed by more grooming and an afternoon feed of leaves. At dusk, within just a few minutes the chimpanzees construct fresh nests in the canopy by the same fig tree, expertly weaving branches and leaves to form comfy beds for the night.

The males especially need the sleep, because chimpanzee politics are brutal and enervating. This constant tension, constrained by social conventions within the group, can burst into homicidal attacks when the males meet stranger chimpanzees. They show no mercy. Richard, who has spent two decades observing the Kanyawara clan, says, "When strange groups meet an individual from another group they'll try to kill him. So the males spend a lot of time trying to stay in big groups for safety, but in big groups there's more competition among themselves.

"Males compete for rank, and it's very clear who is the alpha male. In our group Imoso has been alpha male for twelve years. The lower-ranking males have problems being able to mate with the females when higher-ranking males are around. But if they are friends with Imoso then they get more matings. So, there is a payoff being a friend of a high-ranking male."

The jungle seems quiet on the next day, and I see only the chimpanzees and the black-and-white colobus monkeys. But Francis says there are plenty of animals all about us and you can find them if you know where to search. The forests are being plundered by villagers poaching for bushmeat, but he adds, "there are still duikers, bushbuck, bush pigs, elephants, and giant forest hogs around."

5

On the sixth day that I am with the group, the apparently peaceful chimpanzees erupt in violence. We are walking along a jungle path with a group of seven including the dominant female, Outamba, her infant, Omusisa, and her daughter Tenkere, as well as a brawny male named Kakama. Another female and two males round out the group.

Kakama's mother was antisocial, a loner, and he grew up with few allies among the males. He is now the second-ranking male and yearns to be alpha male. "One day he'll challenge Imoso, but for now Imoso has more allies among the males, and Kakama knows he'd lose the fight," Francis whispers. "But I believe that one day Kakama will succeed." He was down near the bottom of the totem pole, ranked number 19 of 21 among the males in 2001, and so his rise has been spectacular.

At mid-morning, the chimpanzees scream and bang their feet against trees to see if there are any other chimps around. Through the jungle thicket come pant hoots, loud calls that rise and fall in pitch, and foot drumming. The group hurries in the direction of the sounds, ready to fight or greet. They meet a female, Rosa, her brother Makuku, and Johnny, members of their group coming to find them. To reinforce their family bonds they pat and kiss each other and then settle into mutual grooming, each chimp choosing a partner and using nimble fingers to sift through their partner's fur. When they find a tick, they eat it. The chimp being groomed swivels around to allow its entire body to be carefully examined, and then they change positions, the groomer to be groomed.

The chimps lay down on the forest floor for a nap that lasts about an hour. Then Kakama begins to lead Outamba away, but she is unwilling to follow and instead takes her family in the opposite direction with Johnny, who is less bulky than Kakama. "Here's trouble," says Francis.

Kakama explodes in a rage. Screaming with teeth bared, he leaps onto Outamba's back and pummels her with his clenched fists. The other chimps scream and bare their teeth as they watch, but do not join in or try to stop him. The beating lasts about a minute and then Out-amba, chastised, follows Kakama in a different direction, accompanied by her daughter and with her infant riding on her back. "Outamba likes Johnny and wanted to go with him, and that made Kakama angry," Francis explains. "He beat her to force her to follow him. But if Imoso were here Kakama would not have dared attack, because Johnny is one of the dominant male's most senior allies."

Had he been there, Imoso would have cowed Kakama in a domi-nance display. One eyewitness described such a display: "Male chim-panzees proclaim their dominance with spectacular charging displays during which they slap their hands, stamp their feet, drag branches as they run, or hurl rocks. In doing so they make themselves loom as big and dangerous as they possibly can, and indeed may eventually intimi-date a higher-ranking individual without having to fight."

Chimpanzees generally like to put on a show: "The grin of fear seen in frightened chimpanzees may be similar to nervous smiles given by humans when tense or in stressful situations. When angry, chimpan-zees may stand upright, swagger, wave their arms, throw branches or rocks—with bristling hair and often while screaming or bunched in ferocious scowls."

Dave Greer once told me of a trick chimpanzee trainers use for the movies: "They made the chimpanzees frightened, and they responded with fear on their faces that looks like an open-mouth grin to humans." Another seemingly cute sight you see in films and TV is the pout face, where the chimpanzee pushes the lips forward into an oval shape and opens the eyes wide. The chimpanzee is actually attempting to com-municate anxiety or frustration, such as when an offer of grooming is rejected or following an attack.

Richard Wrangham told me that the males' violence also plays a major role in mating: "In relationships between males and females, it's the males using violence against the females who are persuading them to mate with them readily when later the females are going to be in a sexual condition. Males who mate more with particular females are the males who've beaten up more females." The same seduction technique was used by Marlon Brando in the film *A Streetcar Named Desire*, the brawny male displaying his impressive chest muscles and constantly raging and roaring at his female mate.

It is not only the males who have to fight for rank and privileges. "Females migrate to stranger groups to avoid inbreeding at the age of sexual maturity, at seven or eight years old, and they often have to fight their way into a stranger group," Richard said. "The other females don't like them being there, and so the newcomers sometimes form coalitions to defend their right to be members of this new group. The males help them. It can take them three or four years to establish themselves and have a baby, and once that happens they stay there for the rest of their lives."

But as with the mountain gorillas, the female chimpanzees in a group never form the same close friendships as the males. When the females are together in a group, the mood is frequently less relaxed than among the males at certain times when they are together. Goodall found the adult females she observed "more vindictive than the males." Over thousands of hours of watching their behavior, she discovered that the males were just as temperamental as the females but calmed down quicker after an embrace, a kiss, or a caring pat. They do not bear grudges for long. In contrast, Goodall observed that the females "harbor malice over long periods of time."

This may be because the male chimpanzees never leave the natal group. They have been with each other through infancy, adolescence, and maturity. Richard told me the males' affiliation and cooperation are critical to the group's success because they have to defend the group's feeding range, and this is crucial to the reproductive success of the females. In contrast, the females enter the group as stranger adolescents and have to find their rank among unfamiliar females.

In a study of the Kanyawara group that lasted 10½ years, Richard and three other researchers found that the so-called resident females are inevitably hostile to the newcomer females and will frequently attack them and even kill their babies. The vulnerable young migrants suffer the highest rates of intrasexual aggression among the group, and urine analysis shows they have high cortisol levels, enduring much higher levels of physiological stress than resident females.

Richard believes the resident females attack the new arrivals because they cause increased competition among the females for feeding space for the best food, such as ripe fruit. This high-value food is essential for prompting efficient ovarian cycles and for the successful rearing of young.

But adult males protect immigrant females because they offer increased mating opportunities. In almost all attacks by resident females, the males who intervened to stop them took the side of the young migrants, even siding against their own mothers. The beatings the newcomers endure from the resident females when the males are not around can be so severe that they have left the group to find another. This is not in the males' interests. Once the males step in, an attack ends abruptly. The migrant females try to stay close to the males for protection and rarely take part in all-female foraging until they have established themselves, and that can take years. Then, they themselves persecute newly arriving females.

Francis and I continue to follow Kakama, Outamba, feisty little Tenkere, and the infant Omusisa, and I am surprised about an hour later when the two adults sit on the grass and begin to groom each other with what looks like care and affection. Each session ends with Outamba gripping Kakama's left hand with her right hand and then thrusting both hands high into the air. The chimps use their free hand to inspect their exposed armpits for minuscule intruders with a grim intensity. When the search is complete they change hands, gobbling down any ticks. They do not even blink when a forest elephant trumpets nearby.

But Francis is wary and leads me away in the opposite direction. "The elephants are very dangerous; they don't like people and have killed villagers," he explains. The elephants have good reason. African forest villagers have been killing and eating them for thousands of years. Though elephants' eyesight is not good, they know us from our human smell. They move about the jungle alone or in small herds, in contrast to the elephants that inhabit the great wilderness plains of the Masai Mara and the Serengeti in Kenya and Tanzania. I have seen the latter, huge creatures, gathered in herds of more than 100, while the much smaller forest elephants, because of their dense habitat, are either solitary or move around in groups of up to four.

About 30 minutes later we spy a pile of elephant dung on the track. Francis bends down and peers at it. It is still soft and smelly. "The elephant was here about half an hour ago. We'd better hurry back to the bungalows."

Late that afternoon I meet Paco Bertolani, an Italian primatologist who has come to Kibale for a year to study the Kanyawara clan. We sit on the porch of his bungalow. The furniture is basic, a couple of wooden chairs and a wooden table plus a wooden bed—there are few creature comforts. He cooks his own food, buying it in the Fort Portal market. The primatologists and anthropologists that I have met in the field rarely enjoy comfortable living, and Paco is no exception. Years in the field have carved his body to an edgy thinness. Paco has extensive tattoos on one leg, and another tattooed line that runs from his jaw on one side up across his forehead to the jaw on the other. "I got them while living with an Indonesian tribe," he says.

He smiles as he points back inside the utilitarian bungalow that he will be living in for many more months. "This is luxury; I'm more used to living in a mud hut."

Paco has come here in connection with a study of chimpanzees of two very different habitats, the forests and the savannah. "My major study was of savannah chimps in Senegal, and so I'm now studying forest chimps here," he tells me as we sip cups of strong coffee. "It's believed our human ancestors moved from the African rainforests to the woodland savannah millions of years ago, and so the difference between the two types of chimpanzees might give us clues to our own evolution. For instance, the rainforest chimpanzees spend most of their time in the trees, but the savannah chimpanzees are on the ground for most of the day. There are not so many trees in their habitat, and most are low, although they do build nests in the trees each night like other chimps. They even rest up in caves."

Almost all chimpanzees avoid going near water—indeed the chimps at the outdoor compound at Sydney's Taronga Park Zoo are stopped from escaping by a moat filled with water. They will venture to the water's edge, but no farther. But the savannah chimps enjoy sitting waist-deep or lolling about in the water of the many swamps where they live, Paco says. The savannah chimps also greet a storm by performing their own version of a rain dance, a wild rhythmic display.

The earliest ancestors of humans, bipedal apes, walked the earth in the Miocene era, more than 5 million years ago. The extreme drying then created enormous stretches of grassland. Primates at the edges of the tropical forests lost their access to a wide variety of fruits as well as lakes and streams. They had to adapt, journeying much farther for water and food.

Some of the chimpanzee clans came up with creative responses. Along with primatologist Jill Pruetz of Iowa State University, Paco caused a stir among primatologists when he recently reported evidence of chimpanzees at Fongoli in the Senegal savannah fashioning rudimentary spears to hunt bush babies. Nocturnal primates the size of squirrels, they secrete themselves in tree-trunk hollows to sleep during the day. It was the first time great apes had been observed making tools to hunt other vertebrates.

"We witnessed the chimpanzees stripping branches of their leaves and then using their incisors to fashion a sharp point at one end, making a primitive spear," Paco tells me. "They then jabbed these into tree hollows, skewering any bush babies inside and eating them. It was mostly adolescent female chimpanzees who did this, although we saw adolescent males also doing it. We documented twenty-two such spearings."

"This clearly is learned behavior, a separate culture, because we've never seen it at Kibale," Richard Wrangham told me. Paco adds that adult males are the slowest to pick up any innovation such as spear hunting, saying, "They're more set in their ways."

Does this innovation indicate a superior intelligence? A *National Geographic* article noted: "Ecological intelligence is the name of the theory that some primates, including those of our lineage, have evolved larger, more complex brains because it helped them adapt to the challenges of surviving a less than giving habitat." According to primatologist Craig Stanford, the first push toward a larger brain may have resulted from a patchily distributed, high-quality meat diet and the cognitive mapping capabilities that accompany it. Eating meat may have been important in the primate brain evolving to become more sophisticated. The less energy you expend on digestion, the more energy you can use to power a bigger brain.

A few months into his stay at Kibale, Paco has come across a baffling display of that expanded brainpower in the meat-eating chimpanzees. I tell him about the encounter with Kakama and Outamba earlier in the day, the male willing to use violence to force the female to accompany him. He responds, "We're witnessing behavior with those two we've never seen anywhere before among chimpanzees, and if you like I'll take you to see it tomorrow. We expect Kakama to challenge Imoso as dominant male, and this could be part of his strategy."

*   *   *

The next morning in the predawn murk, Francis leads me in the opposite direction from our usual morning treks. The trees are less dense, and there seem to be fewer types of trees, and that means less comfortable nests and less choice of food. "This area was logged years ago, and so this is patchy secondary forest," he explains. It is a pale imitation of the jungle in the opposite direction. Rob Muir, the conservationist in the Congo, had told me that he once did a comparison between a primary and a secondary forest that had been logged out in Madagascar, saying, "The primary forest had over three thousand species of trees and plants, while the secondary forest had just a handful."

Francis and I settle under the trees and are soon joined by Paco. Above us are three nests in three scraggy trees—those of Kakama, Outamba, and her juvenile daughter, Tenkere. Outamba's infant, Omusisa, sleeps with her mother. Richard Wrangham had told me that chimpanzee mothers "give birth in tree-nests. Females are sometimes infanticidal toward very young newborns, so tree-nests may be the safest place to escape from chimps as well as from predators."

Just after 6:00 a.m. Kakama climbs down from his nest, but instead of going as usual to a nearby fig tree for breakfast, he sits below Outamba's tree and stares upward. The minutes pass. There is a rustling of leaves above as Outamba departs her nest, but she surprises me by not climbing down and going to the fig tree. She must be hungry, but for almost an hour she and her daughters stay in the canopy near her nest while Kakama waits below, staring up at them. Finally, she climbs down the tree and dutifully follows Kakama to the fig tree with Omusisa on her back and Tenkere trailing her.

"This happens when a female is in estrus," says Paco. "The alpha male will take her away from the others and guard her day and night to mate with her. We call this consortship. But, as you can see, Outamaba is not in estrus [signaled by a massive pink swelling of her nether parts as a come-on to the males]. We've never seen this before anywhere. Outamba clearly doesn't want to stay with Kakama, and so she stayed up in and around her nest for so long until she and her daughters had to come down. It seems Kakama is making it clear to her that when she comes into estrus she'll have to go with him as his consort or suffer the consequences."

I stay with the chimps all morning as they feed, snooze, and groom each other, and am intrigued by their complete silence. Somehow Kakama has communicated to Outamba and her brood that there will be no pant hooting, no tree stamping to let the other chimps know where they are. While Kakama, Outamba, and Tenkere settle into a solid bout of feeding on leaves, the two-year-old has a great time swinging nonstop on the branches near her mother. Omusisa is still suckling, and will do so for up to another year, and can play as much as she wants while her mother eats a hearty breakfast. Sometimes she dangles on one arm high in the trees, swinging backward and forward, and I bite my lip for fear she might fall, but I have seen countless little chimpanzees, bonobos, orangutans, and gorillas do the same in the wild and have never seen one slip.

Little Omusisa has a lighter-colored face than the other three chimps and a prominent white tuft of hair at her rear end. This is a signal to the group that she is still learning the rules of chimpanzee society and so she has special status. Generally, she won't be punished if she breaks the rules. But her face will grow darker and the tuft will darken by the time she is about four years old, and then Omusisa will have to behave properly or suffer the consequences.

A chimpanzee infant is born almost helpless, and is held securely by its mother's grip when they are together. But it does possess a grasping reflex and in a few days is able to cling to its mother's fur without any help. It is soon riding on her back, and like Omusisa and Tenkere, will stay by its mother's side for up to seven years.

Having studied the Kanyawara chimpanzees for two decades, Richard Wrangham says they all have strong, distinct personalities: "Imoso became alpha at the age of twenty-five; he's now thirty-seven, and he's had the support of the males because he's very good at establishing relationships with those that can help him. He's got a very good eye for balancing his level of power, he is individually dominant, he can fight anyone and win. But he doesn't exert that fighting power, he uses it to make the others a little afraid of him, and so they come to him to be friendly, and he's friendly back to them. So he's very good at manipulating others.

"At the same time he's very power hungry. He grew up with Johnny, they're both about the same age, but Johnny is a much more relaxed individual, he's never really had much interest in trying to vie for high rank, he's remained around ranks four or five. He's a loner, he doesn't have any strong links with males his own age, and so this raises fascinating questions about the power hungry and the lower-ranked relaxed males. You'd think it would pay for everyone to vie for the number-one status. There's a trade-off. If you're like Johnny you live a lot longer, and you can sneak matings while the other males are fighting over the females.

"Kakama is really interesting because he's very friendly with females, and he's frustrated by the fact that he's next in line to Imoso and all these years Imoso has been ahead of him, protected by a group of senior males. Kakama has tried to attack Imoso on his own, but Imoso is much too smart for him. One interesting example occurred three years ago, when Kakama and Imoso were hunting. Kakama was being attacked by red colobus males, Imoso came and joined them, and instead of doing what you'd expect to happen with Imoso joining Kakama and defending him against the monkeys, Imoso went up alongside them and he and the colobus attacked Kakama together. Kakama took off.

"Kakama has shown intense interest in status, and he's always displaying and showing off in a way that's not particularly of interest to the males, but it seems to appeal to the females. He's spent a lot of time currying favor with the females. So, maybe he's preparing for the day when he will become an alpha male, and hopes that he'll carry the common populace.

"Outamba has always been incredibly relaxed, amazingly calm, when the males are displaying around her, which is incredible because she's very frequently the victim of male aggression. Once a male attacked her for eight minutes on end, which is the type of beating that would kill a human. She was clubbed several times. The male had a stick in his hand and beat her with it. But she remains emotionally unreactive, and so when males display she doesn't shout and scream like some of the other females do. She pays a lot of attention to her offspring. She has a very large family; every three years she produces a new offspring, and they tend to survive. She's a very good mother, very calm, very tolerant, perhaps because she likes the males quite a lot and spends a lot of time with them.

"After she came into the Kanyawara group from outside, she formed an alliance with Tongo, who arrived a year after her, to help establish themselves, and that was successful because they gave birth and became central mothers in the group. But Tongo never became as relaxed as Outamba, she was more likely to spend time on her own, and she doesn't like the males as much because she's nervous, more flighty."

Richard's latest project is the study of how young chimpanzees play. In original research he has discovered that "the juvenile females cradle dolls in the form of sticks, and the juvenile males use the same sticks as weapons in play. Everyone has access to the same sticks. With humans there are different toys being used by boys and girls. Most people think this is caused by social learning, cultural influences, and so it is striking in the chimpanzees that what we find is that juvenile females carry the sticks as if they were dolls, but the juvenile males sometimes use them to hit other males, using them as clubs or throwing weapons. This is before adolescence, and so it could be a case of children's culture being passed on by the children themselves."

He has also discovered that young chimpanzee males learn important lessons in other forms of play: "The males play with each other, and a pair of them will often spend half an hour in a slow chase going around a tree, smiling, laughing, touching each other's feet. Some are more playful than others. This is contributing to building important alliances."

Kakama showed his maverick side early in life. Researchers noticed that when his mother was pregnant, the young chimp would cradle small logs as if they were babies, in the same way as the young females. Richard says the clearest sign that Kakama treated the log as an imaginary baby was that he would make a small nest next to his own and place the log in it.

In western Africa, more than a thousand miles away, young chimpanzee females have also been observed playing with "dolls." Tetsuro Matsuzawa is a researcher with the Primate Research Institute at Kyoto University and at his study site in a Guinean jungle noted: "At 5 p.m. we heard loud screams in the forest. A male chimpanzee, Yoro, caught a live hyrax [a plump rodent-looking creature distantly related to the elephant], which elicited much general excitement among the group. Sounds of fighting continued to be heard for half an hour. Finally peace returned and when we next saw the hyrax it was dead—we spotted Vuavua, eight years old at the time, climbing up a tree with the lifeless animal in her grip."

Vuavua carried the dead hyrax around for the remainder of the day, and as night approached she made a nest and then lay in it with the hyrax in her arms. "She started to groom the body with her fingers and lips, and held it up in the air with her hands and feet [just as chimp and bonobo mothers do with their babies]. "We continued to observe her until late in the evening. When, at last, she went to sleep, she did so while holding the hyrax." Vuavua abandoned her "toy" at noon the next day.

Matsuzawa observed another example of toy play among the same group. Ja, another juvenile female, was the same age as Vuavua. "She was walking with her mother who was carrying Ja's two-and-a-half year-old younger sibling. Ja wasn't empty handed either: following her mother's example, she carried a large branch with her, about 50 cm in length and 10 cm in diameter. She held the branch doll at her side with her arm, as if it were an imaginary infant."

These were not isolated incidents. "Young female chimpanzees treated objects—like a branch or a dead animal—as if they were handling real infants. In a way, such doll play behavior is good practice for the future, bringing up your own young."

In contrast, almost all the male chimpanzees grow up learning to make war from their fathers. Some chimps are born warriors, and others could not care less. "When males hunt, some of the juveniles are much more interested in taking part than others," Richard Wrangham told me. "Chimpanzees hunt monkeys to kill and eat them. But, the pure joy of hunting for some seems more important than consuming the meat. I've seen Johnny doing this, going up a tree and killing a monkey, eating a few bites, and then going up and killing another one. He never came back to the first one. Two individuals, Imoso and Johnny, are responsible for initiating almost all the hunts. This is a temperamental feature; they are more willing to take risks."

Jane Goodall, Richard's mentor, has made a decades-long study of male dominance strategies among the chimpanzees at Gombe, and in a novel twist offers advice in workshops for chief executives who want to understand the common imperatives that drive power-hungry chimpanzees and humans. The money she raises goes toward her Jane Goodall Institute, which supports chimpanzee research and education.

On a recent visit to Australia, Goodall told the national paper, *The Australian*: "I have observed chimpanzees for many years, and it's clear that we share many traits. I'm now very conscious of things like body language and people's eyes, and it does become easier to predict how somebody might behave." She explained that the most successful chimp groups had a leader who was not aggressive. "The best ones are . . . self-confident, smart, and have alliances. And as soon as you lose the alpha male you have absolute chaos."

She listed the behavior of adult chimpanzee males in their never-ending power struggles:

Chimpanzees have a complex social structure where everybody knows their place.

They have a hierarchy, most often with a male at the top.

The alpha male demands the respect of others in the group, and also the right of access to food, and the sexually desirable females.

Others are submissive to the alpha.

Not all males want to be the alpha.

The alpha's rise is based on superior intelligence, superior force or both.

The alpha must be able to manage a coalition.

As a male becomes more powerful, he brings his supporters through the ranks with him.

A displaced alpha male can quickly slide in the rankings.

Young males follow the leader and imitate his behavior.

Goodall's co-promoter, Andrew O'Keeffe, arranged for the workshops to be held at the Taronga Park Zoo in Sydney, Australia. "Business leaders who attend the workshops, almost always see someone they recognize," he told *The Australian*.

At Kibale, Francis Mugurusi had told me that a clan of pygmies lived not too far away. Traditionally, pygmies eat chimpanzee and gorilla meat, and I want to see whether they are still killing the great apes. Four hours' drive from Kibale are the legendary Mountains of the Moon, named so by the famed Roman-Egyptian geographer Claudius Ptolemy in the second century A.D. They soar as high as 16,763 feet above sea level, straddle the equator, and form the border between western Uganda and eastern Congo. The tropical forests in the valleys far below on both sides of the border have long been the home of the Batwa, among the most numerous of the pygmy tribes. One of the first pygmies ever to visit the United States was a Batwa, a young man named Ota Benga.

In 1904, Benga was among several pygmies brought to the United States from the Congo to live in the anthropology exhibit at the St. Louis World's Fair. Two years later, he was housed for a short time at New York City's American Museum of Natural History, and then briefly at the Bronx Zoo, where he was put on exhibition each morning in a cage with an orangutan.

On the Ugandan side of the border, our Land Cruiser trundles along a dirt road high in the foothills. They have long been stripped of trees, but their slopes plunge to lush green valleys, a vast forest designated as a Ugandan national park. This action doomed the pygmies' traditional lifestyle here. At a pygmy encampment nudged against a town populated by Bantu (as the taller Africans are called here), I find about 30 pygmies slumped against their mud huts, looking bored and dispirited. Their leader, Nzito, at four foot three, is the smallest pygmy I have ever seen. He tells me that their ancestors migrated from the Ituri forests across the Congo border more than a century ago and came to live in a land of plenty.

"In 1993, President Museveni forced us from our forests, and never gave us compensation or new land," Nzito tells me. The forests were declared a national park, and the clan had to move to borrowed land by the Bantu town. They know they can never again live in their forests and are banned from hunting there. I ask Nzito whether his people still eat chimpanzees. A wistful look clouds his eyes as he says, "We can't hunt chimpanzees anymore, but when I was a young boy, the men hunted chimpanzees and I still remember that they tasted very good."

The pygmies here now eat pork, fish, and beef. At the back of the huts, Nzito shows me about half an acre of marijuana plants to explain where they get the money to buy food in the Bantu market. The sale and use of marijuana in Uganda—even a small amount—is punished by stiff prison terms, and yet the pygmies' crop, hundreds of flourishing plants, is in the open for all to see. The Bantu police must pass it every day. "We sell the marijuana to the Bantu, but the police never bother us," Nzito explains with a smile. "We do what we want without their interference. I think they're afraid we'll cast magic spells on them if they upset us."

Like the pygmies, the chimpanzees are facing an intense challenge to their existence in their jungles, and the crucial question is how long will it take for those habitats to be destroyed or made uninhabitable by humans. Already the chimps at Kibale and Gombe and elsewhere in equatorial Africa, their homeland, are under threat. "Although it's a national park, poachers are hunting bushmeat here, and several chimps have been killed," Francis Mugurusi said back at Kibale.

Rob Muir, the conservationist at Goma, told me that because of overpopulation, habitat loss, poaching, and illegal charcoal trading, he believed great apes in the wild will largely disappear over the next 50 years, a view also common among many prominent primatologists such as Richard Wrangham. "All we'll have left will be great apes in much smaller areas in the wild protected by armed rangers, as well as zoos and sanctuaries," Rob predicted.

To visit one of the best of the many sanctuaries for chimpanzees in Africa, I drive back to Lake Victoria near Entebbe in Uganda and clamber aboard a small boat with the manager of Ngamba sanctuary, veterinarian Lawrence Mugisha. The island sanctuary is funded largely by the Jane Goodall Institute. A wind whips up the waves as we use a GPS to plot the exact position of the equator. As we cross the imaginary line, we travel from the northern hemisphere to the southern hemisphere in an eye-blink. Sixteen miles from the shore, ahead of us is a small, button-shaped island mostly swathed in forest. "The sanctuary takes up the entire island of one hundred acres and so there's no way off it for the chimpanzees," Lawrence tells me. Sharing the island and its shoreline with the chimpanzees are fruit bats, monitor lizards, a trio of hippos, dozens of bird species including fish eagles, and crocodiles.

"All forty-five chimpanzees on the island are orphans, rescued from poachers and people who kept them as pets," Lawrence adds. "They've come from as far away as the Congo, and have suffered terribly. To get an infant the poachers usually have to shoot its mother. They chop up her carcass in front of the infant and sell it as bushmeat, and then sell the baby as a pet." The profit makes the slaughter and capture worthwhile. In the United States, a baby chimpanzee fetches up to $60,000 in the pet trade according to Steve Ross, chair of the Chimpanzee Species Survival Plan for U.S. zoos, the chimpanzee species coordinator in the U.S.

At Ngamba, sweet-faced Bili has a typical story. About 10 years old, she was rescued when still an infant from poachers in the Democratic Republic of Congo. Arriving at Ngamba, she had scabies all over her body and most of her hair had fallen out. "We bathed her in lavender oil and rubbed vitamin E cream into her skin," Lawrence says as we leave the boat and stroll toward the chimpanzees' jungle compound. "Being with other chimps turned her into a happy girl once again. All the orphans have suffered similar experiences."

As we approach the night cages near the landing, a major problem facing chimpanzee and also gorilla sanctuaries becomes evident. In a large cage, a stocky adult male chimpanzee is screaming in anger and throwing himself about the bars in frustration. The cage is about 10 yards high and he swings up, down, and around the cage with a mania-cal ferocity. He looks to me as if he is going mad. "That's Mawa," Lawrence tells me. "He's challenged the dominant male twice on the island, but the dominant male and his allies attacked him. Mawa fled, bursting through the electric fence both times. We've since had to keep him in this cage for two years, and don't know what to do with him."

I go close to the bars and Mawa, who has good reasons to hate humans, screams, bares his big canines, and charges me on all fours with all his hair standing on end to make him appear bigger. Even though the bars protect me, it is a terrifying sight. Goodall must have been blessed with supreme courage, because she endured many such ferocious charges in the wild unprotected when she was getting the chimpanzees at Gombe used to her presence.

"No hard feelings, mate," I murmur as Mawa flings dung at me and then squats by the bars and scowls. "I'd do the same if I were you."

The remaining chimpanzees are separated from visitors by the elec-trified fence. From a raised walkway I watch them being fed breakfast—bananas, pineapples, papaya, and other fresh tropical fruit. Even in a sanctuary of orphans, the chimpanzees have formed themselves into a typical group with a dominant adult male, 12-year-old Mika, who grabs the choicest fruit. Seated by him is his close ally, potbellied Tumbo. "We called him that because it means 'big belly' in our language, and he's very greedy," says Lawrence. "He has no ambition to lead the group, content to be number two to Mika." After the feeding the chimps groom each other, take a nap, or wander into the thicket.

Lawrence tells me, "Mika came to us from the Congo eight years ago. He was tied at the waist, led by a boy, and was owned by an army officer. He was high-ranking even among the juvenile group here and formed alliances among the males that later helped him become the dominant male. The females like him a lot."

Lawrence points out Afrika, a one-year-old female who is happily playing with other infants in the grass. "She looks fine now, but she was a mess when she arrived. Poachers had placed her in a tiny wooden cage that was so small she couldn't reach out. It was piled with her own filth. When we rescued her she was dehydrated and covered in sores. She was clearly depressed, but we fixed up her health problems and she's now playing with the other infants."

One of the small juveniles has a plaster cast on his arm, and has a tough, determined look despite his disability. Lawrence says, "That's Rambo; he's very aggressive with the other juveniles. So, recently the high-ranking adult males decided to teach him a lesson. They beat him up, and in doing so broke his arm."

The most recent arrival, two-year-old Mac, was rescued a month before my arrival. Two villagers illegally had him on sale in a booming black market. Like the other orphaned chimps, he most probably witnessed his mother being slaughtered by poachers. Ugandan police joining with Jane Goodall Institute workers found him in a villager's dark hut, quivering with fear. I was unable to see Mac because he is undergoing a three-month quarantine so that he doesn't pass on any human diseases before being released among the other chimps.

Clearly, chimpanzee sanctuaries are no answer to the looming species crisis because they represent a very expensive way to care for a relatively few chimps, and many have the females on birth control measures so as to avoid inbreeding and to limit their numbers. And there are many sanctuaries where the chimps live in terrible conditions. One of them, the Limbe Wildlife Centre, on the other side of the continent on the Cameroonian west coast, has an excellent international reputation, but when I visit, I find it to be a hellhole for the 53 chimps of the rare central subspecies and the Nigerian-Cameroon subspecies kept here. An information board at the entrance lists its U.S. donors, including Disney, Del Monte, and also Brigitte Bardot's animal foundation. Given sanctuary inside are orphaned chimpanzees, gorillas, drills, pygmy crocodiles, and other threatened species.

The area has one of the highest amounts of rainfall in the world and hosts an enormous biodiversity of species, so I expected much more than what I found. Across the road from the sanctuary is a century-old botanical garden, a lush hill swathed in tropical trees and plants, ideal chimpanzee territory. But the Limbe Wildlife Centre houses its orphans in shocking conditions. The chimps are mostly victims of poachers, who kill their mothers for bushmeat and try to sell the orphans on the pet-trade black market. They are kept in hardscrabble yards of bare earth scattered with stones and with scant grass. A keeper tells me the chimpanzees frequently throw stones at visitors, and no wonder.

At Limbe you hear the same pitiful tales as at Ngamba. Two-year-old Nanga Eboko was found in a Cameroonian village, suffering from gunshot wounds and a fracture. An X-ray showed there were several pellets in the little chimpanzee's body. Limbe's devoted keepers nursed him back to health, but he still will not grow up anything like a normal chimp. Most of the few trees in the chimpanzee compounds are protected by electric wire so that the chimps cannot climb them. I witness little interplay among the chimps, rarely even grooming in contrast to Ngamba. Many of the apes seem to suffer from mental problems, pacing restlessly up and down their compound, rocking to and fro, and ignoring each other.

I spent a few hours at the sanctuary and was glad to leave, feeling very sorry for the chimps. Funds for sanctuaries are always in short supply, and they are forever seeking money, mostly from foreign donors. However, my impression was that Limbe was such a disaster not because of any shortage of money, but because of bad planning. At Ngamba, the conditions were kept as close as possible to those of the chimpanzees' natural habitat, in contrast to the horrendous chimpanzee enclosures at Limbe. It cannot be that difficult to grow grass in the enclosures when you have such an abundant rainfall.

A year after my visit, the Limbe sanctuary opened a new grassy enclosure for the chimpanzees that is far better than their previous accommodation.

Many African zoos are far worse than Limbe sanctuary before it opened the new enclosure. In Niamey, capital of impoverished Niger, near the edge of the Sahara, the zoo is a tumbling-down relic from the former French colonial days. A handful of chimpanzees are kept there in the most horrible conditions. They live in tiny, dark concrete cells that resemble caves fronted by narrow enclosures that barely allow them a hop, skip, and jump. I stayed there for more than an hour on a recent visit, hoping to see whether they were healthy, but not once did the chimpanzees leave their caves. There were no other visitors while I was there. I spotted several sets of dark eyes gleaming in the darkness as I sat patiently by the bars, hoping to earn their trust. The ill treatment that might have lasted for most of their lives won out. I must have been just another pale-skinned great ape to be feared and avoided.

Also shocking were the conditions suffered by the chimpanzees at the Central Zoo in Pyongyang, the reclusive capital of North Korea. I visited the zoo in 1974, and apparently the revolting conditions there had gotten no better three decades later when Jane Goodall went there in 2004. The chimpanzees' plight moved her to tears.

Goodall's biographer, Dale Peterson, described the place where the chimps were kept as "a stark enclosure entirely devoid of vegetation, or even furniture except for a strange pole-and-cable erection." The keeper led out by the hand a pair of emaciated chimps who looked more like scarecrows. On command, the female climbed the pole and slowly and awkwardly walked a 30-foot tightrope. The keeper ordered the male to mount a bicycle, but he was so weak from lack of food that he nearly fell off. The keeper then made him do one-arm pull-ups on a crossbar, even though it hurt him because one of his arms was injured. He held out an arm in the chimpanzee-pleading gesture, which Jane knew well, but the keeper ignored it.

North Korea is an impoverished dictatorship where hundreds of thousands of "enemies of the people"—men, women, and even children—are hidden away in gulags, prison camps where they are kept at starvation level and threatened with summary execution, merely for the crime of criticizing its megalomaniac ruler, Kim Jong Il. Across the country, hundreds of thousands of youngsters suffer from malnutrition, so the chimpanzees in Pyongyang's zoo can expect no better than their shocking treatment. Goodall later mailed the zoo notes on the proper diet for chimpanzees, but it is doubtful that they were ever used.

6

A world away from the cruelty of the Pyongyang zoo, the most impressive chimpanzee sanctuary I saw is in the United States. Save the Chimps, in Florida, the world's largest sanctuary for rescued chimpanzees, is set up as a retirement home for chimps who have been subjected for decades to cruel, invasive experiments by NASA and medical laboratories using chimpanzees as human substitutes. Some of them are descendants of chimpanzees captured in the wild in Africa for NASA's "chimpanaut" project.

From the Miami airport I take a bus for the four-hour ride north through the Florida flatlands, nudging the ocean shoreline, to Fort Pierce, south of Orlando. A high gate bars the outside world. Inside, a dedicated staff of 50 keepers in neat shorts and sanctuary-branded T-shirts scurries around in golf carts looking after the 150 chimps settled on seven 3-acre artificial islands built on the 200 acres of an old orange grove. A further five islands await more than 100 other chimps being kept at a former medical laboratory in New Mexico until they can be moved to the sanctuary.

The sanctuary's director, Dr. Carole Noon, a slim 58-year-old veterinarian, founded Save the Chimps in 1997 when the U.S. Air Force abandoned experiments using chimpanzees, labeling the ones it still had as "surplus equipment." The Air Force dispatched the chimps to the Coulston Foundation in Alamogordo, New Mexico, a biomedical laboratory. "It had the worst record of any biomedical lab in the history of the Animal Welfare Act," Carole tells me. "Chimps were kept there in tiny windowless concrete cells for decades, never seeing other chimpanzees. They were subjected to experiments that included constant blood-taking and injecting them with diseases such as AIDS." Steve Ross, chair of the Chimpanzee Species Survival Plan for U.S. zoos, agrees: "The Coulston facility was the Auschwitz of research labs for chimps."

Appalled at the conditions, Carole fought for five years in the courts to gain custody of the NASA and Coulston chimps, successfully suing the U.S. government for their release. Since then, she has progressively relocated these grizzled veterans to their Florida retirement home. "We've moved just over half," she tells me. "I've had to carefully group them in chimpanzee families, each with its own island, and that's why it's taking so long, and why we still have so many chimps to bring here."

After a good night's sleep to shake off my jet lag, I join Carole in her regular breakfast-for-the-chimps routine. Although their cage doors are open day and night to allow entry onto the islands, most of the chimps choose to sleep in the cages rather than out on the grass. Maybe that is because each has chimp-sized hammocks hooked near the ceiling of the cages to mimic the wild chimpanzees' nests in the forest trees, even though none has ever been near a forest. Carole bangs on a cookhouse-style iron triangle to signal the day's first meal, and the chimps are pressed against the bars as we arrive, eagerly eyeing breakfast.

Many of the chimps are in their fourth or fifth decade of life, and truly look like residents of an old people's home as Carole feeds each one individual medication for heart disease, arthritis, and other ailments they share with elderly humans. The medicine is concealed in cups of fruit juice, and they lap it up. Then she gives them oatmeal studded with raisins. The final course is a tasty spread of fruits and vegetables. Most then wander out onto their grassy islands to enjoy the jungle gyms strung with ropes, log posts that substitute for climbing trees, raised walkways, and other delights for the acrobatic chimps. Some lope on all fours along the walkways, while others clamber up the vertical logs for a chimp's-eye view of their world or swing from the ropes on the jungle gyms.

Some chimpanzees, kept in their lab cells for decades, cannot make the transition from concrete to grass. At one island, all the chimps are gamboling in the grass save for 25-year-old Alice, who grips the cyclone wire at the edge of the concrete strip leading to the island. Carole looks at her with sympathy. "Alice has been here two years, but doesn't ever dare go out onto the grass," she says. "It still frightens her."

Riding in the golf cart with us is imp-faced Melody, an 11-month-old chimpanzee born at the sanctuary. Carole tells me, "Her mother was a first-timer, and abandoned Melody at birth. We've hand-raised her, and a human caregiver is with her twenty-four hours a day, but in the next couple of months we'll put her into a family group." We stop at the cage of Ron and April, two more old-timers enjoying the Florida sunshine. "We showed Melody to all the families, and Ron and April seemed attracted to her the most. We'll put her with them and monitor them carefully."

Melody reaches out to grab me with her long, slender pink fingers that look just like a human's. She turns her sparkling dark eyes on me as she pulls at my polo shirt and, though it may be my imagination, there seems a hint of a cheeky smile on her lips. As they reach adulthood, chimpanzees' fingers—and their toes, which are just like another set of fingers—thicken and grow coarse and tough. Melody keeps reaching out to me, wanting to touch my hands, and for me to touch her hands. "Melody likes human males a lot," says Carole with a smile. "She can tell the difference between male and female humans."

A truck arrives, decorated with paintings of chimpanzees and custom-built to carry 10 chimps at a time from New Mexico. Suddenly, the chimps on the islands start screaming, and it sounds more like terror than anger. After the tranquility, the noise shocks me. Moments later, most of the still screaming chimps flee back along the walkways and into the security of their cages. "That's the warning signal chimps use when there's danger about," says Carole. "Maybe they think the truck has come to take them back to the lab in New Mexico."

Food alone for the chimps here costs half a million dollars a year, and Carole's critics claim the money should go to feed starving children in developing countries rather than a bunch of aging chimpanzees in the United States. "Feeding starving children would not solve the bigger issues," she says. "The world is not divided into starving children and captive chimpanzees—it's not either–or. As they are our next of kin, I feel an affinity to help them.

"Furthermore, our budget for the chimps is four million dollars a year. Recreational fishing in Florida is a two-hundred-million-dollar-a-year business. Americans spend billions of dollars a year on weight-loss products, billions more each year on their pets, and how can anyone justify these luxuries in the face of starving children in poor countries? By this logic people who criticize me should not go to the movies or the opera, play golf, eat at restaurants, have pets or any of the many luxuries in our lives while children go hungry."

On my final night at Save the Chimps, I walk in the darkness to the moat facing one of the islands. Soon after, a burly chimpanzee male knuckle-walks out on the grass and climbs onto the walkway. He clambers up a log post, perches there, and stares up at the silvery moon. For decades, the boundaries of his world were a medical lab cell's concrete walls pressing against him. Now, he can breathe fresh air and can see all the way to the stars.

For how much longer can his kin in Africa do the same?

Just when we are beginning to unravel and comprehend the full complexities of the chimpanzee society in the wild, and how it relates to our own evolution from our common ancestor, the ape-man, it seems we are witnessing the beginning of the end for most free chimpanzees. We humans are both the cause of their demise and the losers.

We should also consider the fate of captive great apes, including the thousands of chimpanzees who are still kept in shocking conditions in medical laboratories and used as performers in TV and print advertisements and movies. Zoos are also not blameless. They pose as our closest kin's saviors and do produce useful research on captive great apes. However, realistically, the lifestyle of zoo great apes is extremely limited when compared with the fulsome and unrestrained behavior of their everyday life in the wild. Zoo chimpanzees can never enjoy that freedom, but we can keep them in conditions that do not subject them to daily stress.

In 2008, the Spanish government took the lead, legislating to protect what it termed the rights of great apes, banning their use in films and TV commercials and outlawing their use in medical laboratories. However, Spanish zoos were still allowed to put great apes on display. This limited stand-alone action was prompted by the beliefs of Peter Singer, professor of ethics at Princeton University, who co-founded the Great

Ape Project (GAP) in 1993. It was, in his words, "a proposal to grant rights to life, liberty, and protection from torture to our closest non-human relatives: chimpanzees, bonobos, gorillas, and orangutans."

Singer claims that we have made great apes and other animals in captivity our "slaves," to use "as we wish, whether to pull our carts, be models of human diseases for research, or produce eggs, milk, or flesh for us to eat." His demand for animal rights is too radical for most, though his advocacy for the great apes has gained wider acceptance, riding on his assertion that "great apes are intelligent beings with strong emotions that in many ways resemble our own." Singer coined an ungainly word, "speciesist," to define what he considers the prejudice of human beings directed against other creatures.

In the United States, Steve Ross, the supervisor of behavioral and cognitive research at the Lester E. Fisher Center for the Study and Conservation of Apes at the Lincoln Park Zoo in Chicago, is pressing for a ban on great apes in films and advertisements. In an op-ed feature for the *New York Times* in 2008, "Chimps Aren't Chumps," he wrote:

You see it on greeting cards and in countless TV programs and commercials: the exaggerated grin on the face of a young chimpanzee, often one that's wearing sunglasses or a grass skirt. It's about as common a ploy for laughs as a pie in the face. Generations have been amused by the antics of Bonzo, J. Fred Muggs, Zippy and, more recently, the business-suited chimps of Careerbuilder.com. A chimpanzee covering its eyes in embarrassment? What's not to love?

But this picture, harmless as it might appear, is giving the public the mistaken and even dangerous impression that chimpanzees have a safe and comfortable existence—and nothing could be farther from the truth.

A survey that I and several colleagues conducted in 2005 found that one in three visitors to the Lincoln Park Zoo assumed that chimpanzees are not endangered. Yet more than 90 percent of these same visitors understood that gorillas and orangutans face serious threats to their survival. And many of those who imagined chimpanzees to be safe reported that they based their thinking on the prevalence of chimps in advertisements, on television and in the movies.

Ross went on to detail the drastic loss of chimpanzees in the wild and concluded:

> *A progressive society should weigh the moral costs and benefits of practices like these. Misrepresentations of chimpanzees may not be as repugnant as racism, bigotry or sexism. But they can still serve as a benchmark for our society's moral progress. The good news is that a growing number of companies, including Honda, Puma and Subaru, have pledged to stop the use of primates in advertisements. The journal* Science *recently stopped its promotional campaign featuring chimpanzees in hats reading the magazine. That two consecutive Super Bowls have gone by without a major ad campaign featuring a chimpanzee is reason for optimism. Sometimes, success has to be measured in small increments.*

This is a worthy ambition, but his comments could just as well apply to chimpanzees held in zoo cages for the amusement and entertainment of paying customers. When I visit him at the Lincoln Park Zoo, I am troubled by the small enclosures for the chimpanzees and gorillas. Daily, hundreds of people file past to gape at them.

When I ask Steve for an explanation, he retreats to a wary defensive tone. "If a hundred chimpanzees in the wild need a few square miles to live in, who is to say that seven chimpanzees don't find enough room in a zoo enclosure?" he counters. "They are free from predators and we provide our great apes with a secure environment and enrichment programs to keep their minds stimulated."

I witnessed Imoso, Outamba, Kakama, and other chimpanzees in the wild at Kibale form small groups and roam over many miles foraging, grooming, squabbling, and taking naps. The blunt reality is that the great ape zoo cages I have seen across the Western world are mostly far too small for the chimpanzees as well as the orangutans, gorillas, and bonobos held captive in them. None are anywhere near as bad as the Pyongyang and Niamey zoos, but even the most modern zoos still hold great apes in barbaric conditions. They cannot be kept in forests, but surely they can be held in safari-style zoos such as San Diego's excellent Wild Animal Park, which houses eight bonobos.

*   *   *

The Congo calls once more, and I steel myself for the journey to a remote province ravaged over the past decade by brutal civil war. There, the bonobos, the chimpanzees' closest relatives, live in their native jungle habitat. Many primatologists claim that the bonobos are the smartest, sexiest, and most peaceful of all the great apes.

The Bonobos

Our Long-Lost Cousins

1

For most people, chimpanzees, orangutans, gorillas, and, for the purists, human beings have long been the only known great apes. But during the 1990s, people became captivated by media reports of another great ape, the bonobo. More commonly known as the pygmy chimpanzee, it resembles Cheeta and his kin, though the bonobo's face, skull, and teeth are smaller and its limbs are more in proportion to the body like ours, in contrast to the hulking chimpanzee. It also has a more upright stance when it walks on two legs.

The wild bonobos live in dense rainforests in the remote northeastern Congo, an area about the size of Montana. The species is so rare that there are only about 160 in captivity in the United States and Mexico and about 100 more in the Congo and Europe. The last of the great apes to be discovered, the bonobo is so threatened that it could be the first to become extinct in the wild. It has the misfortune to inhabit some of Africa's most dangerous jungles, and its numbers have plunged from more than 100,000 three decades ago to fewer than 20,000 today.

The gregarious, super-intelligent bonobo is known as "the sexy ape," and is the only great ape whose society is claimed to be matriarchal. While the orangutan is a loner, and chimpanzees and gorillas often settle disputes by fierce and deadly fighting, researchers claim that the bonobo commonly makes peace by means of feverish sex orgies. Even the youngsters join in.

Largely based on studies done on bonobos in captivity and at provisioned food dumps in their native habitat, the bonobo was tagged by the media as the hippie great ape because researchers found that, in contrast to the brutish chimpanzees, it preferred to make love, not war. It had a rich promiscuous bisexual sex life that seemed straight from the pages of an advanced sex manual, and its clans were said to be dominated by females who formed sisterhood alliances to rule the males.

Many American feminists came forward to claim the bonobo. Until the release of this research, the chimpanzees had been taken as the primate model for the creature from which we humans evolved. Their bellicose nature, and eagerness to kill stranger chimpanzees they catch in their territory, supposedly explained why mankind, the chimps' evolutionary cousin, has been making war across the globe for as long as our historical and tribal memories go back.

But now a great ape species had been found that seemed to prove a significant strand of our nature had evolved from a gentle, tolerant, peace-loving ancestor. It explained our yin-and-yang dual nature, Mother Teresa and Adolf Hitler. The *New Yorker*, tongue-in-cheek, described this "pop image of the bonobo—equal parts dolphin, Dalai Lama and Warren Beatty."

This puzzled me, because from pictures I had seen, the wild bonobo males were brawnier than the females, and they lived in jungles haunted by leopards and giant pythons. How was it then possible for female bonobos, alone in the great ape world, to somehow come to dominate the males?

A leader in the research of this new kind of great ape is primatologist Frans de Waal, a Dutchman who is director of the Living Links Center at Emory University in Atlanta, Georgia. De Waal has written a number of books on the bonobo, but he has never studied bonobos at length in their native habitat in the jungles of the Democratic Republic of Congo, a country that seems forever embroiled in bloody civil wars.

Bonobo territory is now relatively secure, though still risky, but toward the end of the 1990s, the Congo civil war had entered a frenzied stage, people in the millions were being slaughtered or dying of war-related disease and starvation, and no foreign researchers would dare journey to the bonobo habitat. I planned to go there, but, while waiting for the Congo to calm down, decided to fly halfway around the world to the Frankfurt Zoo in Germany.

In 1998, on a chilly November morning, Carsten Knott, the keeper of the great apes at the Frankfurt Zoo, led me into the ape house, where our primate relatives were sheltering in their heated cages from the snow and ice of a bleak German winter. Far from their hot and humid equatorial African and Southeast Asian homelands, they had deserted their outdoor play areas. Here, Carsten, in his early 30s, had found his heaven on Earth. "I wanted to be keeper of the great apes at Frankfurt Zoo ever since I was a little boy and went to the zoo," he told me as we walked around the large covered building. It housed lowland gorillas, chimpanzees, orangutans, and a clan of rare bonobos in their respective cages.

This was my first-ever sight of bonobos, and prompted by their other, erroneous name, pygmy chimpanzee, I expected to see a dwarfish version of the chimpanzee, with the same swagger and strut of the male and timorous fealty of the female that I had witnessed in the jungles of Africa. Instead, the egalitarian bonobos sat calmly together in a large cage, and they charmed me because there was no apparent antagonism between the sexes. Their black hair fell in a neat part down the middle of their high foreheads, above dark, expressive faces. Their lips had a pinkish hue and they were less bulky than chimpanzees, especially the males.

If male chimpanzees resemble a great ape version of Arnold Schwarzenegger in his body-building youth, the male bonobos I saw that day had the elegant appearance of a great ape Rudolf Nureyev. With their humanlike long arms and legs, the bonobos resembled *Australopithecus afarensis*, the missing link in our evolution, the bipedal hominid ape-man who strode the African savannah 3 million years ago. The diminutive apelike creature named Lucy by paleontologists is the best example we have found of the *Australopithecus*. Her ancient bones were discovered in an Ethiopian highland in 1974. When scientists constructed an image of Lucy based on her bones, she bore a startling resemblance to a bonobo.

Although swathed in black hair, the bonobos walking peacefully about the enclosure looked eerily human with their upright bipedal gait, long, slim arms and legs, slender neck, and a torso whose proportions more closely resemble ours. Along with chimpanzees, they are our closest evolutionary cousins, sharing about 98.4 percent of DNA with us. They are even closer to the chimpanzees, sharing 99.3 percent of DNA. Gorillas share 97.7 percent of DNA with us, while the loner enigmatic orangutans in far-off Southeast Asia share 96.4 percent DNA with humans, gorillas, chimpanzees, and bonobos. In contrast we share about 40 of percent DNA with earthworms, 60 percent with chickens, and 80 percent with mice.

Primatologists who have worked with both species say bonobos seem innately smarter than chimpanzees. In 1923, pioneer American primatologist Robert Yerkes of Yale University was captivated by a bright young chimpanzee captured in the wild and named him Prince Chim. Comparing him with other chimpanzees he was studying, Yerkes considered him an intellectual genius. Bonobos had yet to be discovered as a separate species, but we now know from photos and his bones that Prince Chim was a bonobo.

Yerkes's son, also Robert, wrote to me from his home near Washington, D.C., that his father brought home to the family farm what he thought were two chimpanzees: "Panzee was a common female chimp, not very smart and bad-tempered, but Chim was bright as a button, a joy to be with, my playmate. He became my close friend, and we'd wander off together across the fields and have a great time. He was very smart."

Carsten Knott agreed: "I tell new keepers that if you throw a screwdriver in with the gorillas, they wouldn't notice it for weeks on end unless they sat on it. The chimpanzees would use it to destroy something within minutes, but the bonobos and orangutans, within thirty minutes, would figure out how to use it to unlock the cage door and escape."

As we talked, an adolescent bonobo female named Elindi began to groom me, her long fingers tenderly searching through my hair. Satisfied that I was clean of bugs, she offered her back for me to groom in return. After grooming her for a few minutes I left to pay respect to the clan's dominant female, who was seated with the other apes and communicating with them in a constant stream of the bonobos' unique high-pitched squeaks. Elindi's eyes seemed to catch fire with an intense, focused glowing, but minutes later she drew me back with a sweet gaze. She looked at me with what seemed deep affection, then suddenly threw into my face a pile of wood shavings she had been hiding behind her back and flounced away.

Science has known about the bonobos for less than a century. In 1928, Harvard zoologist Harold Coolidge was searching through a collection of what he thought were chimpanzee bones at a colonial Belgian museum near Brussels when he made an important discovery. Looking at the skull of what had been identified as a juvenile chimpanzee brought from the Belgian Congo, he noticed that the skull bones had joined together, indicating that the head belonged to an adult. Yet it was significantly smaller than an adult chimpanzee's skull.

Rumors had long been circulating among colonial officials that mysterious human-like chimpanzees lived in jungles south of the Congo River, then under the rule of the Belgian king. Coolidge passed on this information to a German anatomist, Ernst Schwarz, who swiftly concluded that the skull belonged to a separate subspecies of great ape and wrote a paper naming the creature *Pan satyrus paniscus*, or pygmy chimpanzee.

Coolidge proved that it was actually a new species of great ape and renamed it *Pan paniscus*—Pan for the Greek forest god and *paniscus* meaning "diminutive." After analyzing the bones, Coolidge concluded that the bonobo, whose name could have come from a misspelling of a village named Bolobo in the Congo, "may approach more closely the common ancestors of chimpanzees and man than does any chimpanzee hitherto discovered."

Scientists found that this new species lived only in a large chunk of dense, swampy equatorial forest on the left-hand side of the Congo River in what is now known as the Democratic Republic of Congo, Joseph Conrad's *Heart of Darkness*. In the late 1990s I was eager to go there, but the country was still very dangerous with back-to-back civil wars, and so foreign researchers stayed away from bonoboland, which was suffering some of the fiercest fighting.

A survey by the New York–based International Rescue Committee estimated that the Congo civil war was the world's deadliest crisis over the past 60 years, with troops of five African nations and numerous Congolese warlords battling in the country's jungles and cities and along its waterways. The fighting caused the deaths of more than 4 million Congolese.

The lure for dictators such as Yoweri Museveni of Uganda, Robert Mugabe of Zimbabwe, and Paul Kagame of Rwanda was the Congo's billions of dollars' worth of mineral resources such as diamonds, copper, and gold and the world's biggest deposits of the rare coltan, an essential mineral in the manufacture of cell phones. They sent thousands of troops storming across the Congo's porous borders to do battle in the civil war so as to grab a share of the plunder in the carve-up.

To learn as much as I could about bonobos, just after my visit to the Frankfurt Zoo in 1998, I flew to Japan to meet Takayoshi Kano, the first researcher to study bonobos in their native habitat for an extended period. His on-site research lasted almost two decades. When I met him at Kyoto University's Primate Research Institute, Kano told me he decided to study bonobos primarily because of their similarities to human beings. The DRC was a relatively peaceful place when he went there in 1974 to seek out the mysterious great apes. Around a campfire near the bonobos' remote jungle habitat in Wamba, just over 600 miles northeast of Kinshasa, the DRC's capital, a wizened elder told him that bonobos and villagers once shared a common life.

"But when humans began wearing clothes, the hairy bonobos spurned their smooth-skinned cousins and returned to the jungle," Kano went on. "Because bonobos were so like humans, since ancient times, the villagers observed a taboo on killing them, and called them brothers and sisters. The elder told me killing a bonobo was like killing a human."

Kano spent months trudging through Wamba's forests before he finally saw his first bonobos up close. It was a foraging party of 10 adults, and he was intrigued by their contrast to chimpanzees. "The bonobos were a little smaller than chimpanzees, but far more graceful and slender, with higher foreheads and limbs shaped more like ours," he told me. "When the bonobos walked upright with their straight backs, a very human posture, I got a chill down my spine realizing how closely they resembled the ape-men, our common ancestors."

Bonobos were arboreal, or tree-dwellers, and hard to follow through the dense, moist forest. In desperation, Kano tempted them down from the trees by provisioning, planting a field of sugarcane deep in their territory, a controversial technique that was also used by Jane Goodall at Gombe. Kano set up sugarcane caches to tempt the wild bonobos close enough to where he could observe the social interplay between individuals. "Because provisioning quickly eliminated an animal's apprehension of humans, it is an effective method of investigation if used cautiously," he wrote.

Kano waited patiently until weeks later he spied a bonobo clan, 40 strong, feasting on the tasty food. "Seeing them so close, they seemed more than animals, more a reflection of ourselves, as if they were fairies of the forest."

Expecting the bonobos to display the aggressive behavior of chimpanzees, with fierce adult males terrorizing anyone who defied their heavy-handed rule, he was startled to witness what seemed like females ruling the roost. They sat amiably by the sugarcane as they groomed each other, and snacked or chatted in squeaks and squeals, like the grand dames of a Parisian salon, allowing favored males to sit with them. Whenever a male, shrieking with anger and sometimes dragging a tree bough for noisy effect, made a rare charge against females seated together, Kano told me they either ignored his boorish display or chased him into the jungle.

In comparing adult female bonobos and chimpanzees, Kano noted that the chimpanzee females did not mix easily with each other. In contrast, the bonobo females were extremely sociable, "and always feed, travel, and rest in clusters. Social interaction between them is also lively."

Overall, the relationships between male and female bonobos is much friendlier than between the chimpanzees. Chimpanzee males daily engage in ritualized subordinate gestures towards high-ranking males, behavior that Kano never saw among bonobos. If a low-ranking bonobo male is attacked by a higher-ranked male, he "grovels and shrieks violently," but Kano saw these as signs of distress rather than appeasement. After it is attacked, a low-ranking bonobo will make a "cry-face," but it is not offered a calming embrace the way the chimps are.

Through daily observations, Kano found that each group had up to 150 bonobos—adult males and females, adolescents, juveniles, and infants—but each day they split into smaller foraging parties. Like the chimpanzees and gorillas, when a female became sexually mature she migrated to another group, and so avoided inbreeding, but the males remained with their birth group throughout their lives.

In Wamba, Kano made a remarkable discovery. Whereas chimpanzee males move away from their mothers emotionally as they grow up, spending most of their time with the other adult males, bonobo males remain bonded to their mothers throughout their life. That lessened the bond between males in bonobo groups and made the mother "the core of bonobo society, and the males lead a life following their mothers."

He concluded: "Thus, although the bonobo differs completely from the gorilla in that the female bonobo is central to the formation of parties or groups, the chimpanzee resembles the gorilla in that males form the nucleus of the unit and females are not friendly with each other."

Kano's observations were startling news for primatologists familiar with chimpanzee behavior. "Among chimpanzees, every female of
whatever rank is usually subordinate to every male of whatever rank,"
Richard Wrangham had told me.

Even more enthusiastic—once Kano's findings filtered beyond the
tight world of primatologists—were many American feminists who
claimed the bonobo females for their own, the great ape world's feminists, a sisterhood in the jungle. "Females . . . form alliances against
males and as a consequence, male bonobos do not dominate females
or attempt to coerce them sexually," wrote University of Michigan psychologist Barbara Smuts.

More salacious were Kano's sexual behavior observations. Bonobos
enthusiastically indulged in group sex with male on female, female on
female, male with male, and even the juveniles and infants joining in
the fun. This showed that the bonobo had a keen imagination. Chimpanzees and gorillas generally have sex in the one position, the male
mounting from the back, but bonobos use many positions including
females rubbing their genitals together, which Kano called GG, asserting, "No chimpanzees, gorillas or orangutans have ever been observed
indulging in GG."

2

By November 2006 the Congolese civil war had largely come to an end,
spluttering on in border conflicts such as the rebellion in the east near
Goma, and bonobo researchers had begun returning to their habitat.
I was invited on one expedition organized by the Bonobo Conservation Initiative, a Washington-based nonprofit organization dedicated to
saving the wild bonobos. Its president, Sally Jewell Coxe, told me when
I was in Washington that "bonobos live in peace, love and harmony,
and are a reflection of the better side of our own nature. Chimps are
from Mars, bonobos are from Venus."

Flying into Kinshasa is not for the easily scared. The terminal is unlit,
and its dark, dingy halls seem under attack by arriving passengers. They

rush toward the immigration booths, shouting, shoving, and thrusting each other aside to get to the front quickly. Fights break out, the participants sleek, dark men clad in Parisian-style suits, brandishing expensive gold watches and carrying laptops. They use their free hand to punch and push each other.

Having survived the melee, I board a dilapidated taxi for the trip into town. As in so many African countries, the streets of the capital city are potholed, and I pass by run-down skyscrapers with broken windows that look to be on the edge of rusting away or tumbling down. The streets are thronged with people. Armored personnel carriers rush by, manned by grim-faced Indian soldiers wearing the blue helmet of United Nations peacekeeping troops. They stand in the backs of the vehicles, gripping heavy machine guns mounted to the floor. With Indian troops also based in Goma on the other side of the country as UN peacekeepers and rebellions constantly breaking out across Africa, UN service seems a growth industry for the Indian army. It had proven its prowess in several wars against its neighbor Pakistan.

Outside UN headquarters in the city center, coils of razor-sharp barbed wire are strung along the high walls, and at the front gate troops peering out from behind sandbag emplacements warn off passersby. To intensify the threat, a tough-eyed Indian soldier perches at the turret of a tank parked directly in front of the barred gate. These precautions are necessary because murderous fights frequently break out in the streets between supporters of the president, Joseph Kabila, aided by the brutish police, and his major opponent, then vice president Jean-Pierre Bemba. The VP's stronghold in the northeast takes in bonoboland, where I am headed.

The day after arriving I drive to Lola Ya Bonobo, "Bonobo Paradise" in the local Lingala language, a sanctuary for orphan bonobos set in fertile hills about 20 miles from Kinshasa. Within 10 miles, the capital's down-at-the-heel boulevards give way to narrow, crumbling dirt tracks that thread together mud-hut villages and open-air marketplaces. Lola Ya Bonobo is set on several grassy hills enclosed by high-wire fences, and I am met by its founder, Claudine Andre, a Frenchwoman raised in the Congo. An attractive redhead, Claudine looks more like the middle-aged manager of a classy Parisian bar than a savior of the bonobos.

She shows me around the 35-acre sanctuary, which has several large jungle-like enclosures for the bonobos, as well as decorative ponds, a clinic staffed by a veterinarian, and an outdoor nursery that together house 44 bonobos, from infants to adults. All have been confiscated from humans who kept them as pets, usually after their mothers had been slaughtered for bushmeat. They play, eat, doze, and have sex by day in the enclosures, but retreat at night to barred cages to prevent poachers from stealing them.

One of Claudine's favorites is Mbali, a three-year-old once kept for months in a basket by what she describes as "a witch doctor." He chopped off one of the little bonobo's finger joints every so often to use in *ju-ju*, or black magic rituals. Most of the fingers of one of Mbali's hands are missing. "Congolese believe that if you put a bonobo bone into a baby's first bath, the child will grow up strong," Claudine tells me. "The witch doctor even chopped off the tip of Mbali's penis to use in rituals."

On seeing Claudine, Mbali lurches over and gives her a huge hug. He then covers her face with sloppy kisses. Bonobo researchers have noticed that kissing, as with humans, bonds individuals and can reduce tension. Leonore Tiefer of the Kinsey Institute found that the research "shows most clearly the constant use of kisses by [bonobos] to reduce tension, to reassure in any situation of fear or competition. Every bonobo, female, male, infant, high or low status, seeks and responds to kisses."

"The bonobos even French kiss," Claudine says with a throaty laugh. "They've done it to me many times."

Claudine puts me up in one of the bungalows she keeps for visitors. The following morning, at one of the enclosures, I watch as eight bonobos grip their night cage bars and stare greedily as attendants pile up their tasty breakfast—papaya, lettuce, sugarcane, pineapple, and other fruits. Let loose, they immediately launch into an orgy of frenzied group sex. Even the youngsters join in, and the air is almost torn apart by their excited squeals. After 15 minutes of indulging their lust, they settle down side by side to eat peacefully. "Bonobos use sex to deflate tension, and competition for the best food could cause a fight, so they defuse it by having sex first," Claudine tells me.

Here, I see another intriguing side of bonobo sexuality. Tshamboli, also known as Tshi Tshi, a 20-year-old female with thoughtful brown eyes, had been rescued by Claudine two years earlier after spending 18 years confined in a cage as a pet in a Kinshasa biological laboratory where scientists carried out HIV research on chimpanzees. She had never seen a tree and never rested on anything softer than the concrete floor of her cell in all those years until Claudine brought her to Bonobo Paradise.

Following breakfast Tshi Tshi allows her favorite, Api, a juvenile male, to mount her and have sex. She looks the other way when the dominant male, Makali, clearly indicates with an erection of his long, thin pink-tinged penis that he, too, wants to mate. She lies on her side, swollen bottom pointed provocatively at him, and stares with studied indifference into space. Makali sits patiently by her side waiting for an invitation and, when it does not come, wanders away. "With chimps and gorillas, a dominant male would have had sex with the female whether she wanted to or not," says Claudine. "This shows, in contrast, the power of the bonobo female."

Despite Mbali's good manners, I witness the bonobo male's innate aggressiveness for the first time. At the open-air nursery set on the side of a steep hill with plenty of climbing branches and vines for practice, a pair of three-year-old males, their little faces tightened with anger, swagger toward me. They leap at me in a coordinated attack, punching me in the stomach, on the back, and on the head. When I seek the dubious sanctuary of a chair, one of the young bonobos leaps onto my shoulders and knocks me onto the ground.

I am not going to give in to a couple of bonobo youngsters, and I grit my teeth and climb back onto the chair. The pair affect a look of disinterest, toying with the skins of bananas they had eaten, but the moment I take my eyes off them they hurl themselves onto my back and pound me with their fists.

At an enclosure, a heavy-shouldered adult male repeatedly drags a tree bough along the ground at high speed and slams it into the fence at me in a power display. His face is a mask of rage as he bares his impressive canines. Kano observed that male bonobos at Wamba do the same in the wild when flaunting their power in front of males from another troop. "I never let male attendants into the enclosures because the male bonobos would attack them," Claudine reveals. So much for the bonobo's much vaunted peace, love, and brotherhood.

Over the two days I visit Tshi Tshi, she plays a clever game of tease with me. At first she offers me a thumb-sized nut and when I reach to take it, she waits until the last moment and then with perfect timing pulls it back, a delighted smile on her face. The game goes on for a couple of hours, and she never tires of putting this curious-looking great ape male in his place. "I think she's fallen in love with you," Claudine says. Tshi Tshi's favorite adolescent male, Api, also seems to think so. He comes close to me, digs up a pile of dirt with his sharp fingertips, and angrily flings it into my face.

On the following day, Tshi Tshi offers me a yard-long length of rope she had found and the game of tease continues, only this time she chooses to play it on a slope. Each time she teasingly pulls back the rope at the bottom of the slope, she then races to the top with a big grin on her face as she watches me stumble up to try yet again to capture the rope. Up the slope I run, down the slope I run, up the slope I run, until I am panting with exhaustion. The more I pant, the wider Tshi Tshi smiles.

Her intelligence prompts her to improvise on the game when I feign disinterest. Time and again she throws the rope to me when I pretend that I am no longer interested in seizing it. To test her intelligence, I tie the rope around a cage bar in a complicated double-knot. Tshi Tshi stares at it for about five seconds, unraveling the puzzle in her mind, and then she confidently unties it without a moment's hesitation. Contrast this with the mountain gorillas who cannot comprehend how to unloosen a wire loop when they get their hand or foot caught in a snare.

An infant bonobo arrives at the sanctuary the next morning. He has been confiscated by police from a trader who was trying to sell him to foreigners at a deluxe hotel in Kinshasa for $1,000, a fortune for most Congolese villagers. He is emaciated and looks heartsick, his little body limp, and with no apparent interest in life, battered by the emotional trauma he has repeatedly suffered.

"The poachers would have killed his mother in front of him a few weeks ago, then stuffed him in a basket for the canoe journey downriver to Kinshasa, and that can take weeks," Claudine tells me. He clings to her with closed eyes, and each time he opens them he stares at us, sighs, and closes them again, as if that could shut the strange, pale creatures forever out of his life. "He's refusing food and water, and he's in danger of dying in a day or two."

Claudine employs local village women to act as surrogate mothers for the little bonobos, each assigned to a particular infant so they bond, and they stay with them all day and sometimes through the night. They carry them on their back, like their mothers did in the forest, feed them, and even romp with them.

The new infant's surrogate mother arrives in the afternoon, a buxom middle-aged woman with a lovely smile. She has raised several children of her own. She stays with the little bonobo all night, and by the next afternoon I see a heartwarming transformation. She has brought a little sweater to keep him warm and holds him in her arms. His eyes are bright and stay open as he constantly looks around him. He seems full of energy as he sucks happily on a chunk of pineapple.

"It has happened like this often," Claudine tells me as we smile at the transformation. "The little orphans thrive once I put them into the care of local women, who treat them with all the love and devotion they give to their own children."

A day later I fly to Mbandaka, capital of Equateur Province, a run-down potholed town by the Congo River a touch south of the equator. Although it has more than 100,000 people, the civil war damage was so severe that the city still has no water or electricity supply. Mass graves of civilians executed by soldiers during the civil war have been found on the city's outskirts.

Hordes of girls have tramped in from the impoverished villages and are offering their charms all over town, in the discos, in the open-air marketplace, by the river, and even in the bakery where I eat breakfast each day on the terrace. To earn a few coins, they wait on the tables. Croissant, coffee, and me!

The most beautiful of the girls, Corinne, is barely 16. She latches on to me and flashes a scintillating smile even when I am not alone. As she serves me she maneuvers her spectacular breasts, barely contained by a skin-tight T-shirt, as close as she decently can to my face and throws me a suggestive glance. My white skin, meaning rich man, is the lure.

My journey is stalled in Mbandaka for several days. I am traveling with Michael Hurley, executive director of the Bonobo Conservation Initiative and Sally Coxe's partner. Tall and solidly built, he has a bluff Irish face topped by a leonine mop of fair hair. He carries a big knife strapped to his waist, and with his khaki bush pants and short-sleeved khaki shirt, he seems to have modeled himself on the 1950s movie hero Jungle Jim.

The BCI people at Mbandaka had secured enough gasoline for the outboard engines to take us up to the bonobo habitat and back, a 10-day journey along the river, but one of Jean-Pierre Bemba's opposition politicians had just arrived in town on the same plane on which we'd flown. Because this is Bemba's stronghold the politician is met with a military guard on the tarmac, and a band that plays the national anthem comically out of tune.

Without a flicker of conscience, the politician confiscates our gas to use for his SUV tour of the villages in his constituency. "*C'est la Congo*," Michael says. "We'll have to wait until we can get more brought up by barge from Kinshasa."

In Mbandaka I hear the first rumblings that Kano's bonobo research, using provisioned sugarcane dumps, had skewed some results to give a false impression of the wild bonobos' true nature because he observed them in unnatural conditions. Dr. Mikwaya Yamba-Yamba, a botanist with CREF, the Congo government's science research body attached to the Ministry of Scientific Research and Technology, has just returned from a year studying the Wamba bonobos' food, observing them daily. Over tasty croissants, served by Corinne with fluttering eyelids and wasted breast-waggling, he tells me that they spend most of their lives in trees laden with edible leaves and fruits:

"By enticing bonobos to the ground to eat sugarcane together, Kano introduced competition for food where that usually doesn't exist in the wild. This exaggerated and even altered their natural behavior. The notion that they have frequent sex is also a myth. That might be so for captured bonobos, who have little else to do, but when they're in the wild much of their day is spent foraging for food and resting. When you're there you'll see this for yourself."

Yamba-Yamba says bonobo males are not the dream primates of feminists. They lead the groups and protect them from deadly predators such as leopards and pythons. He tells me, "The dominant male, Tawashi, is the leader of a troop I study, though the dominant female, Kiku, does have some say in the decisions, which is unique among the great ape species in Africa. Tawashi wakes the troop at dawn, leads a foraging party through the day, chooses where they eat, what they do, and where they sleep at night." This of course resembles the daily routine of the mountain gorilla silverback patriarch.

In their native habitat, Yamba-Yamba says, the bonobos choose from a wide variety of food—plants, fruit, insects, and meat. They live mostly on fruit and plant food—seeds, leaves, bark, roots, pith, stems, and mushrooms—but they also raid wild beehives for honey and bird nests for eggs. They are especially fond of the larvae of a butterfly as well as green caterpillars and earthworms, tastes they share with the local villagers.

The bonobos also eat meat from small mammals they capture such as flying squirrels, rodents, and duiker. But when Kano left remains of small animals and meat at his food dumps, the bonobos spurned them, preferring the sugarcane, pineapple, and other fruits piled there. The thrill of the chase seemed to spur their appetites for fresh bloodied flesh, and lifeless lumps of meat did not have the same primeval attraction.

When a fresh supply of gas arrives to get us to bonobo territory, Michael and I embark on a six-day trip upriver from Mbandaka by a giant pirogue—three canoes hacked from huge tree trunks, roped together, and propelled by two outboard motors totaling 200 horsepower. We have five boatmen aboard and a young woman who lives in a village near Kokolopori. She will cook our food in exchange for a ride home. Months can pass before a boat travels from Mbandaka to Kokolopori, because the civil war destroyed much of the commerce along the river.

We depart Mbandaka at mid-morning, speeding along the Congo, one of the world's longest rivers—2,900 miles from source to sea, and in places 10 miles wide. This uncrossable geographic barrier kept chimpanzees and gorillas in the jungles on the Congo's right side and the bonobos on the left. That is why the DRC is the only place where bonobos are found in the wild. "We call bonobos the left-bank great ape," Michael jokes.

Many crafts chug along the Congo River, the most bizarre being steamers crammed with hundreds of passengers and towing floating huts strung together like giant straw wings, up to a dozen on each side, their occupants hitching a ride to Kinshasa. As darkness drops a velvet curtain along the great waterway, we turn off to the right, entering the Maringa, a tributary river that stretches for up to a mile from bank to bank here and cuts deep into the heart of the Congo.

The river twists and turns like a giant snake on the move, and is guarded on both banks by towering walls of trees woven together by dense vegetation. In *Heart of Darkness,* Joseph Conrad described these as "an exuberant and tangled mass of trunks, branches, leaves, boughs, festoons motionless in the moonlight."

By day, hornbills, herons, kingfishers, and fish-eagles perch by the fast-flowing muddy water while locals pole canoes to market with freshly caught fish or home to riverside straw huts. At night, the darkened riverbanks echo with the urgent thump of unseen drums and raucous tribal singing.

Breakfast is catfish caught in the river by our boatmen, cooked with oil and eaten with rice and spinach purchased at Mbandaka. Lunch is catfish caught in the river by our boatmen, cooked with oil and eaten with rice and spinach. Dinner is catfish caught in the river by our boatmen, cooked with oil and eaten with rice and spinach. Bottled water slakes our thirst.

At mid-evening, two days after our departure from Mbandaka, the pirogue chugs along the darkened Maringa close to the riverbank with the high trees in silhouette looming over us. I am sitting alone near the prow enjoying the cool breeze, luxuriating in the contrast to the sultry daylight heat. We swing around a bend in the river and suddenly I see, a few yards ahead, a huge tree that has toppled over into the water. Its trunk rests at the surface and its many branches rear out of the water like a natural barricade. It is too late to stop the pirogue and it crashes into the tree.

A heavy branch thumps into my chest and other branches attack me from all angles. The biggest danger is to my eyes—a branch can take them out as easily as pitting olives. I throw up my hands in front of my face and branches bang against my arms. I save my eyes, but one branch whacks into my mouth. I spit blood into the water.

One of my upper teeth, a capped premolar, wobbles loose. My tongue moves it to and fro. As the boatman slowly reverses the pirogue, untangling it from the branches, Michael comes to the prow. "Are you okay?" he asks. I wipe blood from my mouth. "All in a day's work in the Congo," I reply with a grim smile. My tooth falls out a few days later.

Rugendo, the mountain gorilla silverback, who created world headlines in 2007 when he was shot dead on the slopes of the Virunga volcano chain in the Democratic Republic of Congo. Also killed were his four adult females.
(Photo: Paul Raffaele)

Humba, the brother of Rugendo, heads a family of mountain gorillas who inhabit the Congo Virungas' slopes.
(Photo: Paul Raffaele)

Lubutu, a female mountain gorilla, left Humba's family to join Rugendo's group after the silverback was killed.
(Photo: Paul Raffaele)

Makumba, the western lowland gorilla silverback studied for several years by primatologists at Bai Hokou in the Central African Republic.
(Photo: Chloe Cipolletta)

Makumba in a pensive mood. (Photo: Angelique Todd/WWF)

At Rwanda's Ruhengeri orphan gorilla sanctuary, a young eastern lowland gorilla. There are only a few thousand eastern lowland gorillas left and their habitat in the Democratic Republic of Congo is constantly under threat by rebels and illegal miners. (Photo: Paul Raffaele)

Noel, the playful infant son of Rugendo—the doomed Congo mountain gorilla silverback. (Photo: Paul Raffaele)

American David Greer, head of the antipoaching patrol at Dzanga-Ndoki National Park, with some of his rangers and a pygmy tracker, holding confiscated poachers' snares. (Photo: Paul Raffaele)

Jean-Rene Sangha, once the most notorious gorilla and elephant poacher at Dzanga-Ndoki National Park. He's killed hundreds of gorillas and elephants. The sign marks the boundary of the national park. (Photo: Paul Raffaele)

A dominant male orangutan in Sarawak, Borneo. They grow enormous cheek pads and weigh up to 300 pounds. (Photo: Sarawak Convention Board)

Outamba, the dominant female in a group of wild chimpanzees studied by researchers in Uganda's Kibale Forest National Park. (Photo: Paco Bertolani)

Kanzi, the super-smart bonobo, with human—ape communication researcher Sue Savage-Rumbaugh at the Iowa Great Ape Trust. They're using a lexigram, with symbols for words, to "talk" to each other. (Photo: Great Ape Trust of Iowa)

Claudine Andre with bonobo orphans at her sanctuary near Kinshasa, capital of the Democratic Republic of Congo. (Photo: Paul Raffaele)

Imoso, the dominant male of the Kanyawara chimpanzee group. Imoso holds his position not only by brute strength and fighting ability but also by guile and by knowing how to co-opt most of the other adult males into becoming his allies and helping him hold on to his leadership. (Photo: Paco Bertolani)

Imoso being groomed by Kakama. (Photo: Paco Bertolani)

Wild bonobos near Kokolopori in the Democratic Republic of Congo.
Bonobos spend most of their time in forest trees. (Photo: Michael Hurley)

* * *

On the third morning we pull in at Basankusu, a riverside town with a military base, where we have to show our permits to travel farther up the river. It is the heartland of opposition to the DRC's young ruler, Joseph Kabila, and strangers are treated with suspicion. There is an eerie quiet despite crowds of people swarming along the waterfront and at the hilltop outdoor market. Not one of the hundreds of people I see returns my smile. They still seem in the grip of some terrible trauma. Fierce battles occurred here during the late 1990s, and sunken barges still lie rusting in the shallows. Riverside buildings have been abandoned, their walls shattered by rocket fire.

According to the French-based relief agency Doctors Without Borders, over a 12-month period during the civil war, in 1998, when the town was held by rebels, 10 percent of Basankusu's population perished. There is a brooding menace here, and a sense that a wrong word or movement could spark a sudden explosion of violence. As our pirogue prepares to leave, 100 soldiers charge down the hill by the river, chanting war cries, led by witch doctors clad in leafy headdresses and grass skirts. "It's their morning exercise," a local assures me.

The military is all-powerful in this distant place. Two days later, the army commander at Basankusu confiscates a barge owned by the U.S.-run African Wildlife Foundation to move troops down to Mbandaka. The barge was carrying agricultural products and humanitarian supplies for the region's impoverished people.

All along the river I spy grim testaments to the chaotic fighting. Much of the DRC's pre-war export income came from rubber, timber, and coffee plantations strung along the Maringa, but the riverside residences and warehouses are deserted and crumbling and, like the waterfront buildings at Basankasu, are mangled by artillery fire and pockmarked by bullets. "The military and the rebels looted everything along the river, even light sockets, and it'll take a long time to return to normal," Michael tells me as we pass yet another abandoned plantation building. As I will hear later, the military and the rebels also played a significant role in causing the bonobos' numbers in the wild to plunge during the civil war.

By the fourth day, the river has narrowed and the riverside villages have almost disappeared. We seem to be reliving scenes from the movie *Apocalypse Now*, which was loosely based on *Heart of Darkness*. An unnerving silence has fallen across the river. A day later we pass Wamba, where Kano lived for years while researching bonobos. Our destination, Kokolopori, is a further 35 miles to the east along the river.

The mighty Maringa is now at times just 20 yards across, with jungle trees on either side towering over us. We slow to a crawl, the boatman weaving through an obstacle course of spiky trees that have toppled into the shallow water. I do not want to think of the consequences should we get entangled with a fallen tree beneath the water. Darkness falls swiftly here because we are right on the equator, and a ghostly mist settles on the river as we tie the pirogue to sturdy reeds by the riverbank. We leave soon after dawn when the mist rises.

At mid-morning, more than 600 miles upriver from Mbandaka, the boatman ties the pirogue to a big tree on the riverbank, which is thronged with villagers. They have come to help carry our supplies balanced on their heads on the two-hour walk through the jungle to Kokolopori.

I ask Michael how they knew we would be here on this day, because Kokolopori has no telephone connection to the outside. "Africa." He shrugs. "We got word to them a few weeks ago that we'd be here about now, but we were delayed at Mbandaka, and the journey's taken a couple of days longer than I expected. So, they've probably been here on the banks for days waiting for us to arrive."

The jungle, home of the bonobos, beckons.

3

We climb a narrow path up a slope, leap across fallen tree trunks, and plunge into knee-high streams. A sudden rainstorm drenches us, spearing through the canopy like a shower of darts, reminding me that I have entered one of the world's biggest and most important rainforests. It spreads for thousands of square miles, linking up to the Congo Basin.

An hour later we emerge from the jungle and head up a slope dotted on either side with mud-hut villages. Children run from their schools and stop playing soccer on bare earth patches with spindly sticks for goalposts to peer at the *mindele*, or white men. Michael and I are more like red men, our faces flushed from the effort and soaked with sweat. They surround us with cheers and pleas for pencils and exercise books. This is a pleasing contrast to the children near the Virunga volcanoes who begged me for money.

Perched on a cleared rising, my new home is one of eight mud huts in a compound used by BCI. We are greeted by Albert Lokasola, a roly-poly middle-aged man with a gleaming smile. Albert's father is paramount chief of this area, but Albert went away to school in distant Kinshasa, stayed there to fashion a life as a top-level bureaucrat, and became secretary-general of the DRC's Red Cross. He has only recently come home on a personal mission to save the bonobos. He heads Vie Sauvage, a local village-based organization he set up dedicated to protecting the estimated 1,500 bonobos inhabiting the surrounding 2,000 square miles of jungle of his native land.

Bofenge Bombanga, a powerful shaman clad in a loincloth and a headdress made from many dried hornbill beaks, leads the welcoming dance in our honor. Dozens of women scream, yell, and fling their arms into the air as they stomp their feet to the throb of drums. Bofenge shuffles about, keeping perfect time to the beat with his aging, gnarled bare feet. He motions for me to join the dance, and I leave a hilarious image that might last through the generations, to be told and retold at the hearth, of the beefy white man who pranced in a frenzy about the dance ground as if he had ants up his pants.

We sit down to take swigs from a bottle of whiskey we have brought with us to be shared with our hosts. Bofenge tells me that long ago, a village elder near here was trapped up a high tree after his climbing vine fell loose. A passing bonobo helped him down. Since then it has been taboo for villagers to kill a bonobo.

Two days later, in a trek through the jungle, we again scramble over fallen trees and balance along slippery logs thrown across chest-high streams, and tread along a carpet of moldy leaves that crunch with every footstep. Leonard Nkanga Lolima, the head tracker, leads me to a clan of bonobos known as the Hali-Hali group.

Bonobos live what primatologists call a fission-fusion life, the group sleeping in adjacent trees each night but breaking into foraging parties during the day. Leonard has been tracking this group of nine bono-bos since dawn. Vie Sauvage employs 36 trackers from local villages to follow five bonobo clans daily, to protect them from poachers. Albert pays each man $20 a month. That is a large sum here, where people are subsistence slash-and-burn farmers, fishermen, and bushmeat hunters.

The money flows from Vie Sauvage's funding agency, BCI. "It costs $250,000 yearly to fund Kokolopori, and it comes mostly from Conservation International in Washington," Michael explains. BCI has brought in many bicycles to transport goods along the narrow jungle tracks, as well as crop seed and sewing machines, all to stimulate a cash economy and persuade the local people to stop hunting forest animals for bushmeat. It also provides much-needed basic medicine at village clinics because the nearest doctor is at Kisangani, a bumpy four-day ride along jungle paths on Albert's motorcycle.

In the forest, Leonard stoops in the murky light to sniff a fallen leaf. "Bonobo urine," he murmurs. The trail plunges into a gloomy rain-soaked tunnel of tall, leafy trees. Minutes later, Leonard raises a hand to halt us and points up at the trees. High above I see a large, dark creature propped between the trunk and bough of a hardwood tree. "Bonobo," he whispers. "The dominant male. He's sleeping. Keep quiet, because it means there are bonobos all around us." We creep toward the tree and crouch beneath it, and for the next 30 minutes I try to ignore the fiery bite of large ants crawling over my arms and legs and the constant sting of mosquitoes.

Suddenly, dung splatters the ground around us. It is the big male. "He's just woken, and he's angry we're here," Leonard says quietly.

The male screams a *waaa* warning to the other bonobos and they respond, shaking the leaves with their shrill cries, so loud they bang at my ears. Through binoculars, I see dark eyes, spookily human-like, peering down at me, alive with intelligence. A baby grips its mother's belly as she balances on a branch while a youngster leaps up and down in imitation of a power display. Moments later, with an aerial skill that would shame Olympic gymnasts, the bonobos are gone, swinging and leaping from branch to branch, tree to tree, led across the rainforest canopy by the big male.

We follow the bonobos and reach them about 15 minutes later, only because they have settled once more in the trees above us. I am cap-

tivated by the bonobos' jungle-toughened athletic build. At Frankfurt Zoo the males had the slim, elegant cast of ballet dancers, but
the males here are broad-shouldered with bulging muscles, developed
from daylong, yearlong workouts as they swing from tree to tree. Even
the females are bulky. An adult male in the wild weighs on average 86
pounds, while an average-sized female weighs about 68 pounds. That
is about 84.5 percent of the average size of an adult male and female
chimpanzee in the wild.

The wild bonobos' coats are much denser than those of bonobos in
captivity. Zoo bonobos are often almost hairless because they pluck hair
from their chests, arms, legs, and even heads through boredom and the
sort of zoo madness many great apes behind bars exhibit when they are
prevented from foraging and the pleasure of wandering. Out here, a
bonobo does not have time to get bored.

Leonard whispers that the dominant male's name is Raphael, given
to him by a researcher. The bonobo impresses me with his dignity and
strength as he sits high on a limb yanking fistfuls of leaves and munching on them. This is no subordinate male, dominated by females, but a
big, strong primate wielding considerable power. All around us bonobos, young and old, settle in their own trees and feast. A juvenile male
lies in the crook of a tree, like a teenager snacking on a sofa by a wall,
one leg dangling down into space while the other rests at a right angle
up the trunk, as he strips and eats leaves from branches close at hand. I
see through binoculars that a bonobo's feet are an almost identical copy
of its hands, like a chimpanzee's, and just as dexterous, with opposable
big toes for gripping.

Two females stop eating for a few moments to rub together their
swollen genitals, GG. Researchers have spotlighted this behavior as
evidence that the bonobos are bisexual. If so, the female bonobos are
swiftly and easily pleased. It looks to me more like a sexy hello than a
sexual tryst. The GG is often initiated by a lower-ranking female with
a female that has a higher status within the group, a pleasant way of
paying homage.

Moments later my heart stops as a youngster casually steps off a
high branch and plunges toward the forest floor, crashing through
the branches and leaves. About three yards from the bottom, and just
a moment or two from crashing into the ground, he grabs a passing

branch and swings onto it. Leonard tells me this is a favorite death-defying game of young bonobos, and invariably produces a wide grin at the end of the fall.

Pairs of bonobos groom each other, searching for ticks, much as I had seen chimpanzees in the wild at Kibale do. Then the dominant male puckers his pink lips and lets loose a piercing scream, a signal for the troop to move again. The others join in with excited squeals. Raphael leads the way, hurtling from tree to tree just below the canopy, followed by the others with another big male in the rear guard. The bonobos leap more often than the chimpanzees when traveling through the trees, and those I witness travel faster.

I stumble behind, my head banging into low branches and tripping on vines and tree roots that spread across the forest floor like a tangle of bulging veins. Soon after, the bonobos settle into another clump of trees. After greeting this new feeding spot with another chorus of excited screams, they begin stripping the branches of leaves by the fistful and eating them. Their daily pattern is similar to that of the chimpanzees—a cycle of feeding, resting, travel, feeding, and resting.

Kano recorded that in a typical day, the bonobos in the jungle spend 43 percent of their time resting, 20 percent foraging, 20 percent feeding in the trees, 13 percent traveling, and 13 percent "other." The total of 109 percent is caused by an overlap between feeding and foraging.

At noon, drowsy-eyed, the bonobos slip into a siesta and we crouch below for two hours as high above, the bonobos sleep off their morning repast. They awake to a loud *waaaa* call by Raphael, and after a few minutes of excited "chatter," the bonobos slither down onto the ground in search of plants, mushrooms, flowers, honey, truffles, earthworms, and algae in a nearby swamp. They move so swiftly through the forest that we mostly see them as momentary blurs of dark fur.

I spy a female who slips down a tree and walks upright on two legs across a moss-covered log, her long arms held high in the air for balance like a tightrope walker. Her upright stance, so unlike the chimpanzees' crouch-walk, brings a lump to my throat. As the setting sun paints the rainforest gold, high in a tree above me, Raphael, the dominant male, seemingly deep in thought, sits on a branch and swings his human-like legs to and fro as he stares at the sun slipping below the canopy rim.

* * *

Michael is a friendly companion here in the Congo wilderness, but for a few days he does not seem to be able to get the trackers together again to lead me to the bonobos, even though they go out every day to monitor the apes. There is always something else to do, and he is thwarted somewhat by a Congolese attitude that places little value on punctuality or reliance on a calendar count-off of the days. He says, "I've been here so many times that maybe I'm beginning to think like a Congolese."

This gives me time to ponder the comparison between bonobos and chimpanzees in the wild. The UN's *World Atlas of Great Apes and Their Conservation* states:

> *Among chimpanzees, males associate closely with one another, grooming one another frequently and cooperating in hunting, in patrolling borders, in stalking and sometimes killing chimpanzees from neighboring communities, and in guarding and mating with swollen (in heat) females.*
>
> *Among bonobos, grooming between individuals of the opposite sex is more frequent and occurs for longer periods of time than grooming between females or males only. Bonobo males are much more peaceful than chimpanzee males, interact less, compete less for copulation opportunities, are not as territorial, are less aggressive with males of other groups and do not hunt other large mammals.*

Another significant difference is sex and the availability of females. Chimpanzee females come into heat for only a few days a month, and so competition for them among the males can be fierce, with the dominant male granting more mating rights to his allies. But bonobo females are receptive to the males for most of each month, and that means there is hardly any fighting by the males for their favors.

The *World Atlas of Great Apes and Their Conservation* concludes:

> *As a result, at any given time in a bonobo community there are on average many more females interested in mating than there are in a chimpanzee community. In these circumstances, it would be much more difficult for a high status male to monopolize mating opportunities, so the male status is less important to individuals. Male bono-*

bos are rarely observed to compete or fight over access to females. It is the female that determines whether copulation occurs.

I smile as I read this, remembering Tshi Tshi back at Lola Ya Bonobo in Kinshasa, turning her back on the dominant male when he wanted to mate with her. He waited with great expectation, his passion clearly on show, but ambled away when Tshi Tshi showed she was not interested.

A day later, a most welcome visitor is Lingomo Bongoli, a Congolese bonobo researcher who worked closely with Kano for years at Wamba and wrote a scientific paper with the Japanese professor.

He dismisses talk that locals do not eat bonobos: "That was once true, but since the war outsiders have come here, and they tell our young people that bonobo meat gives you strength. Too many believe them." Lingomo took an extensive survey among the people in and around Iyondji, his village, and found that 27 percent had eaten bonobo meat. "It's the same at Wamba, and is probably more, because it's forbidden by law to kill bonobos and so people probably hide their involvement."

Among the biggest killers were soldiers, rebel and government, based here or passing by during the civil war. "They destroyed or stole everything from the plantations and villages—our goats, chickens, radios, sewing machines, generators, even the electricity wire," he tells me, his lined features bunching in a grimace. "And they killed bonobos for bushmeat. The soldiers shot dead the wife of Koi, the man who first led Kano to the bonobos, because he wouldn't show them where they were in the forest. Koi wouldn't tell them even when they slashed his head with a machete, but others did."

A couple of days after I speak with Lingomo, two Congolese CREF researchers from Wamba make the 35-mile journey by bicycle along jungle paths to talk with me. "There are still plenty of bonobos at Wamba because they were deep in the forest, far from the soldiers," says Mola Ihomi. He and Kumugo Yangozene spend the entire year at Wamba, collecting bonobo data to share with Japanese researchers from the Kyoto University's Primate Research Institute, who journey there for two-month stretches a few times a year. "We track and monitor three groups," Mola tells me as we shelter from a fierce noon-day sun in an open-air hut, poles holding up a roof of thatched banana leaves.

Mola disputes Kano's finding that coalitions of females dominate bonobo society, claiming like Yamba-Yamba that by using sugarcane dumps the Japanese primatologist altered normal bonobo behavior in the wild. In the jungle, the responsibility for ensuring a happy and prosperous life seems shared. "The dominant male is usually in charge and leads the group," Mola tells me, "but if the dominant female doesn't want to follow him, she sits down and then the rest of the group follows her lead and doesn't move. She always has the last say. It's like the dominant male is the general, and the dominant female is the queen."

What is it, then, in zoos and provisioning food dumps that prompts the adult females to form coalitions and so dominate the adult males? Is this behavior a core part of the bonobo nature? Much more research is needed.

Mola is also skeptical of claims by Western researchers that when strange groups of bonobos meet in the jungle they do not battle it out like chimpanzees, but defuse the tension by intense sexual activity between almost all members. "Of three groups we're studying, two were once joined together but they split, and now when they reunite in the forest, they do have sex," he explains. "But when either of them bump into the other group, which isn't often, they display fiercely to defend their territory. Males and females scream, and throw dung and sticks at each other. They even fight, sometimes inflicting serious bite wounds."

A member of Kano's research team also reported such conflict. When two separate groups met, "a violent fight occurred, leaving several individuals injured." But Mola offers an important qualifier that suggests bonobos are a uniquely peaceful great ape: "No one has ever seen a bonobo kill another bonobo."

However, they do kill other creatures. Gottfried Hohmann, a German bonobo researcher at the Max Planck Institute for Evolutionary Anthropology in Leipzig, has visited bonobos in the Congo many times since 1989, and told the *New Yorker* that he had witnessed bonobos several times killing and eating tiny antelopes called duiker. The females nearly always do the dismembering, Hohmann said. He wrote, "Bonobos start with the abdomen; they eat the intestines first, in a process that can leave the duiker alive for a long while after it has been captured."

Findings from a milestone bonobo study released in 2008 by Hohmann revealed that, like chimpanzees, the bonobos also hunted the
red colobus monkey. However, unlike chimpanzee females, who have
rarely been seen hunting monkeys, some female bonobos are enthusiastic hunters. Of five observed hunts, two of three where bonobos
succeeded in catching and killing red colobus monkeys were carried out
by females.

Mola tells me that a bonobo mother gives birth to a single infant
only about every five to six years. She weans it at three years but carries it around for another two years, sharing her nest and protecting it.
Bonobos, like chimpanzees, live to about 40 years of age in the wild if
they manage to sidestep or cope with the jungle's many dangers including leopards, pythons, and disease.

Bonobo researchers at Wamba abandoned the sugarcane dump
method more than a decade ago. Mola says that as bonobos are up
near the canopy for much of the day, it is now harder to observe and
understand the full spectrum of the relationships between individual
members of a group.

However, Mola says that some of Kano's important findings match
his own. As the months stretched into years for Kano at Wamba, he
came to recognize 150 individuals, and the Japanese researcher told
me in Kyoto that his excitement grew as he noticed the close attachment between certain females and males. A highly ranked female he
had named Kame was always accompanied by a pair of males, Ibo and
Mon. She groomed them, fed side by side with them, but never mated
with them. Kano realized he was watching a mother and her two sons
bonded by familial affection. Mola nods when I tell him this.

"I saw other mothers and sons stay together, and realized that mothers were the core of bonobo society, holding the group together," Kano
told me. "They even pushed their sons' status by encouraging them to
mate with other females in their social circle, because the more females
a male can mate with, the higher is his status. And if a male dared attack
another male, his mother would marshal her female allies to defend
him."

Patient observation over many years convinced Kano that male bonobos bonded with their mothers for life, a finding that was probably not
affected by studying them at the sugarcane dump. That contrasts with

chimpanzee males, who rarely have close contact with their mothers after they grow up, instead joining other males in never-ending tussles for dominance. Jane Goodall had found that a chimp will occasionally display fear of her adult male son, but this has never been observed among the bonobos.

"This was a major revelation because it proved the chimpanzee model was not the only one to point to our origins, that another primate close to us had developed a social structure mirroring our own mother-son bonding," wrote primatologist Frans de Waal.

On my most intense encounter with the Hali-Hali group, I follow the bonobos for a full 24 hours and see no trading of sexual favors, and much less sexual activity than at Kano's sugarcane dump or at Bonobo Paradise in Kinshasa. That is probably because the wild bonobos have abundant food available in their habitat. They spend much of the day feeding or dozing.

As the sky darkens, I watch as the smaller foraging groups arrive, fusing together as a big group, and settle in a clump of trees high in the canopy. They swiftly build their springy night nests, yanking leafy branches together and weaving them into a comfy resting place up to five feet across. They are the most elaborate of the great apes' nests and resemble the nests of giant birds, with many leafy branches woven together. For about an hour they chatter nonstop in their squeaky voices, but the noise drifts away by 6:00 p.m., and as the light swiftly drains from the sky, each bonobo has settled out of sight in its leafy bed.

We set up camp nearby, and I sit by a fire with the trackers, sharing my tinned meat and coffee with them for our evening meal. The fire is both to cook the food and warn off prowling leopards. Suddenly, Leonard, the head tracker, stands up, murmurs something in an angry voice, and walks away. The other trackers shrug. He does not return. I plan to ask Albert to seek an explanation from Leonard when I return to Kokolopori.

At 5:00 the next morning, I crouch with the remaining trackers beneath the trees as the bonobos wake, stretch, and immediately begin munching on leaves and fruits next to their nests. It is breakfast in bed bonobo style. A female swings to the next tree to rub genitals with another female for a few moments, squealing perhaps in pleasure. Nearby, a male and a female, balanced on a thick bough, mate with obvious pleasure in the frontal position, her legs wrapped around his waist.

Kano was amazed when he first saw bonobos having sex in the face-to-face position, a favorite way. No other African great apes generally do this, though an inventive pair of Congo western lowland gorillas were once spotted mating face to face. For centuries humans believed that only we mated this way, an intimate position involving eye contact that dramatically increases the emotional bond between partners.

Two youngsters watch the pair intently and then one climbs between the male and the female, but they ignore the intrusion. Kano noted that this is normal bonobo behavior. The adults tolerate curious youngsters jumping onto their backs during mating or pushing their little bodies between the amatory couple during the act.

Male bonobos are more precocious than female bonobos and begin to act sexually before they are about one year old. A male bonobo's penis can become erect just six months after birth. Kano wrote: "When a mother finishes GG rubbing, her male infant clings to her female partner and inserts his erect penis into her." Sometimes, a mother takes her infant's penis herself and inserts it into her GG partner.

As the infants grow into juveniles, their sexual activity becomes more regular: "When the juvenile encounters his mother or other adults engaging in copulation and GG rubbing, he immediately runs and clings to either one's stomach or back, and screams. Then when the adults conclude their activity, they embrace the juvenile and practice similar behaviors with him. Other juveniles, attracted by the sight, will come and sit patiently waiting their turn. The female will accept them."

Adult males are sometimes less tolerant with the youngsters and push them away when they try to intrude during mating. But the adults usually cooperate in their youngsters' awakening sexual desires. Kano has seen adult males invite juveniles to join in.

In contrast to the randy juvenile males, female juvenile bonobos rarely take part in GG rubbing with other females until they are adolescents. Because of their "small openings," Kano told me they rarely engage in copulation until they are adolescents and their genitalia are fully developed.

Adolescence is a difficult time for a female bonobo, similar to adolescence in chimpanzees and gorillas. She has to leave her natal group and migrate to a stranger group. During her settling in she remains reserved, stays away from disputes, rarely shows aggression, and if attacked by one of the new group restricts her response to screaming. To establish and cement social bonds, she spends much more time grooming than being groomed.

Based on thousands of hours of observation by himself and fellow Japanese primatologists at Wamba, Kano divided a female's adolescence into three stages. At age seven to eight, she leaves the natal group, but her sexual organs are small and she is unlikely to be penetrated. During the middle stage, from nine to 12 years old, her sexual organs swell and she begins to enjoy GG rubbing with females and copulation with males. In the final stage, at age 13 to 14, she conceives, and then nine months later gives birth for the first time.

Kano told me that the bonobos' rich sex life is a pointer to their superior intelligence when compared with the one-dimensional sex life of chimpanzees and gorillas. Bonobos have been observed in the wild having novelty sex, with both partners dangling precariously from tree boughs as they mated. Bonobo females are in estrus for most of their 46-day cycle, signaled by swollen pink rumps, and are eager to mate almost all the time. Chimpanzee females, in contrast, are interested in sex for only the few days of each monthly cycle when they are in estrus.

What makes face-to-face mating possible for the bonobos, in contrast to other African great apes, is that the female genitalia, like that of humans, is frontally inclined. The clitoris is relatively much bigger, which may also explain the females' liking for GG. Like the chimpanzees, bonobos are promiscuous, with each ape mating with many partners, and Kano even saw young females soliciting juicy sugarcane stalks from adult males, offering sex in return.

The bonobos have at least 20 gestures and calls that signal willingness to have sex such as feverish hooting, the displaying of their bodily charms, and food proffering. De Waal also saw this tradeoff among zoo bonobos, observing a young female approach a male who held an orange in each hand and offer to mate with him. When they finished, she took one of the oranges and loped away.

These are skewed observations of bonobo life, marked by artificial conditions at the Wamba sugar dump and in zoos. I saw for myself that bonobos in the wild have a feast of food all around them, and each usually feeds in a separate tree or on separate branches of a tree. So, a female desiring a tasty tidbit does not have to beg from a male in the wild, offering sex in return; she has only to leap across the branches and indulge herself.

An hour after the bonobos woke, most are leisurely feeding and some are having sex. But I also see some of the adult males playing with juvenile males, lying back on branches as the younger ones launch mock attacks on them. The juvenile bonobo is usually the aggressor, gently biting a foot or grabbing an arm or leg, all the while play panting, a rapid-fire *hat hat hat*. A female in another tree dangles her tiny baby by the arms above her raised feet, playing "airplane" just like human mothers.

Adult males also like to carry infants on their backs. Kano noted that to do this, the male "approaches a female carrying an infant and stares at the infant while extending his hand." The infant can jump on his back, but if it clings to his stomach, the male might move away from the mother. "The mother will tolerate a separation of about five meters," Kano observed, "but if they go further or the infant begins to whimper, the mother will hurry to take it back."

In all my time with the bonobos, I never see an adult strike an infant or a juvenile. Kano says the adults' behavior toward the young (two to four years) can be summed up in one word—tolerance. "There are no reports of severe scolding or violent attacks directed at juveniles," he asserts.

Just after 6:30 a.m., led by Raphael, the bonobos signal with a few minutes of excited squeaking and squealing their readiness to abandon the trees where they had slept. Then they swing off for another day's foraging. There are about 40 bonobos in the Hali-Hali clan, but as they head into the forest along their aerial highways, they split into a handful of parties. They will next meet up at dusk at some other place to prepare their night nests.

Bonobos vocalize far more than any of the other great apes I have been with in the wild, and they seem to be able to communicate with each other, mostly with an almost constant stream of squeaks, squeals, and hoots. How else can the foraging parties range over the forest in different places all day and then meet up at a different nesting site at night? They have a mighty hoot that carries across the jungle, but sometimes the foraging parties can be more than a mile apart.

The bonobos may also be able to pass on direction messages using marked vegetation, much as I had seen the pygmies do many times in the Congo jungles. Renowned bonobo researcher Sue Savage-Rumbaugh once told me that she noticed this when she and a tracker were following a group of bonobos in a Congo forest: "We came to a fork in the trail and weren't sure which way the bonobos, who were traveling on the ground, had gone. The tracker said they'd taken the path to the right, but I noticed that a plant had been deliberately bent, which indicated to me that the bonobos we were following had done it to show other bonobos that they'd taken the path to the left. The tracker insisted that he was correct, and so we followed the way he felt they had gone but found no trail. We returned to the fork, then took the left-hand path and soon after found the bonobos."

After a good night's sleep at the huts I hope once again to go back in the jungle to be with the bonobos, but Albert tells me Leonard has refused to take me, and the other porters have joined his boycott. Albert shakes his head when I ask why and says, "I don't know why he's doing this. I'll get someone to fetch him from his village and ask."

Leonard dare not disobey the son of the paramount chief and the man who pays his wages. He arrives at the compound with scowling features and turns away when I look at him. I sit patiently as Albert questions the head tracker for more than 30 minutes. Then he turns to me and says, "Leonard says he believes you are a witch doctor and have cast a spell on him. He feels you want to kill him with the spell."

"Are you joking?"

Albert shakes his head. "I know it sounds unbelievable, but that's the way our people think. They really fear the supernatural. Leonard says you put a curse on him while you were in the forest the other night, and he refuses to go with you again."

Albert keeps questioning Leonard, who frequently shakes his head as if saying no. I offer him $20 to change his mind. Leonard refuses the offer, claiming no amount of money will make him change because he fears I want to kill him with black magic. But when I increase it to $50 he reluctantly agrees to take me to the bonobos on the following day, our last day at Kokolopori.

At mid-morning the following day, as Michael joins the porters carrying our bags back down to the three pirogues roped together and tied to a tree at the river, Albert and I wait for Leonard and the trackers. Thirty minutes later, three of them appear and explain that Leonard still believes me to be a witch doctor and refuses to come. "I'll take you," Albert exclaims. "The trackers will follow my orders."

Despite his bulk and a bulging belly, Albert is agile as he leads me through the gloomy forest, jumping over fallen trees and even balancing like a bonobo, with arms extended, as he crosses a surging stream on a large tree trunk that has fallen across the water. "I loved being in the forest when I was a kid, and come back whenever I can," he says.

Three hours later we still have not found any bonobos. Then, Leonard unexpectedly steps from behind a tree about 20 yards ahead. He must have known along which path the trackers planned to take us. I smile, but he glowers back at me. "I've found a group of bonobos about twenty minutes' walk from here," he tells Albert. "They're from a group we rarely see, and so they're not habituated."

My heart leaps. Such bonobos are an extremely rare sight. Leonard's 20 minutes is an hour of hard trekking for me. We find the bonobos on the ground, eating ants from a nest, and I see them as flashes of fur through the undergrowth. Then they take to the trees and we follow them, me in a by-now-familiar stumbling role, almost bursting my lungs to keep up with Albert and the trackers. I fall so far behind that they are out of sight, and when I turn a corner, I find a young bonobo male hugging a tree at my eye level, about five yards away. He must have heard me coming, and froze in fear.

We stare at each other in surprise, then confusion, because both he and I do not know what to do next. The moment freezes, and then he snaps out of the spell and leaps to the ground and scurries away. About 10 yards ahead he jumps onto a tree and scrambles up onto a high branch. From the safety of his perch, he stares back at me.

He utters a *waaa* warning screech and I hear rustling above. A big male bonobo appears high in a nearby tree and stares down at me. I fix him with my binoculars and see that his face is taut with anger. He breaks a bough from the tree and throws it at me, but it either must be a warning or his aim is wild, as it misses by 10 yards. I stand my ground. The bonobo tears off another big branch and throws it at me, under-hand in a looping motion. It soars through the air, and I jump aside as it crashes into the forest floor about a yard from me. Had it hit, it would have brained me. I turn to walk away and see Leonard, almost hidden in the thick bush, watching with a wicked smile.

At the river, the porters are still loading our equipment and I sit by a tree with Michael and eat a chunk of fresh pineapple, its copious, tasty juice dribbling down my chin. Suddenly, I feel as if a knife has just been thrust into my lower lip. The pain is intense. My lip begins to swell like a balloon. "It was a wasp, I saw it," says Michael. "It was on the other side of the pineapple, and as you went to eat it the wasp stung you."

"All in a day's work in the Congo," I reply with a forced smile. I take an antihistamine tablet and rub antihistamine cream on the lip, know-ing that the damage, a lip swollen about three times normal size, looks worse than it is. The swelling and the pain will be gone by the evening.

We wave good-bye to the villagers, board the pirogue, and push off. We are now traveling downstream, and the rush of water shoves the pirogue this way and that. About 30 minutes later, as we round a bend, Michael shouts a warning. A few yards ahead a tree that had fallen into the river, spiky with branches, pushes up through the surface of the surging water. It is too late to alter course. The bow smashes into the tree trunk, and the impact's force sends the pirogue out of control and veering sharply to the left.

The bow plunges into the bank with a thud, and the canvas canopy smashes sideways into a big wasps' nest on a tree by the river. A horde of striped wasps pours out, buzzing with anger and seeking the intruder. They spy us. "Get under a blanket," Michael shouts at me. I peer out as he and the head boatman jump into the waist-deep water, ignoring the wasps attacking their bare heads and arms, as they shove, pull, and push at the pirogue's prow, trying to extricate the pirogue from where it had dug into the soft earth of the riverbank.

The pirogue comes loose, and they jump aboard as the heaving water grabs the craft and flings it downriver away from the wasps. Both Mi-chael and the boatman have many red swollen stings on their bare arms and faces. Michael smiles as he takes a swig of whiskey. "All in a day's work in the Congo," he says to me with a wink.

4

Bonobos are among the cleverest of the great apes, and the person who probably knows their minds and cognitive ability best is Sue Savage-Rumbaugh, a world-famed expert in ape–human communication. A decade earlier, my visit to Takayoshi Kano in Japan coincided with an International Primatological Society conference being held there. I attended the sessions and met Richard Wrangham as well as Sue for the first time. Sue had worked for two decades at the Georgia State University Language Research Center (LRC) in Atlanta studying the bonobo's ability to comprehend language, most famously with a male bonobo, Kanzi, who has been featured on the cover of *Newsweek* and been described as the smartest known non-human on the planet.

So, to find out more about the depth of bonobo intelligence, which researchers claim is reputedly at the level of a three-year-old child, one evening during the conference I invited Sue to a traditional Japanese meal. As she and I settled on tatami mats to enjoy raw fish, shabu-shabu, and a few tumblers of sake, she began to tell me about Kanzi.

Sue saw her first great apes, chimpanzees, when she was eight years old. She attended a monkey show in St. Louis. The chimps, clad in human clothes, walked about on stilts and rode motorcycles through flaming hoops. It got her wondering how animals felt about things, given that they could not talk to us.

The next time she saw chimpanzees was in 1970 at the University of Nevada, as a graduate teaching assistant in psychology. A newly arrived lecturer brought along a chimp named Booee whom he had trained to use the American Sign Language, used by deaf people. Sue was astonished when the chimp employed it to make the correct signs for objects such as a shoe, a hat, and a ring of keys. She wondered whether the chimp really understood the meaning of the words he was signing. "I had heard little of the so-called ape language studies," she recalled, "and I thought of apes as smart dogs with hands and goofy faces." But this ape was different, and she was impressed that he made the correct sign every time he was asked.

To learn more, she began working as a volunteer at a chimp farm owned by the psychologist near Norman, Oklahoma. There, she helped Booee practice the signs by holding up an object and then taking hold of the chimp's hands and moving them into position. Whenever he got the sign right, he got a reward of food. Sue was intrigued and decided to study human–ape communication, believing that it could lead to a better understanding of "the essence of humanity itself."

She read of the famous English diarist Samuel Pepys's prophetic reaction on seeing a chimpanzee in London in 1661. Pepys noted how human-like the chimpanzee was and added, "I do believe it already understands much English; and I am of the mind that it might be taught to speak to make signs."

Pepys's extraordinary insight was a forerunner to Robert Yerkes, who wrote in 1925: "I am inclined to conclude that from various evidences that the great apes have plenty to talk about, but no gift for the use of sounds to represent individual as contrasted with facial, feelings or ideas. Perhaps they can be taught with their fingers, somewhat as does the deaf and dumb person and thus helped to acquire a simple, non-vocal, sign language." In reading this, Sue was inspired to study the ability of great apes to understand human language and respond to it.

Pioneers in the field used chimpanzees and gorillas to test the ability of great apes to communicate with humans, most famously a female chimpanzee named Washoe, and so Sue started with a chimp named Lucy. But she changed to bonobos for her experiments, because she considered them "more human-like than common chimps in many ways, including being more vocal and more communicative and having extremely expressive human-like faces." She ignored the scorn of linguists who regarded ape–human communication as little more than a circus trick.

In 1975, Sue received a post-doctoral fellowship at the Yerkes Regional Primate Research Center in Atlanta. Three bonobos caught in the wild were brought to the center for study. Interest in the bonobo had been stirred by Robert Yerkes's book *Almost Human*, in which Yerkes noted that the bonobo Chim had a pronounced laugh and once even picked a handful of flowers and presented them to "a lady attendant." As we know, Yerkes believed Chim to be a super-smart chimpanzee, but he proved to be a bonobo.

"It is impossible to spend more than a few hours around the two species [chimpanzee and bonobo] without being overwhelmed by the differences between them and the uncanny echoes of humanity one constantly experiences with the bonobo," Sue has written.

The bonobos Matata, Lolekelma, and Bosondjo, two females and a male, arrived in Georgia from the Congo at the end of 1975. Sue soon noticed the difference between the two species. The inventive bonobos were given a plastic feeding pail in their cage, but it soon became much more than that. They used it for holding drinking water, inverted it as a seat, used it as a repository for urine, placed it over the head as a blind, carried it on the stomach as if it were an infant, and played with it as a toy. The neighboring chimpanzees were able to observe all these activities, and Sue wondered whether they would imitate the bonobos' antics if they were given a pail. They did not. Instead, they used the pails as props in aggressive displays, shaking them in the air, slamming them against the cage sides, and kicking them across the floor.

Sue says Kanzi's name means "buried treasure" in the East African language Swahili. He was born in October 1980 when his mother, Lorel, was on loan to the Yerkes Center from the San Diego Zoo for breeding. He was her first infant, but she had been reared in the zoo nursery and not by her mother. Sue says that baby Kanzi was "a small ball of black fur with spindly arms and legs."

After giving birth, Lorel was nervous, paced the floor, and looked bewildered. Matata, a wild-born female who must have seen births among her jungle clan, gently took hold of Kanzi and placed him on her stomach. Lorel tried to get him back by tugging on his tiny leg, but Matata kept hold of the baby bonobo and would not let go, day after day. She adopted Kanzi and was allowed to keep him. Lorel returned to San Diego.

Matata and Kanzi arrived at the Language Research Center when he was six months old, and Sue began to teach Matata how to communicate with a lexigram, an electronic board containing 384 separate lexigrams, complicated geometric symbols as well as a few words such as *jump* and *burrito*, and also some numerals. Each geometric symbol depicted nouns and verbs that relate to everyday life such as *orange, go, here, river, yogurt, dog, would, do, backpack, dessert, cold, draw, Dan, nook, Sue, slap, swimming pool, Jell-O, towel*, and many more.

By pressing the symbols on the lexigram, Matata began to "talk" with Sue and her team of researchers in a very basic way, making her the first bonobo to be language trained. But Matata took a long time learning each symbol on the board. As she did, Kanzi acted like a bored kid. He ran around the space, leaping on his mother in play. Sue said that Matata was extremely indulgent of Kanzi's antics, like all bonobo mothers, and did not like Sue trying to discipline him to make him behave. Sue hoped that Kanzi could be brought into the program one day, but it was too early yet.

Although Sue had not noticed, Kanzi had been observing his mother's lessons. At 14 months he occasionally pressed one of the symbols on the board, and then ran to the machine that dispensed food when the right answer was given on the lexigram. By the time he was two years old, he had already begun selecting the symbol for *chase* and then trying to get Sue to chase him. It was then, and still is, his favorite game. Matata plodded along, and after two years of training had mastered only six symbols.

When Kanzi was 30 months old, the Yerkes Field Station decided that Matata should conceive under controlled conditions. They separated the pair. The day after Matata left for the field station, Sue introduced Kanzi formally to the lexigram. She remembered that the experiment was spectacularly successful. The little bonobo used it spontaneously, without any specific, food-reinforced, training more than 120 times that day. He pressed the symbols for words such as *outdoors*, *swing*, *bite*, and *tickle*. He already knew the meaning of some of the symbols, even though his mother was able to learn only a handful.

Kanzi pushed the symbols for *apple* and *chase*, then looked at Sue in a meaningful way and ran away with a big smile on his face. He pressed keys for food items such as *banana*, *raisin*, and *peanuts*, and amazed Sue when she took him to the refrigerator and he unfailingly chose the food he had just signaled. She had toiled for two years trying to teach Matata with hardly any success, and yet Kanzi had shown he knew the meaning of eight of the symbols, achieving more than his mother without Sue giving him a single lesson. He was learning language as a child does, without training, and Sue said Kanzi was "thus the first [ape] truly to understand a spoken human language."

For Sue it was "one of the most intriguing times of my professional life. I recognized that if what we thought we had accomplished proved indeed to be accurate, it could revolutionize our understanding of the nature of language acquisition, indeed perhaps of all learning processes." She realized that it could also challenge a central tenet of our knowledge of ourselves—that among all the millions of species on the planet, the human mind was unique. Sue broadened her tests with Kanzi, setting up 17 food locations in the forest near the LRC, 55 acres of pristine trees, to see whether Kanzi could associate them with the appropriate lexigram symbol. He did.

With Matata gone, Kanzi was keen to communicate with his new caregiver, Sue. After four months, his lexigram vocabulary had expanded from the original self-taught eight to more than 20 signs, pinpointing types of food, locations in the forest, and people. This let him tell Sue where he wanted to go into the forest to find a particular type of food she had placed there such as cheese, blackberries, bananas, or oranges. Another important symbol he learned and understood was the one for *now*, as opposed to the symbol for *later*.

He progressed steadily. A photo of Kanzi at three shows him confidently pressing one of the lexigram keys on a portable keyboard Sue had devised, laminated paper sheets imprinted with columns of the word symbols, because he loved going out into the forest. It provided a replica for Kanzi of a typical day for young bonobos in the wild, with Sue leading Kanzi to forage for food, rest, and play.

Another picture shows Kanzi with an apple that he had just signaled on the sheet after one of Sue's researchers had taken it out of a backpack in the forest. At four, Kanzi began using the lexigram to talk to himself and a picture shows him, by now a big, beefy adolescent, hunched over the lexigram pressing symbols, oblivious to those watching him. At six years of age he had a vocabulary of 200 words.

Kanzi was a great fan of TV and had his favorite programs, which he chose by pointing to photos pasted onto the casing of videotapes. His favorite films were *Greystoke: The Legend of Tarzan, Lord of the Apes,* and *Quest for Fire*, in which actors played the parts of costumed great apes and primitive humans.

Kanzi enjoyed painting, but he could not produce recognizable images, a trait Sue says he shares "with most children under the age of three." His utterances became spontaneous, and he started to put words together—the symbol for *command*, the one for *ice water*, and the one for *go*, indicating that he wanted someone to get him ice water. Sue said that Kanzi was able to string words together to form basic sentences with meaning, and that he was able to invent a syntax of sorts, adding, "The fact that Kanzi is able to invent such rules is strong evidence for the continuity theory—that is, the idea that the mind of man differs in degree from that of the ape, but not in kind." In one videoed example of his linguistic prowess, Kanzi was asked by voice: "Give the dog a shot." Kanzi took hold of a syringe that was in front of him, took off its cap, and injected the needle into a stuffed toy dog.

But many linguists scoffed at Sue's claims. Famed linguist Noam Chomsky of the Massachusetts Institute of Technology, skeptical of great ape comprehension of language, emailed me soon after my meeting with Sue: "If an animal had a capacity as biologically sophisticated as language but somehow hadn't used it until now, it would be an evolutionary miracle." Chomsky believes language is pre-wired only in human brains, a unique neural circuitry evolved from our pre-human ancestors and shown in the common patterns evident in the grammars of all human speech. He claimed, "This research is just some kind of fanaticism."

Another prominent American linguist, Stephen Pinker, was also a skeptic. A cognitive scientist, he studied how children acquired language. Pinker told the *New York Times*: "In my mind this kind of research is more analogous to the bears at the Moscow circus who are trained to ride bicycles. You can train animals to do all kinds of things."

Sue countered that as humans and bonobos share 98.4 percent of DNA, there was good reason to expect that there was some sharing of language capacity. She contended, "How do we know that bonobos are not using these abilities in complex ways in the wild, ways that are not yet transparent to us? The more we learn about them, the more sophisticated their communicative abilities appear."

Clive Wynne, a psychology professor at the University of Florida and author of an authoritative textbook on the subject of animal cognition, was impressed when he put Kanzi's achievements to the test. Writing in the magazine *Skeptic*, Wynne noted, "What other nonhuman can convey so much to his caregivers, or understand so much of what they say to him?" He was disappointed, however, by what he felt was Kanzi's lack of understanding of grammar.

One day by the river at the center in Georgia, Sue asked Kanzi using words, "Can you throw your ball into the river?" She said, "He'd never done such a thing," and indeed Sue had tried to keep all his toys out of the river. A red ball was and still is his favorite toy. "He promptly tossed the ball into the river." In another test she asked Kanzi to give his half sister, Panbanisha, an onion. They shared the same mother, Matata, who had returned to the Language Center after giving birth to Panbanisha. Kanzi knew where to find onions because they were his favorite food. He searched about for an onion and presented Panbanisha with one.

Sue began to experiment with complex verbal sentences. When she asked Kanzi if he would swap a monster mask of which he was fond for some tasty cereal being eaten by another of her bonobos, he readily carried out the trade. At the end of a nine-month trial, Sue said, Kanzi had correctly answered or acted upon 74 percent of the spoken sentences directed at him.

On an outing in the forest attached to the language center, Kanzi used a long, hairy finger to touch the symbols for *marshmallow, make,* and *fire.* Given marshmallows and matches by Sue, Kanzi snapped sticks to build a fire and grilled the marshmallows skewered on a stick.

Sue also claimed, controversially, that her bonobos know the meaning of up to 3,000 spoken English words. Wary of claims by her critics that Kanzi was responding to facial expressions and gestures, like those that circus animal trainers used, as clues to what he should do, Sue and her researchers began wearing masks in their language experiments. In one of many experiments to prove Kanzi's comprehension of unfamiliar spoken English, she gave him a rubber snake and a stuffed collie dog. It was the first time he had seen the toys, but he knew what they represented, shown real dogs and snakes on forest walks.

"Can you make the snake bite the doggy?" she asked, her face concealed by a mask so that her expression gave no hint of meaning. In response, she told me, Kanzi sank the snake's fangs into the dog's rump. She then asked, "Can you make the doggy bite the snake?" Kanzi put the snake's head in the toy dog's mouth and pushed the dog's mouth on it.

Bonobos are far more vocal than chimpanzees, a difference that is clear to anyone who has witnessed both species in the wild. Sue raised Kanzi's half sister Panbanisha with a chimpanzee named Panzee. She said, "She never fully matched Panbanisha's skills, though, either in production or comprehension."

Sue noted: "The apes I know behave every living, breathing moment as though they have minds that are much like my own. They may not think about as many things or in the depth I do, and they may not plan as far ahead as I do. Apes make tools and coordinate their actions when hunting prey, such as monkeys. But no ape has ever been observed to plan far enough to combine the skills of construction and hunting for common purpose. Such activities are a prime factor in the early lives of hominids. Fortunately for those working with apes, they sense the world much as we do. Their vision, hearing, sense of smell, and so on are very much like our own."

The University of Minnesota's Center for Cognitive Sciences, in its Millennium Project, named Sue's monograph *Language Comprehension in the Ape and Child* as one of the top 100 most influential works in cognitive science in the twentieth century.

After many years of watching and studying bonobos, Sue says that she "still cannot help but sense that I am in the presence of the emergence of the human mind, the dawn of our peculiarly human perspective feeling."

Some hours later, at the end of the Japanese meal, Sue said to me, "Please come to Des Moines sometime. I'd like Kanzi to meet you."

5

That chance to meet the world's most famous bonobo, Kanzi, whom Sue says can converse in a meaningful way with humans, comes directly after my visit to the wild bonobos at Kokolopori. From the Congo I journey to the most unlikely place for great apes, America's heartland, Des Moines, with the temperature dropping from a steamy 90 degrees Fahrenheit in the African jungles to minus 10 degrees with a deep layer of snow on the ground in Iowa.

By the time I have flown from the Congo to transit in Johannesburg, then on to New York and Des Moines, jet lag has blanked my mind and I am in automatic mode, though I know from long experience that a good sleep will put the zip back into my brain. I book into the venerable Hotel Fort Des Moines, one of the most famous sites in the American political landscape. In a presidential election, one of the most important primaries takes place in Iowa, and contenders flock to Des Moines, the capital, with many staying in this hotel.

The hotel is almost a century old, and on the wintry night I arrive it looks almost deserted. As I wander down an empty, chilly, ghostly corridor on the way to my room, I seem to be reliving a Jack Nicholson scene from the horror movie *The Shining.*

In 2005, Kanzi, his foster mother, Matata, his half sister Panbanisha, her whip-smart young son Nyota, and four more bonobos moved to Des Moines to their own $10 million, 18-room house and lab complex on the grounds of the Great Ape Trust of Iowa. A benevolent Iowan billionaire, Ted Townsend, provided the funding to study the intelligence and language abilities of great apes at the cutting edge. Townsend's great wealth comes from his father, Ray, who invented a machine named the Townsend Model 27 Pork Skinner in 1945. Nine years later, Ray invented a combination of machines that turned a pig's carcass into sausages. Iowa is the home of the U.S. porker, and most of the nation's hot dogs are made using Townsend's machines.

His son, Ted, made many trips to Africa to see wildlife and was stunned to find that the magnificent great apes he encountered in the wild were facing extinction. He had heard about Kanzi and flew to Atlanta to meet the famous bonobo. Kanzi hit the keys for . . . *Kanzi . . . wants . . . grape . . . juice . . .* on his lexigram, and Townsend fetched the drink for him. *Thank . . . you . . .* Kanzi tapped out.

Charmed, Townsend decided to form the Great Ape Trust of Iowa and put down an initial $20 million to provide the trust with its own facilities in his hometown, Des Moines, dedicated to the study of the great ape mind. The trust also hosts a pair of orangutans in their own building on site, and when complete, the grounds will house all four species of great apes, including western lowland gorillas and chimpanzees.

The morning after my arrival, a taxi takes me five miles from Des Moines's business center through suburban America to the "paradise"

Townsend set up for some of the world's foremost ape language researchers.

A high-wire fence surrounds the center's 230 acres, which has wetlands, ponds, and an 80-acre riverine forest, all fashioned from an old quarry. It is North America's largest great ape sanctuary. The bonobo blockhouse boasts a 13,000-square-foot lab, with closed-circuit cameras watching the apes' every move. There is a kitchen where the bonobos can cook food in a microwave and get snacks from a vending machine, and there are drinking fountains. Outdoors, they have seven acres to themselves to ramble and roam, including three outdoor play yards by a pond and their own cave.

Sue meets me inside the huge bonobo home. She explains that Kanzi and his family "have all grown up in a bicultural group consisting of humans and bonobos of varied ages and sexes. They are the only captive apes reared in this way and the only reproductively competent, captive apes who have acquired human language and stone-age manufacturing skills. They make up the only linguistically competent group of apes who continue to experience direct and intimate contact and communicative interaction with multiple adults."

Through a glass window I see several bonobos, adults and adolescents, playing a vigorous game of chase in a large sunroom, their features gleaming with delight. Sue points out Kanzi sitting by a wall watching them. Now middle-aged at 28, he has the mien of an aging senator—balding and paunchy, with serious deep-set eyes and heaps of gravitas.

She tells me that housing Kanzi and his kin at this ultramodern facility has a serious purpose. The center conducts cutting-edge experiments testing great ape comprehension of human language. "Kanzi truly understands spoken English," she says. Through a glass panel, I watch as she asks Kanzi, the clan's dominant male, if he will allow me to enter. She beckons him with a hand, and he comes to the window and peers at me. "Kanzi, this is Paul," Sue says. "Can he come and visit you?"

The bonobos control who comes into their quarters. Kanzi squeals apparent agreement and pushes a button activating a hydraulic door that rumbles open, allowing us inside. A see-through wire barrier separates us from the bonobos. Despite his age, "Kanzi is extraordinarily strong and can cause you serious damage if he wants," Sue explains. He is short-tempered, has intense emotions, is highly sensitive and impulsive, and wants to dominate all that he sees.

The other bonobos peer at me for a few minutes and then resume their games. Sue enters a small room, and I watch outside through a glass panel as she and Kanzi embrace and then roll onto the floor. She enjoys a vigorous game of play-tickle with the roly-poly Kanzi. Each attempts to tickle the other under the arm and along the chest, and they both chuckle with delight.

An international study of great apes showed that they all enjoy a good tickle. Davila Ross, a primatologist at the University of Portsmouth, enlisted keepers at seven European zoos to tickle their great apes on their palms, feet, necks, and armpits. She analyzed more than 800 recordings of the resulting apes' laughter. Ross concluded that human laughter was closest to the laughter of chimpanzees and bonobos. Orangutan laughter was the most distant, while, "gorillas took up an intermediate position."

On this day at the Great Ape Trust, there are no language experiments and so I provide the entertainment for the bonobos. Sue has remembered that during our time together in Japan, I mentioned that I knew how to perform the Maori *haka*, the ferocious New Zealand war dance made famous by the national rugby team; now Sue asks me to perform it for the bonobos. It will be the first time they have ever seen a male human displaying his power in this way.

I shout out the traditional challenge, a short, sharp fierce burst of words, to get the bonobos' attention. They stop their games to peer at me and then at each other, not sure what to do. Kanzi stares at me. I take a deep breath, put on a maniacally bellicose face in the Maori way, and start the war dance. I punch my chest, slap my thighs, and holler a challenge to the bonobos at the top of my voice.

All but Kanzi sit shocked for a few moments, and then they snap into an angry frenzy, the noise deafening as they challenge me back, screaming, hooting, and charging me right up to the wire barrier between us. I intensify my shouting and my body slaps as I glare at them and thrust out my tongue and roll my eyes in the proper way.

The traditional words I yell claim that I am going to tear out their guts, ravish their daughters, and put their homes to the torch. They scream back and bang their fists on the floor. A big male grabs a plastic stool, rushes at me, and slams it into the barrier just inches from my face.

As the tumult bounces around the walls with a fearful echo, only Kanzi remains calm. He draws Sue to him with a waved hand and lets loose a torrent of squeaks and squeals. I stop the *haka* when Sue calls me over. "Kanzi says he knows you're not threatening them but they're scared, and so he'd like you to do it again just for him, in a room out the back," she tells me.

I nod but keep my skepticism to myself until I see what happens. I follow Sue out through the hydraulic door, out of Kanzi's sight, and am amazed when we find him, alone at the back of the building in a small room that cannot be seen by the other bonobos. As I enter, he stands upright behind protective bars and slaps his chest and thighs, imitating me. I take this to mean that he wants an encore.

I respond with renewed vigor, launching into the *haka*, and in return Kanzi intensifies his war dance. He then presses against the bars, trying to reach me with his brawny arms. His brown eyes glow, and I know he does not want to harm me because he is flourishing an impressive erection at me.

Once he calms down, Kanzi squeaks something at Sue and then leaves the room. She tells me he wants to show me his new lexigram, and we return to the outer corridor and peer into another small room. Kanzi is already there and seated before an electronic lexigram touch-pad. It is connected to a computer monitor that shows the words as Kanzi selects them, and even has an American male voice pronouncing each word. But the electronic lexigram is having teething problems and the words Kanzi punches in get garbled, because his finger constantly slips off the keys and touches other words. "We're trying to solve this problem," Sue says.

Sue believes bonobos use a type of language to communicate with each other in the same way in the jungle. In a blind-test experiment to judge the extent of bonobo vocalization, she placed Kanzi and Panbanisha in separate rooms; they could hear each other but not see each other. Kanzi was told he was to be given yogurt by using the appropriate symbol on a silent lexigram. He was verbally asked to announce this to Panbanisha. "Kanzi vocalized, then Panbanisha vocalized in return and selected 'yogurt' on the silent keyboard in front of her," Sue says.

She mentioned different foods to Kanzi such as peanut, tomato, egg, and banana, and unfailingly Panbanisha recognized his bonobo vocalization for the food type chosen and pressed the correct lexigram. With these and other ape-language experiments, Sue claims, "the mythology of human uniqueness is coming under challenge. If apes can learn language, which we once thought unique to humans, then it suggests that ability is not innate in just us."

Many primatologists believe that Kanzi and his clan do have an extraordinary grasp of the English language. They join several chimpanzees, gorillas, and orangutans, who have a simpler understanding after being taught to use a few hundred words of American Sign Language, the language of the deaf. The jury is still out among other scientists, especially linguists, who do not believe that great apes, however smart, can comprehend language structure, but rather that Kanzi and the others are moved by the desire for treats to learn simple language-based tricks. That may be so, but I have no explanation as to how Kanzi indicated to Sue that he would meet us in one of the back rooms where he wanted to see a repeat of the war dance.

Juan Carlos Gomez, author of the book *Apes, Monkeys and Children: The Growth of the Mind,* is on Sue's side. He claims that with many ape language experiments, most of the time, "what they understand is the tone of the voice, or perhaps they have learned to associate the sound of a word or sentence to a particular situation (going for a walk, getting food, or being reprimanded and so forth)." But Gomez admits that the results of rigorous tests Sue devised, comparing Kanzi's comprehension of spoken English against that of a two-year-old human, "were so spectacular, Kanzi, at 8 years of age, performed as well or even better than the infant in almost all types of sentences, revealing a capacity to understand spoken English equal to a two-year-old."

As the end of the day nears, Kanzi and the other bonobos sprawl like couch potatoes on the floor of a room facing the outside corridor. I see them through the glass panel snacking on M&M's as they watch a DVD, one of Kanzi's longtime favorites, the Greystoke Tarzan movie. He had selected it by pressing a still picture from the movie on a computer screen attached to a DVD machine. Another favorite is Clint Eastwood's *Every Which Way But Loose.* "The bonobos' most loved films star great apes being friendly with humans," Sue tells me.

I sleep well at the hotel and the next day arrive just after noon. Kanzi seems to recognize me, because he leaps up and down as he spies me through the glass panel and presses the button to let Sue and me in without being asked. Although we are still separated by the barrier, he does not object when Sue takes me into an adjacent room he can see through the barrier and brings in Panbanisha's eight-year-old son, Nyota. His name means "star" in a Congolese language, Sue tells me, and adds that Nyota is "the new Kanzi."

Nyota, though a juvenile, is still much stronger than I am and can do me considerable damage if he wishes. But he initiates a game where I sit on the floor while he charges at me, avoiding me with a neat sidestep at the last moment. He leaps to the ceiling and swings on bars there, and then drops almost on top of me with a throaty chuckle. Nyota's acrobatic skill is as impressive as that of the wild bonobos. Not once does he hit me like the little terrors back at Bonobo Paradise, the bonobo sanctuary in Kinshasa. After 30 minutes, he stops the game, lopes across the floor to embrace me, and plants a wet kiss on my cheek.

"Nyota likes you," Sue says. "Let's take him for a ride around the grounds; he loves that."

We get into her car, with Sue driving and me in the front seat with Nyota seated on my lap with his arms around my neck. He puckers his lips and kisses me several times as the car drives slowly around the trust's perimeter skirting the high fence.

In another intriguing experiment, Sue put to the test Kanzi's ability to make crude stone-cutting tools in the manner of our pre-human ancestors in the Paleolithic era. By smashing rocks together, they produced sharp-edged flakes to be used as crude knives. This is a considerable advance on the chimpanzees' ability to fashion twigs and grass stems to reach inside termite and ant nests and extract the insects for food. In his classic book from 1949, *Man the Tool-Maker*, anthropologist Kenneth P. Oakley stated that humans were unique because "Possession of a great capacity for conceptual thought, in contrast to the mainly perceptual thinking of apes and other primates, is now generally regarded by comparative psychologists as distinctive of man."

Nicholas Toth, an American anthropologist, agreed to try to teach Kanzi to make stone-cutting tools and show him how to gain access into a food box that could be opened only by cutting a string attached to it. Toth is a professor of anthropology at Indiana University and co-director of the Stone Age Institute there. He is an expert at making stone-cutting tools like those of our 2-million-year-old ancestors *Homo habilis*. Examples of these tools, made from basalt, were found at the Olduvai Gorge in Tanzania and consisted of hammer stones, scrapers, choppers, and small, sharp flake knives.

With Kanzi watching keenly, Toth struck one rock with another to produce flakes. He selected a sharp flake and used it to cut the string that held closed the box lid. He opened the lid and gave Kanzi the treat inside the box, establishing the connection: bang together two rocks and use a resulting stone flake to cut the string, open the box, and get the food inside.

Kanzi copied Toth, knocking a pair of rocks together, but did not produce any flakes. As days passed, he was still not successful, and so Toth gave flakes he had produced to Kanzi, who used trial and error to choose the sharpest flakes from duller-edged ones by using his eyes and lips.

Despite his strength—two or three times that of a human male adult—Kanzi pulled back from hitting the rocks hard against each other. Sue wondered if he was afraid of hitting and injuring his fingers. But one afternoon, Toth heard a BANG followed by more bangs. He found Kanzi making his first flakes, actually tiny chips. He was still not hitting the rocks hard enough together. Then one day in the fourth month of the experiment, Toth saw Kanzi looking at the two rocks thoughtfully. The bonobo suddenly stood up and threw one of the rocks onto the ground. It shattered into flakes. Kanzi grabbed the sharpest and dashed to cut the string securing the food box.

Copying Toth, Kanzi went back to banging a pair of rocks together. Sue wrote: "So far Kanzi has exhibited a relatively low degree of technological finesse . . . compared to that seen in the early Stone Age record. The amount of force he uses in hard-hammer percussion is normally less than ideal for fracturing these rocks."

Toth concluded that there was a clear difference between Kanzi's stone-cutting tools and those made by ancient men. They had a physiological advantage as well as superior mental ability. Sue pointed out: "The structure of bonobos' arms, wrists, and hands is different from that of humans, and this structure constrains the ability to deliver a sharp blow by snapping the wrist, a movement Nick considers important in effective took-making."

What delighted Sue was Kanzi's thoughtful problem-solving of throwing a rock against the ground to break it into flakes to use for cutting tools. This may have been the technique used by ancient man before he discovered how to manufacture cutting flakes by banging rocks together. In honor of his achievement, in 1991 Kanzi was given the inaugural CRAFT Annual Award for Outstanding Research Pertaining to Human Technological Origins.

Toth remarked that the bonobo received the award "because the work with Kanzi has given us an important insight into Paleolithic technology."

On the following day I see the bonobos' meals prepared in the kitchen, using the best-quality fruit as well as yogurt. I watch them eat with no suggestion of the jealousy I saw at the Kinshasa sanctuary, where the bonobos quelled it with a high-energy orgy. Nyota plays with me again, for an hour bouncing around the barred walls of the room to launch mock attacks but never harming me. Then, I sit by the wire barrier in the largest room and Kanzi lopes across the room to sit beside me on the other side.

The other bonobos, including his sister, ignore us. I peer into Kanzi's gleaming brown eyes, looking into their depths. He resembles many of the wise old men I have encountered on my journeys across the world. I am unable to communicate with Kanzi as effectively as Sue, and so whatever he thinks about me will remain within his mind, and he can only guess at what I feel about him.

As I leave the room and head for the exit for the final time, Kanzi spots me through the glass panel and jumps up and down. Sue gives me a farewell hug. "You're the human Kanzi," she says. I assume that it is a compliment.

Just as we are beginning to plumb the minds of these extraordinary creatures, Kano's "fairies of the jungle," who could have resembled our ancestors 8 million years ago, bonobos in the wild might slip away forever in a few decades. I fly on to San Diego to visit its world-famous zoo, which is home to 18 bonobos. Karen Killmar, the associate curator of mammals, tells me there are probably no more than 20,000 bonobos in the wild, though she adds that it is difficult to come up with an accurate number because research is so scanty. "There's no doubt that their numbers dropped dramatically during the [DRC] civil war," she says.

The UN organization GRASP (Great Apes Survival Partnership), set up in Nairobi in a crisis move to arrest the dramatic decline of the great apes, published an estimate of bonobo numbers in the wild in September 2005. In all nine territories including Wamba and Kokolopori, GRASP reported that the bonobo population was "unknown since war."

Daniel Malonza, a GRASP spokesman, told me by phone, "The threat of renewed civil war, a surging human population, the thriving bushmeat trade, and the destruction of the bonobos' habitat in the DRC is hurrying them toward extinction in the wild."

Even peace in the Congo might not save the bonobos. "International logging companies are poised to exploit the forests here where the bonobos live," Michael Hurley told me at Kokolopori.

The zoo bonobos have a far greater chance of survival. Karen Killmar helps administer the Bonobo Species Survival Plan in the United States, which strives to maintain healthy sustaining populations of captive bonobos. American zoos together hold about 79 bonobos. "We select suitable bonobos and move them between zoos to mate to widen the gene poll," she tells me at the San Diego Zoo.

The zoo populations lack the great genetic and cultural diversity of the wild bonobos. But they are a fail-safe option should anything happen to the wild bonobos in the decades ahead, such as a calamitous epidemic or more widespread hunting for bushmeat and habitat destruction.

Research on the wild bonobos is severely hampered by lack of funds. They have slipped off the media radar, and now hardly anybody outside the primatology world seems to have heard of them, so there is widespread ignorance of their fate. Research into the lifestyle of the wild bonobos is scarce, and compounding this lack is some suspicion and hostility among the nonprofit groups dedicated to saving bonobos. There is little cooperation between them, prompted by the scramble for limited funding.

The most urgent task is to identify the full extent of the bonobo range by transects in likely forests, to tabulate how many are still left in the wild and where they range. Southwest of Kokolopori in the vast Salonga national park, bonobos were once common but are rarely sighted these days. However, gruesome evidence recently gave hope of finding more bonobo groups. On my return to Mbandaka from Kokolopori, Jean-Marie Benishay, BCI's field director, showed me a picture of the stacked skulls and bones of six bonobos on sale at an outdoor market for use in black-magic rituals. "They come from a place where we thought bonobos had disappeared," he said with a grim smile. "This proves bonobos are still out there."

From witnessing one of the rarest of the great ape species in the Congo wild and in captivity, I journey to the Central African Republic. There, I will find the jungle habitat of perhaps the most numerous of the great ape species, the towering western lowland gorilla, the inspiration for King Kong.

King Kong at the Brink

THE WESTERN LOWLAND GORILLAS

1

Living in Sydney on the other side of the Earth from Africa, I almost always have to take a long flight on my way to work. The bonus on this journey is that I must fly to Paris to get a visa. My destination is the Central African Republic, a tiny landlocked military dictatorship infested with rebels and brigands, and forever on the brink of another bloody coup d'état. The Dzanga-Ndoki National Park, in a remote corner of the country, is a principal habitat for one of the last remaining significant stands of the western lowland gorilla, the subspecies you find in zoos worldwide.

A gorilla is a gorilla is a gorilla to most people, but to zoologists it is a species, and there are four subspecies—the mountain gorilla, eastern lowland gorilla, Cross River gorilla, and western lowland gorilla. We have already met the mountain gorilla in East Africa, living at altitudes from about 8,000 feet to 12,000 feet above sea level. The remaining three kinds of gorillas live in the jungles and forests of East, West, and Central Africa, and their habitats range from sea level to about 8,000 feet high.

During a span of about 2 million years, the various gorillas evolved from a common ancestor, all belonging to the same species but having some distinct characteristics. The mountain gorillas are the biggest and are terrestrial, ground dwellers. The three lowland kinds are a little smaller and far less furry, and are semi-arboreal, spending part of their day perched on branches in towering fruit trees.

Western lowland gorillas mostly live in tropical rainforests in Central and West Africa, and were the only kind known outside Africa for centuries. Hanno, an intrepid Carthaginian admiral, sailed down Africa's west coast in the sixth century B.C., and in his account of the journey he refers to a wild, hairy people the translators called "gorilla" living on a lake. Filippo Pigafetta's *A Report of the Kingdom of Congo*, published in 1591, taken from the notes of Duarte Lopez, a Portuguese sailor, noted that "in the Songan country, on the banks of the Zaire, there are multitudes of apes, which can afford great delight to the nobles by imitating human gestures." These may have been chimpanzees or gorillas, or both.

The most numerous of the subspecies, the lowland gorillas' habitats spread across the jungles of equatorial Africa. Though the mountain gorilla was discovered by European explorers two centuries after the western lowland gorilla, due to Dian Fossey's expertise in public relations, the mountain gorilla is now the far bigger star. And since Fossey went to live in its habitat four decades ago, there has been considerable research done on its habits. In contrast, research on the western lowland gorilla has been limited.

According to primatologist Dave Greer, who studied both subspecies in their native habitats, this is because the elusive lowland gorillas are much shyer, having been hunted for millennia by pygmies and other tribes living in or near their habitats. Unhabituated lowland gorillas flee at the first sign of humans. They live in dense forests where the visibility can be just a few yards and are semi-aboreal, spending part of each day high up in the trees out of researchers' sight, snacking and snoozing. As a consequence, the western lowland gorillas are far harder to study than the mountain gorillas, who live most of their life on the ground and range each day over significantly smaller distances.

Unlike almost all the other African great apes, western lowland gorillas do not avoid water and happily sit for hours in swamps and water-logged clearings, munching on water plants. The silverbacks perform dramatic splash displays, slapping the water with their huge hands, much as I did when I was a kid at the beach in Sydney when playing with friends. The big male gorillas leap into the water during these displays to generate waves and spray.

Far from playing, however, the silverbacks use the displays in an attempt to intimidate rival silverbacks and win over females. "If you see a 160-kilogram silverback charging into the river, you are unlikely to lead to the conclusion that he is bathing," commented Richard Parnell, a primatologist at the University of Stirling in Scotland. He and a fellow researcher studied the gorillas at Mbeli Bai in northern Congo, which they found to be a socializing spot as well as a feeding site.

These displays usually take place at a *bai*, a pygmy word meaning "where the animals eat." A *bai* is a swampy forest clearing in the jungle, measuring from 25 to 250 acres, where animals including elephants, buffaloes, and gorillas go to soak up the mineral-laced waters in the ponds and puddles. These are sometimes formed by elephants digging holes in the soil with their tusks, which are then filled by the heavy tropical rainfall.

"The *bai* is a great place to check out the females," Parnell said. Like the mountain gorillas, loner western lowland gorillas hope to pick up a female with a roving eye during an encounter with a group. That, said Parnell, is probably the reason for the boisterous splashing displays. Only males splash. Silverbacks commonly slam their hands against the ground as they prepare to charge a rival, as I have seen myself, and the researchers believe the splashing displays show how the gorillas at Mbeli Bai have adapted their behavior to an aquatic environment.

The University of Stirling researchers wrote: "We anticipate that gorillas, maligned as cognitively poor cousins to the other great apes, will emerge from further *bai* studies as adaptable, innovative and intelligent creatures that exploit a complex environment."

They must like the *bai*, because a study of western lowland gorillas in the Republic of Congo, not far from where I am headed, found that the groups spent time there on 95 percent of 380 observation days. The attraction is probably related to diet. The swamps offer a rich source of aquatic herbs providing carbohydrates and high-quality protein as well as health-beneficial minerals, especially sodium, which is found in higher concentrations in the aquatic rather than terrestrial plants.

A *National Geographic* photo shows a western lowland silverback soaking in the muddy water. He yanks up the roots of swamp herbs and methodically washes the mud from them in the water. Then he happily sucks them down like strands of noodles.

At Mbeli Bai, not far from where I am going, gorillas were observed using tools for the first time. Primatologists define tool use as "the employment of an unattached environmental object to alter more efficiently the form, position, or condition of another object, another organism, or the user itself when the user holds or carries the tool during or just prior to use and is responsible for the proper and effective orientation of the tool."

Researchers from the Bronx Zoo's Wildlife Conservation Society observed an adult female named Leah in Mbeli Bai at the edge of a pool in an area where the gorillas had not been seen for six months. A branch was sticking out of the surface: "[She looked] intently at it in the water in front of her for 1 min[ute] before entering the water. She began to cross the pool walking bipedally, but after her first steps the water quickly became waist deep and she returned to the pool edge." Leah reentered the water, again standing upright, and grabbed the stick. She held it out in front and used it to test the depth of the water as she crossed the pond. On shore her baby started crying, and so Leah returned to the pool edge to get her baby. Then, she "moved around the pool while feeding on aquatic plants."

Leah clearly had an inventive mind because she was the only gorilla, with her silverback partner, George, ever observed mating in the face-to-face position. Gorillas, like chimpanzees, commonly have sex in the front-to-back position.

A female named Efi from another gorilla group at Mbeli Bai was witnessed breaking off a stump of wood and holding onto it for support as she dug for herbs. She then used the stump as a bridge to get across a muddy patch. The footprints of other gorillas in nearby swampy clearings showed that they also used branches as bridges to get across the mud.

Primatologists, cheering for the gorillas, hailed the discoveries as proof that they were not the great ape cousins of Homer Simpson. They argue that gorillas simply do not need tools as much as the other apes do. Chimpanzees use rock hammers to break edible nuts and termite-fish with twigs, but gorillas break the nuts with their huge teeth and smash open the termite mounds with their massive fists. All the great ape species can learn to use tools in captivity, and so it could be that tool-making for them is an environmental adaptation rather than that the chimpanzees possess a higher innate intelligence.

King Kong was essentially a giant version of a western lowland gorilla, characterized by his short, thick, dark fur, stocky body, huge jaw muscles, pronounced brow ridge, large, strong teeth, small eyes, large nostrils, and small ears. Although tiny when compared with the eponymous movie star, the subspecies' mature silverbacks are truly massive. They can stand six feet tall, and each can weigh as much as two sumo wrestlers—about 400 pounds. Like the mountain gorillas, they are dimorphic, with the females weighing less than half as much as the silverbacks. Their size has probably evolved over a very long time, as the females continue to choose the biggest and brawniest of the males.

The lowland gorillas share many similarities with the mountain gorillas. Each group is dominated by a silverback who is generally gentle and amiable unless he or his family is threatened. They can live up to 50 years in the wild, and a little longer in captivity. The oldest known gorilla, a western lowlander named Jenny, died at the Dallas Zoo at the age of 55. She had an inoperable stomach tumor that prompted her to stop drinking and eating. The zoo veterinarians mercifully euthanized her.

Females leave the natal group at about eight years old in search of a silverback who can protect them and their future babies. Based on gorillas kept in captivity, the National Zoo in Washington says that the normally quiet gorillas become "unusually loud" during the act of mating.

Before I depart for the Central African Republic, at a secondhand bookstore in Sydney my daughter finds a book published in the early 1930s, about the same time as the movie *King Kong*. The title gives the plot away: *The Gorilla Hunters, A Tale of the Wilds of Africa*, by R. M. Ballantyne. The book documents an expedition by what were then called white hunters to the Congo Basin in search of western lowland gorillas to shoot. Gorillas were much more plentiful then and the hunters had no trouble finding them. They were very successful in their grim ambition, and the author notes that they shot many young and old gorillas.

Their first kill was a silverback. The author relates:

Shouldering our trusty rifles and buckling tight the belts of our heavy hunting knives, we sallied forth, after the manner of American Indians, in single file, keeping as may well be supposed, a sharp look out as we went along . . . If [we] were to go home and write a book detailing our adventures in these parts, at least half the sportsmen of England would be in Africa next year, and the race of gorillas would probably become extinct.

When the hunters came upon the silverback in the jungle and shot him unseen through the tangle of bushes, the gorilla roared a first and final challenge. The narrator goes on:

I have heard the roar of the lion, and the tiger in all circumstances, and the laugh of the hyena, besides many other hideous sounds, but I never in all of my life listened to anything that in any degree approached in thundering ferocity the appalling roar that burst upon our ears immediately after that shot was fired.

Wound up by his fertile imagination, the author began to approach hysteria:

Apart altogether from its gigantic size, this monster was calculated to strike terror to the hearts of beholders simply by its expression of its visage, which was quite satanic. It seemed as if I were gazing on one of those hideous creatures one beholds when oppressed with nightmare!

Suddenly it rushed at us. "Look out Jack" we cried in alarm. Just as the monster approached to within three yards of him, Jack sent a ball into its chest and the king of the African woods fell dead at our feet! It is impossible to convey in words an idea of the mingled feelings that filled our breasts as we stood beside and gazed down at the huge carcass of our victim. Pity at first dominated in my heart, then I felt like an accomplice to a murder, and then an exulting sensation the joy of having obtained a specimen of one of the rarest animals in the world overwhelmed every other feeling. I shuddered as I looked upon it, for there was something terribly human-like about it despite the brutishness of its aspect.

With this sort of guff propagated in books and magazines, audiences in the 1930s were able to believe that King Kong was but an exaggeration of the real gorilla's true ferocious nature. And yet in this lurid account there are stirrings of some hint of modern feelings toward the great apes in the momentary pity the narrator felt at the death of the monster, prompted by his realization of the gorilla's closeness to humans.

Like the narrator, I head off to the Congo Basin to seek out western lowland gorillas. Up to 100,000 are spread across the rainforests of Africa according to the National Zoo in Washington. They are found only in equatorial Africa—in the Congo Basin, Cameroon, Nigeria,

Gabon, the Central African Republic, and Equatorial Guinea. A large chunk of their habitat is close to the equatorial coast on Africa's midwestern side, where many of the European explorers landed beginning in the sixteenth century, and so they were the first of the gorilla subspecies to be discovered by "whites."

A French missionary in 1847 in what is now Gabon reported encountering enormous manlike creatures in the jungle. Explorers, eager to thrill and chill their readers in distant Europe and the United States, sought the great ape out and branded it with a false reputation as a monster and a man-killer. In contrast, the National Zoo describes western lowland gorillas as "quiet, peaceful and non-aggressive."

Until the 1990s the western lowland gorillas seemed to be doing okay, but an outbreak of Ebola, that terrifying viral disease, has since devastated the subspecies in some places, killing more than 75 percent of the gorillas here over the past decade. Its numbers have also been decimated by the usual culprits—poachers, civil war, the charcoal trade, shrinking habitat, and other diseases. It is now listed as critically endangered by the International Union for Conservation of Nature.

2

Sitting next to me on the flight from Paris to Bangui, the capital of the Central African Republic, is an American diplomatic courier, a bulging briefcase never out of her grip. I am disappointed that it is not handcuffed to her wrist, like in the movies. "You're going to Bangui?" she asks just before the jet descends into N'Djamena, capital of Chad, her destination, in the murky predawn light. "It's on the CIA's list as one of the poorest nations. Be careful, there's been a civil war with a lot of fighting. It's so dangerous that our government has banned all nonessential travel there."

I already knew of the danger. The travel advisory issued by the Australian Foreign Ministry warned: "We strongly advise you not to travel to the CAR at this time due to the activities of rebel groups and rogue security and military forces. We also advise you to reconsider your need to travel to the capital Bangui at this time due to the tense and unstable security situation. Skirmishes between government forces and opposition groups have occurred in the capital. Banditry and crime are common throughout the CAR. Theft and robbery occur regularly in Bangui and armed forces and gangs operate in the outlying residential areas of the city."

The Central African Republic is not officially a war zone, but something considerably worse because there are no front lines behind which you can seek shelter. Violence is liable to erupt without warning anytime, anywhere. The problem is massive corruption. The Big Men grab what they can while they can, plundering the national coffers, stashing their loot in Switzerland, Hong Kong, or increasingly in the Arab kingdom of Dubai while looking behind their back all the time, making every effort to forestall the next coup d'état. That usually means spending an obscene slice of the national budget on the defense forces to keep the soldiers loyal, and throwing opponents into some of the worst jails on Earth to rot, or executing them.

In the meantime, the poor, even in oil-rich countries such as Equatorial Guinea, eke out their lives on the customary one dollar a day per family. The CAR does not have anything worth selling to the outside world, apart from a few diamond mines, and yet its Big Men still plunder what they can. It is near the bottom of the United Nations' Human Development Index, meaning it is close to being the most miserable and poorest place on Earth. Life expectancy is a disastrous 40 years, and the per capita yearly income is $300. I wonder what chance gorillas have in such a place.

The jet takes off from N'Djamena airport and climbs back into a shadowy sky, and about an hour later daylight filters through the gloom. Dense forests and winding rivers appear far below, shimmering with ribbons of drifting mist. We land at Bangui airport, a ramshackle huddle of sheds. Sullen barefoot porters in torn shorts and T-shirts dump our bags on a concrete floor under a tin roof. When all the bags are taken, there are still a handful of us waiting empty-handed. I recognize my fellow business-class passengers.

"Your bags must have been left behind in Paris," a stocky manager from the airline tells us in an offhand manner. "They'll arrive in a couple of days on the next flight."

"How come it's just the business-class luggage that's missing?" I ask.

"It's obvious, they were in the same container." He shrugs, all lips, shoulders, and hands. The CAR was colonized by the French, and it is still tied tightly to Paris by economic, defense, and cultural links. Over the long colonial rule, the locals obviously took a liking to the emotive Gallic body language.

A tall, square-shouldered Frenchman in an elegantly cut dark suit, blue silk tie, and polished black shoes mutters something to the official. Though clad in the uniform of an international businessman, he has the merciless eyes of a battle-hardened soldier. The airline official shrugs once more. The Frenchman peppers him with a machine-gun clatter of short, sharp words. This time the official nervously pulls at his collar and nods.

Bangui, nestled against the great Oubangui River, which forms the border with the Democratic Republic of Congo, is a hot and humid relic of French colonialism littered with run-down buildings, potholed roads, and crumbling monuments, the remnants of former dictators. Many of its buildings are pockmarked by bullets from a recent civil war. "This place is hell on Earth," the taxi driver says in English with a soupy French accent when I ask about all this visible evidence of mayhem. He surprises me by not suggesting a tour of Bangui's delights, often the opening offer by airport taxi drivers in developing countries. Perhaps there are none.

The capital is a tragic and frightened city, its residents cowed by decades of violent coups d'état from a ruthless army beholden only to itself. Armored personnel carriers manned by grim-faced French Foreign Legionnaires roar through the streets, a watchful soldier gripping a machine gun mounted on each carrier. We pass unsmiling soldiers from an African peacekeeping force patrolling in single file and gripping assault rifles, their nervous eyes swiveling from side to side.

Surly local police in ragged uniforms and carrying clubs also prowl the streets. An SUV tears toward us with its siren screaming. It is followed closely by an open-backed truck carrying troops brandishing assault rifles and a long-barreled anti-aircraft gun. The driver quickly pulls to the side of the road like all the vehicles in sight. In the SUV, the occupants are hidden by darkened windows. "It's the president," the taxi driver says. "He's always worried that someone from the army is going to shoot him."

My hotel lobby is the haunt of shady-looking men whispering in Sango, the local language, and French. My room faces the swimming pool, where prostitutes, some as young as 11 and 12, frolic in flimsy underwear, displaying their wares. From room balconies overlooking the pool, several middle-aged European men are watching the children, like spiders eyeing prey. One middle-aged man with a military-style crew cut waves to a wiry little girl whose exposed breasts are still child-like buds, and beckons her to his room. She obeys, clad only in her wet panties. A few minutes later they both appear on the balcony so she can wave in triumph to her friends below in the pool, and then they disappear back into the room. An hour later she returns to the pool, glassy-eyed and clutching a wad of money.

I seek out the hotel manager and remind him that it is a crime of rape if the man had any form of sexual activity with such a young girl, and ask that he call in the police to investigate. He looks at me with disinterest. "How do you know anything like that happened?" he challenges. "Maybe they watched cartoons on TV."

At mid-afternoon the phone rings in my room. It is an airline official advising that my suitcase has been found. At the airport, the Frenchman is waiting in the baggage hall with other passengers from our flight. "They were here all the time," he says as the still-sullen porters carry our bags into the hall.

"How did you discover them?"

The Frenchman smiles, but his eyes do not soften. "I told the airline manager that friends of mine in Bangui would apply not-so-gentle persuasion to him if he didn't find our bags quickly. He got the message. The bastards had put them aside, and were going to loot them when they finished work today."

That evening I see the Frenchman again in my hotel dining room, downing wine with several senior Legionnaire officers in uniform, perhaps the friends who are expert in the art of "persuasion."

The next day I visit the Bangui marketplace, where bulky middle-aged women clad in colorful robes sell piles of smoked wildlife, the bushmeat trade. I ask a few of the women if they are selling gorilla and chimpanzee meat, and they say no, shaking their heads, uttering horrified denials. But Angelique Todd, an English gorilla researcher whose friend peddles meat at the market, tells me at my hotel, "The sale of elephant, gorilla, and chimp meat is banned, but you can still purchase it in the Bangui market. A buyer has to know whom to approach. They don't display the meat, but they can get it for you for a price."

During an afternoon walk, I stop in the main street near my hotel to take some pictures. Suddenly, somebody from behind grabs my shoulder and spins me around. I face a tall young man carrying a club and scowling at me. Two more men with clubs stand by his side. My throat clenches. Robbers! Pedestrians amble by as if nothing is happening. His grip is too tight for me to run. He points to a badge on his breast. "Gendarme!" he snarls. "Tourist!" I reply. It has no effect. He hails a taxi and pushes me into the back seat. He and his two colleagues pile in and we head for the roundabout at the city center.

There, an enormous policewoman in a short-skirt khaki uniform that accentuates every bulge pulls us over to the side of the road. She gives me a ferocious look and snaps something in French. The policemen get out, my arrester dragging me by the collar. They begin to argue, shouting and swinging their arms about with their clubs whirling in the air.

I have no idea what is happening until a middle-aged man steps out of the crowd watching, and explains in faltering English that I am under arrest for taking pictures in the street. "The men are city police and the woman belongs to the national police," he explains. "They are fighting over who has the responsibility of arresting you."

"Is it a crime to take pictures in the street?"

"Maybe yes, maybe no. They think you're a spy. You know, like James Bond, 007. The policewoman saw you passing by in the taxi and she probably thinks she can get a bribe from you, or part of what you pay the men."

I decide that I will not pay any bribe, to see where this ends up. The shouting goes on for another 30 minutes, and by then several hundred people are pushing and shoving each other to get a look at the foreign spy and listen to the argument. I am a little scared because I am carrying several thousand dollars' worth of cameras, and if the police decide to seize them as part of the evidence against me, then I will feel the loss.

My arrester bundles me into the taxi that has been waiting for us and we head off for the central police station. The policewoman hails her own taxi and follows.

Three middle-aged policemen peer over a balcony as I am prodded upstairs to be charged. One of them speaks English. He is slim and clad in a ragged uniform, but he has a calm and confident look that tells of decades of asserting unchallenged authority. He questions the policeman who arrested me as well as the policewoman for about five minutes. Then he asks me in English if it was true that I was taking pictures in the street.

"Yes, I'm a tourist and I wanted to take home pictures of your beautiful city to show my wife and daughter. If I've broken the law, then I sincerely apologize."

He smiles and snaps an order to my arresters, who leave the police station with their heads low. The policewoman follows them down the stairs. "I'm sorry. The boys were so enthusiastic because they thought you were a spy for the rebels, taking pictures to plan another assault on the city."

"I'm simply a tourist."

"Yes, I can tell you mean us no harm, but we see so few foreigners here, besides the foreign legionnaires, and we tell our young men to be vigilant."

He sends a constable to a nearby café to get Cokes for himself, me, and his two colleagues. We sip them at a small table on the balcony discussing European soccer, an obsession of the men in all of the 18 African countries I have visited. As the drinks arrive he says, "I hope you don't mind paying. We're a poor country and we police often don't get paid our salaries."

"Gladly."

Late that afternoon I walk from the hotel back to the city center to meet a friend from Bangui's WWF office for coffee. We linger, and by the time I am ready to return the city is in darkness; there are no workable street lights, and no police in sight. My hotel is just 100 yards away, down a street that looks deserted, but my friend insists I take a taxi. "Robbers prowl the streets after dark and you're an easy target," he explains. I heed his advice and take my shortest taxi ride ever in four decades of roaming the world.

* * *

On the following day the Land Rover I have hired bounces, wheezes, groans, and rattles as we head southwest on a winding dirt road marred with potholes. The farther we travel from Bangui, the worse the road gets. Every 10 miles or so we are stopped by a roadblock manned by hard-eyed soldiers. They are aggressive and unfriendly. In Asia, in surviving two decades of military roadblocks, I honed a way of easing the tension by joking with the soldiers. Usually, they broke into smiles and joked back, even when there had been recent fighting nearby. But I fail to get these soldiers to smile. They aim their guns at me and stare with undisguised hostility.

"The soldiers are not joking people," warns my driver, a Bangui local named Jacob. "If you make them angry they'll shoot you. Maybe they think you're French. They're very afraid of the legionnaires who shoot to kill when there is a disturbance."

He is not exaggerating. During one civil war a battle broke out between government troops and rebels at the state radio station, always a major target during any coup attempt worth its salt. The legionnaires were sent in, and when they stopped firing most of the soldiers, government and rebel, were dead or badly wounded.

I fall asleep not long after the sky darkens into night, undeterred by the roll and rock of the SUV as it bounces over the countless ruts. Just before dawn, 12 hours after we left Bangui, Jacob nudges me awake. "Pygmies," he says. "We've arrived. They'll take you to the gorillas."

For thousands of years the pygmies were masters of equatorial Africa, but today there may be only 100,000 of these seminomadic hunter-gatherers left. There may be only about 200,000 in the CAR. Much of their rainforest realm has been destroyed and turned into grazing land by the Bantu, and as I saw at the Mountains of the Moon in Uganda, they are mostly scorned by their taller neighbors. The pygmies, along with the western lowland gorillas, inhabit the same rain-soaked jungles of the Congo Basin.

So, I am delighted that they will take me to the gorillas in the wild. Mention of the pygmies evokes an alien world in deepest Africa, humming with mystery and adventure. The ancients were dazzled by wanderers' tales of a mysterious race of tiny Africans. Pharaohs and emperors of Rome were captivated when captured pygmies, the world's smallest people, were brought before them. And they still have allure today with modern moguls, Hollywood movie directors, casting them as exotic extras in tales of primitive Africa.

The earliest known reference to a pygmy—a "dancing dwarf of the god from the land of spirits"—is found in a letter written around 2276 B.C. by Pharaoh Pepi II to the leader of an Egyptian trade expedition up the Nile. In the *Iliad*, Homer invoked mythical warfare between pygmies and a flock of cranes to describe the intensity of a charge by the Trojan army. In the fifth century B.C., the Greek historian Herodotus wrote of a Persian explorer who saw "dwarfish people—who used clothing made from the palm tree" at a spot along the West African coast.

More than two millennia passed before the French-American explorer Paul du Chaillu published the first modern account of the pygmies. "Their eyes had an untameable wilderness about them that struck me as very remarkable," he wrote in 1867. *In Darkest Africa*, published in 1890, the explorer Henry Stanley wrote of meeting a pygmy couple: "In him was a mimicked dignity as of Adam, in her the womanliness of a miniature Eve."

The dirt track has narrowed, and the cleared land leading up to the jungle for about 50 yards is lined with huts shaped like beehives. They are waist-high, and I wonder how entire families can fit inside. No creature stirs, and even the camp dogs are still asleep. "The pygmies will wake soon," Jacob says. "They're Africa's greatest hunters, and each morning they go early into the forest to catch animals."

We have reached a singular place. The Dzanga-Sangha Dense Special Forest Reserve at the southwest edge of the country, about 500 miles from Bangui, is the premier wildlife sanctuary in the Western Congo Basin. The rainforest's 1,296 square miles is home to one of the richest and most diverse assemblies of animals, birds, fish, and insects found on Earth, a pristine jungle nestled against the Sangha River. The huge river snakes through the massive rainforest, forming a triangular border with Cameroon and the Republic of Congo not far downstream.

At a bend in the river, about six miles on from the pygmy encampment, we arrive at Bayanga, a Bantu settlement. Bantu is the common name given to sub-Saharan Africans who are not pygmies. As the sun rises over the treetops, Bantu men and women emerge from their wooden huts yawning and rubbing sleep from their eyes and begin to set up market stalls, gossip, or fetch river water. They are a tallish well-built people, mostly immigrants who flocked here when a French timber company, Sylvicole, set up a sawmill to cull the rainforest two decades earlier.

Aside from poachers, logging is the wildlife's greatest enemy here. Logging concessions cover up to 75 percent of the CAR's remaining forests, and up to 10 percent of the forests still left are cleared each year. The northern half of the country is virtually a wasteland, with hardly any forests remaining and almost no wildlife. This is largely a tragic consequence of mostly Sudanese and Chadian poachers, armed with assault rifles and traveling in convoys of trucks carrying the smoked wildlife carcasses stacked in the back to their home markets. But here in the south there are still significant stands of pristine forests remaining where you can find chimpanzees and gorillas.

Jacob takes me to a small lodge, a handful of wooden bungalows perched on poles dug into the riverbank. Even at this early hour, pirogues—canoes carved out of entire logs—slip quietly by, poled by wiry Bantu. The air drips with humidity, and dozens of big mosquitoes spin about me in a frenzy, unable to penetrate the protective lotion I have applied.

A slim, dark-skinned Bantu in his early 20s greets me with a firm handshake. "I'm William Bienvenu, your translator," he says. "I'll go with you and the pygmies to the gorillas. Jacob says you're a Catholic. I spent four years in a seminary, but had to leave in the final year because I liked girls too much."

"That was probably a good thing, William. A good-looking priest with an eye for the girls can be quite a devil."

We drive back to the same pygmy village. It sprawls along a rise skirting the jungle, now dappled with sunlight gleaming through the trees. About 20 beehive-shaped huts are scattered across the slope, made from bent striplings covered with leaves. Smoke spirals into the damp air from several fires. Squatting around them are tiny women clad in sarongs, warming water and cooking food. Outside the huts, miniature men sit on their haunches and chat.

"The village is called Mossapola," says William. "About one hundred pygmies of the tribe called Bayaka live here. As you can see, they are a very primitive people. We call them *babinga*, which means 'subhuman' in our language."

The Bantu indeed believe the pygmies to be subhuman, and have persecuted them for centuries. Until the 1980s, the CAR denied them passports even though the pygmies were here thousands of years before the Bantu arrived. William's remark hints at a deep-rooted prejudice, but just as I am about to reply a dozen pygmy men run down the slope, shouting. When they reach us, their laughter cascades like a waterfall as they start pulling hairs from my bare arms and staring wide-eyed at them. Their own are smooth-skinned.

Parting a path through the children and adults gathering around me, a small man moves to my side. "I'm Wasse," he says, speaking through William.

"Wasse is the Bayaka's greatest hunter and tracker," William says. "He'll take us to the gorillas."

I like Wasse immediately, struck by his keen eyes and calm manner. Like the other pygmies he has a childlike face, his nostrils wider and flatter than William's and Jacob's. The top of his head is in line with my shoulder, and I am short myself at five foot six. The word *pygmy* comes from the Greek for "dwarfish," but the pygmies differ from true dwarfs in that their limbs are in proportion to their bodies. Their children are almost the same size as Bantu children until they are about 11 years old, when they do not have the adolescent growth spurt of most other humans.

"The activity of IGF, the insulin-like growth factor, is decreased in all pygmies tested," wrote Italian geneticist Luigi Luca Cavalli-Sforza, who spent time with the Bayaka pygmies. He explained, "Their small stature is probably an adaptation to the forest. All forest people tend to be smaller, though not as small as the African pygmies."

A tiny woman walks down the slope and stands by Wasse. Her front teeth have been carefully chipped, and they resemble triangular shark dorsal fins. Many other Bayaka women have the same look. "They do it with machetes," shudders William. "She's Jandu, Wasse's wife."

"It makes me look beautiful for Wasse," Jandu tells me.

Wasse touches my arm. "We've talked enough. The sun does not stop moving. Let's go find the *ebobo*, the gorillas."

We are joined by another pygmy. "This is Wunga, my best friend," Wasse says. "He's our greatest hunter of elephants." Wunga is about the same height as Wasse but stockier.

Leaving the dirt road near Mossapola, we head straight into the jungle in the SUV and drive for several miles along a narrow, grassy lumber track until we reach a small clearing. Along the way we disturb thousands of large white butterflies that have settled on the damp, mossy ground but swarm into the air as we rumble by. "We call this place Mongambe, and we should find *ebobo* here," says Wasse.

He tells me that the Bayaka until recently hunted gorillas for their taste and for their massive hauls of meat. "I was the greatest *ebobo* hunter in our tribe, and so all the girls loved me and many slept with me," he says matter-of-factly. Wasse's face lights up at the memory. Pygmy adolescents practice free sex. From the time they reach puberty, they can sleep with as many partners as they can attract, but once they marry, the freewheeling times end. Adultery is forbidden.

"Where did you meet Jandu?"

"At a feast. She loves me because I'm a good hunter who can feed my family. I love her because she's given me three children, and that's how we Bayaka count our riches. When I die, people will still know my name because of my children."

"Do you still hunt gorillas?"

Wasse shakes his head. "Most of us have stopped hunting *ebobo*. We protect them because when we go into the jungle we see they're growing fewer. But some Bayaka secretly help the poachers to track *ebobo* for their meat."

We leave the SUV in the clearing and plunge into Tarzan Land. The jungle presses in from all sides, and overhead, the dense vine-woven canopy echoes with whooping monkeys and screeching parrots. "Watch where you walk, because there are many mamba and cobras in this forest," Wasse warns. The mamba is one of the world's most poisonous snakes, and if one of us is bitten, he might die before we could reach the doctor at Bayanga. The forest elephants are also dangerous. "If we're attacked by an elephant, jump behind a large tree and hide there. The elephant will not see you. The most dangerous is a mother with a calf." As I knew from Kibale in Uganda, Africa's forest elephants inhabit remote equatorial African rainforests and are mysterious creatures, rarely seen.

The heat and humidity are suffocating for me, but they do not bother the pygmies. Wasse peers at the ground as he and Wunga walk splayfooted over the forest mulch, with me and William stumbling a few paces behind. Wasse quickly reveals his peerless tracking skills, and where I see tangles of vine and patches of mud that cling to my trekking boots like glue, Wasse sees a road map of animals that have recently passed by.

The air is so dense and humid that I struggle to breathe, and the sweat runs off me like water, but we press on. We trek for about an hour through the jungle gloom, and then Wasse holds up a hand to halt us. His dark eyes flit from side to side and his nostrils flare as he tests the air for scent. Visibility is just a few yards, and he points toward the undergrowth. I freeze. In the dim light, I see a mother elephant and calf a few steps ahead of us. The mother's huge head and ivory tusks thrust defiantly through the leaves. The calf nestles by her front leg, unafraid, imitating its mother's belligerence. A lack of fear must be your birthright when your mother is one of the biggest and meanest creatures in the jungle.

The leaves rustle as the mother moves a step forward, glaring at us. She pushes her ears out, threatening a charge. "She'll try to kill us if she decides we're after her baby," Wasse whispers.

I look around and select the biggest nearby tree, tensing to dash behind it. Suddenly, the elephant raises her trunk in the air and trumpets a war cry. Wasse turns in an eye-blink and runs for his life, Wunga and William close on his heels. For a few moments I stand rooted to the spot by my astonishment at being abandoned by the pygmies, and then race after them, leaves and saplings slapping me. It is amazing how fast a middle-aged, overweight city dweller can move through the undergrowth when threatened by an angry mother elephant.

I dare not look back to see if the elephant is chasing us. Wasse calls a halt about 300 yards later, and I slump against a tree, gasping for air. "I thought we were supposed to jump behind a tree if an elephant was about to attack," I mutter after recovering my breath. "That's what you told me to do."

Wasse is laughing so much that it takes him a few moments to stop. "I did that to see how fast the white man would react and follow us. No one with any sense stays around when an elephant is angry."

He starts laughing again at the thought of me trembling behind the tree as the mother elephant tries to stab me with her tusks. Pygmies daily face such danger in the jungle, and Wunga and William join in the laughter, and then I do too, until my sides ache.

A few minutes later, with Wasse still smiling at my gullibility, we resume our silent quest. The dense tree canopy has plunged the rainforest into flickering shadow, but Wasse spots a trail of bent grass that weaves a path through the giant trees. He lopes through the forest, and the sweat streams from me as I try to keep up.

At mid-morning Wasse raises a hand. Ahead, the ground is strewn with *bambu*, a pink-hued fruit the size of an apple. He examines the tooth marks. "*Ebobo*," he mouths. Gorillas!

Wasse and Wunga spread out and follow pathways through the jungle that only they can see, whistling like birds to keep in touch. William and I follow Wasse, and I notice that every 20 yards or so he bends a twig from a sapling, marking our way. Halting once more, he cocks his head and listens to a sound I cannot pick up. I hear only insects shrieking, monkeys chattering overhead, and birds warbling, but Wasse whispers that several gorillas are about half a mile away and are hammering termite nests with their giant fists, seeking a tasty snack.

Hampered by thick jungle, it takes us about half an hour to get close, the gorillas' presence signaled by the musky odor of an unseen silverback. Wasse and Wunga fan out once more, communicating with hand movements, signaling the path the gorillas have taken, mouthing silent words to indicate where they are feeding.

Wasse holds up three fingers, meaning three gorillas, and makes a clicking sound by snapping his tongue against his palate. He had already told me this sound lets the gorillas know that we do not intend to harm them. He touches my arm and points through the undergrowth. Moments later, two swarthy faces pop up from behind a tree wreathed in vine leaves about 10 yards ahead. The creatures look like Neanderthals, their brown eyes shining with intelligence. This is one of nature's rarest sights, young male and female western lowland gorillas in the wild, among the most elusive of Africa's creatures. Their coats are typically brownish gray, a contrast to the glossy black of the mountain gorillas.

Gripping a creeper wrapped around a tree trunk, the potbellied female climbs about three yards up the tree for a better look. She stares at us, and then glances at a dense patch of bamboo near a fallen log. Wasse sniffs at the air. "There's a silverback in there, and she's letting him know where we are," he whispers. "He's probably their father. Silverbacks are very good fathers, and they'll fight to the death to protect their children."

Suddenly, the silverback roars a challenge. The fearsome, deep-chested explosion of noise drops us to our knees in homage. We are so close to him that the sound bangs against my ears. Still hidden from view, he beats his massive chest, the *thock, thock, thock* sending a clear warning. "He's telling us to leave," Wasse says softly.

He motions with his hand, asking if I want to get closer. The big male has signaled that he will fight to defend the youngsters—he must worry that we are a threat—but the chance to see a lowland silverback up close in the wild in all his charismatic glory is worth the risk. When I answer with a nod, we crawl on our bellies through the undergrowth until we are about five yards from the log. The silverback suddenly appears from behind it, his enormous face scowling as he glares at the intruders in his realm. He looks massive with his huge shoulders, enormous chest, and very large head. He must weigh over 400 pounds.

We are lucky. Almost all unhabituated gorillas flee at the first sight or sound of humans, because for generations pygmies have hunted and killed them with poison arrows. We are many miles from a World Wildlife Fund project that has habituated gorillas to study them, and so it might be the first time the silverback and his children have seen humans up close. As we look silently at each other, my throat clenches with emotion. Millions of years separate us on the evolution tree, and yet somehow I sense that the silverback is drawn to us, as I am to him, intrigued by some hint of similarity that he cannot comprehend.

Neither the silverback nor I want to break off contact, and this silent peering lasts for several minutes. That is what my watch says, but time drifts during such a supreme experience and I could not have judged whether it was three minutes or 30 minutes. For all this time I keep my eyes averted, switching my glance to the jungle and then momentarily back to the silverback and back to the jungle. But then I lock eyes with him, committing an act of deliberate defiance.

Leaping up, he whacks his barrel chest in anger. He charges from side to side, screaming as he bangs his massive fists against the trees in a terrifying display of his great strength. He tears heavy boughs from the trees and tosses them to the side. To mollify the huge ape, I join Wasse, Wunga, and William in hugging the ground, bending my head and casting my eyes downward, a primate gesture of humility, deference, and respect that the silverback understands, indeed expects.

Wasse had told me that if we try and run away we could provoke the silverback into an attack, just like the mountain gorilla silverbacks. The silverback can run much faster and could throw me to the ground and perhaps slash me with his huge canines. Steeling every nerve in my body, I keep perfectly still. The silverback glares at us for a few minutes, and then, perhaps accepting that we are no threat, he lumbers away with the youngsters scurrying after him. The jungle gloom swallows them.

Wasse hisses with relief and grins at me. "The *ebobo* are like the Bantu. We make them happy by pretending to fear them, and pretending they are better than us. So, we call the Bantu *ebobo* because we know they call us by a bad word, *babinga*."

William repeats this word for word in English without a flicker of emotion. Whatever other virtues he might have abandoned from his time at the seminary, he holds true to the monastic rule of strict obedience to his duty.

Up close the silverback seemed more a bellicose guardian angel than a demon to me, but in the late nineteenth century, hunter and explorer Paul du Chaillu described an encounter with a western lowland gorilla in nightmarish terms. Because his books were so popular, more than anyone else, du Chaillu was responsible for spawning the King Kong monster image of the gorilla. He described an expedition in what is now Gabon when the jungles there teemed with gorillas. Like Ballantyne's account, the outcome seemed to du Chaillu triumphant, but now seems tragic:

During the next few days we observed gorilla tracks, and followed them up. Suddenly, as we were creeping along, the woods were filled with the tremendous roar of the gorilla. Then the underbush swayed rapidly, and before us stood an immense male gorilla, which looked us boldly in the face. He was nearly six feet high, and presented a sight I shall never forget. With immense body, huge chest, great muscular arms, fiercely glaring, large, deep-grey eyes, and a hellish expression of face, it seemed to me some nightmare vision.

This one's eyes began to flash fiercer as we stood motionless on the defensive, and the crest of short hair on his forehead began to twitch rapidly up and down, while his powerful fangs were shown as he again sent forth a thunderous roar. He reminded me of nothing but of some hellish dream-creature, half-man, half-beast, which old artists pictured in representations of the infernal regions. He advanced a few steps, then stopped to utter that hideous roar again, advanced again, and finally stopped at about six yards from us. Just as he began another of his roars, beating his breast in rage, we fired, and with a groan which had something terribly human in it, and yet was full of brutishness, he fell forward on his face. The body shook convulsively, and then death had done its work!

After the gorilla encounter, we trek back through the jungle to the clearing where we left the SUV and set up camp. I keep my shoes on, because the soil in this forest is host to jiggers, worms that burrow through the skin into your feet and itch like hell. William and I will sleep in a tent we raise, while Wasse and Wunga prefer to bed down in the open on mats under a fast darkening sky. Once we have finished dinner we sit on logs by a sparkling fire under a bright sky, the stars like glittering diamonds scattered across a swath of black velvet.

Wasse tells me that the silverback and his youngsters were foraging for food when we found them. Like the mountain gorillas, western lowland gorillas live in a family with a dominant male gathering around him several adult females and their offspring. And like the mountain gorillas, the silverback is the unquestioned boss of the family, waking its members in the morning, leading them to food throughout the day, calling rest and play halts, mediating disputes within his family, and at the end of the day choosing where to build the night nests.

When he moves, they move. When he stops, they stop. The silverback will defend his dominance against any stranger silverback who takes him on, with the fights sometimes ending in the death of one of the giant combatants. The winner takes all.

The pygmies traditionally killed gorillas for food. Wasse has brought along his crossbow to show me and says he always aimed the arrow at the chest, and preferably at a female gorilla because he says their meat is more tender. He points to the green-tinged tip of an arrow. "It's *nadebale*," he says, a poison made from a forest plant known to the pygmies as far back as the Bayaka tribal memory reaches.

"Some of the gorillas ran away, but some climbed trees to escape from us," he says. "The poison on the arrow tip can kill monkeys but not gorillas," because of the small amount. It makes the gorillas dizzy, and the fall from the tree is often enough to kill or at least stun them. Then, Wasse and his clan would finish the job with machetes. He sees me flinch. But, no more, he assures me.

Wunga tells me how he used to hunt forest elephants for the tribe to kill, giving them enough meat for weeks from one kill. He would track a likely target, sometimes for an entire day, until he saw his chance. Gripping a spear whose blade was coated in poison, he crept as close as possible downwind to the elephant and then ran at it from the back and plunged the spear deep into its side. The enraged elephant would turn on him, but he always scampered away.

The entire clan then followed the stricken elephant, sometimes for a few days as the poison weakened it. Then it would collapse and die. Once that happened, the pygmies were efficient butchers. One of their first actions, Wunga says, was to cut a big hole in the side of the elephant and then the pygmies would crawl inside, emerging covered with blood and bearing the elephant's internal organs. They cut up and dragged back to their jungle huts virtually every part of the elephant.

The gorillas' end was relatively merciful. Wasse tells me that he and his fellow hunters would creep up on a family group after tracking them, the same as we did earlier in the day. Sometimes the gorillas heard them coming and fled, but often the pygmies caught them while they were foraging or napping. "We usually killed just one gorilla," Wasse says. "That was enough meat for our people. We roasted the body over a fire and cut it into pieces to share."

There can be as few as two in a gorilla group and as many as 20 have been spotted, but generally the western lowland gorillas live in the smallest groups of any gorilla type, averaging between four and eight family members. Now and then, silverbacks who do not have a family group stay together in what are called bachelor groups. When a silverback dies, his group splits apart and the members search for and join other groups.

*　　*　　*

It rains overnight, and we wake at sunrise to a damp, misty morning. Half an hour by the fire, which had been burning through the night to keep away leopards, and a hot cup of coffee stir the blood, and by 7:00 a.m. we leave camp on our search for more gorillas. At mid-morning, deep in the forest, we come across a small camp used by poachers. Their sleeping mats have been rolled up and stacked in a pile. Bamboo stands hold chunks of smoked meat, mostly duiker. "No gorillas," Wasse says. It is among the least popular bushmeat sold in the Bantu markets.

Wasse squats beside the fire and feels the charred wood. "The poachers were here at sunrise but they've gone hunting." They are probably armed with AK-47s and so we move on quickly, not wishing to be around if they return.

Suddenly we are attacked, not by poachers, but by thousands of small stingless sweat bees eager to suck up anything that is salty and moist— such as our body fluids. They get up my nostrils, in my ears, and dive-bomb my eyes. I drop to the ground and cover my head with my hands as they swarm around me. When I take a peep, I see that the pygmies and William are hardly affected, probably because I sweat more. Wasse hands me leaves to plug my nostrils and ears. I pull my jungle hat low over my eyes and ignore the swarms of bees sucking the sweat from my arms, face, and neck. We move on.

At noon the bees have disappeared, and we sit by a stream and eat lunch—cassava for Wasse and Wungu, and baguettes and cheese quarters wrapped in foil for William and myself. By mid-afternoon we have picked up the trail of a gorilla group, and we follow dung deposits they have scattered liberally along their path. The trail is speckled with their feeding leftovers, grass torn out by the roots, stems stripped, and bamboo ripped open. The two pygmies bend over every so often to search the muddy earth for the gorillas' knuckle and footprints.

We keep a strict silence, and an hour later, Wasse halts us with a raised hand and then points at a tall tree. About 20 yards above us, a female gorilla perches on a bough and feeds on leaves. On a sturdy lower branch is another female with a youngster sitting beside her, who follows her example in stripping saplings of their leaves and eating them.

The female sees us, screams a warning, gathers her youngster on her back, scrambles down the tree, and disappears into the undergrowth. From there comes the drumming sound of a silverback beating his chest. My instinct is to flee, but I hold my ground, shoulder to shoulder with Wasse. We fall to our knees, expecting a charge. The silverback does not disappoint. I see him emerge from the undergrowth to stand upright and beat his chest with cupped fists. Then, he comes storming straight at us on all fours.

I risk glancing up and see he has halted a few yards in front of us. His wide-open mouth reveals huge canines and his eyes glow with anger. A younger male, a blackback, emerges from the undergrowth, beats his broad chest, and rushes at us, also screaming in rage. He reaches the silverback, probably his father, and halts, crouching on all fours behind him.

For at least five minutes the gorillas scream at us, the noise so loud that it hurts my ears, but they come no closer. Then the silverback turns and waddles away into the bushes, followed by the blackback. We retreat in silence, but when we are about 500 yards away, and with dense forest between us and the gorillas, Wasse says, "It was not the same silverback we saw yesterday."

Wasse and Wunga converse excitedly in Bayaka, and then Wasse turns to me. "You are a man," he says with a smile. He pauses and thumps my chest with his fist as he repeats, "You are a man." I assume he offers the compliment because I did not turn and run.

3

Dave Greer, the American primatologist from Rwanda, has moved to Bayanga and now leads the anti-poaching patrols at Dzanga-Sangha. William tells me Dave risks his life daily to save some of Africa's most threatened animals, including western lowland gorillas and forest elephants.

Dave greets me at his spartan hilltop home in Bayanga, clad in shorts, with bare chest and bare feet. With his casual dress, thatch of red hair, and great love of the outdoors, he looks and acts like a grown-up Huck Finn. In Rwanda he had impressed me with his dedication to the goril-

las and his refusal to carry a gun in that very dangerous country, still in
turmoil from the Tutsi genocide. So I am surprised to learn of his new
combative role, deciding to fight the poachers at their own game and
take up the gun. In doing so he was joining a new breed of eco-warriors
who carry weapons, and are not afraid to use them as they go after the
poachers.

"Hop in," he invites, opening the door of a mud-splattered SUV.
"Let's go see some gorillas, and I'll tell you why on the way."

Bayanga perches on the Sangha River a mile from the Dzanga-Sangha
forest complex, which merges with massive contiguous rainforests in the
two neighboring countries of the Republic of Congo and Cameroon to
form a giant wilderness reserve, a single ecological block. It is protected
by the three governments with the financial help of the World Wildlife
Fund and the German Development Agency. Dave is employed by the
WWF as park advisor for anti-poaching.

As we drive through Bayanga, men and women wave to Dave and
smiling children run alongside, calling out, "Darveed." He returns
their greetings in fluent Sango, knowing many of their names. "They
perceive you in a different light if they see you've taken the time to learn
their language," he tells me. "They know I like to live with them and
eat their food, enjoy their culture, and play basketball with them."

On Bayanga's outskirts we enter a dense rain-soaked forest split by
a logging track. Saplings and leaves slap against the SUV's windows.
A sign featuring a painting of a gorilla signals that we are now in the
reserve's crown jewel, the Dzanga-Ndoki National Park, 471 square
miles of pristine rainforest. Dave tells me over the roar of the SUV
that the park is inhabited by 88 species of mammals and 379 bird species, including rare and threatened creatures, many the targets of the
poachers.

"All fishing, gathering, hunting, mineral and forest exploitation are
prohibited in the park, which is a vital reservoir for endangered species,"
he says. "However, the Dzanga-Sangha Dense Forest Special Reserve,
at two thousand square miles, surrounds the park, and there the pygmies can carry out traditional hunting, fishing, and gathering plants.
They're still banned from hunting endangered species such as gorillas
and forest elephants, which are protected by national legislation."

The deeper into the park we go, the wider Dave smiles, but then he believes he was born to a life in the wilderness. I am intrigued by the fate of Amahoro, the giant silverback mountain gorilla at the Virungas. "Did you ever habituate him?"

"Yes, if you persist long enough you can usually habituate gorillas. But I was gone by then. And, as you know, not long after your visit, the *Interahamwe* finally got Fidele. They chopped him to death with machetes merely because he refused their demands that he stop tracking the gorillas for the Fossey researchers. He left behind a wife and eight children."

Dave's fax at the time, sent to me in Sydney, was a hell of a shock. I sat in my study for a long time with the light off thinking of Fidele, wanting to believe Dave was mistaken. I had hoped to return one day to go again with Fidele to the mountain gorillas. I read about war, see it on TV, experience it in real life, but this was the first time a friend had been murdered. I felt heartsick, not only for him but also for his wife and kids.

For some time Dave faced down similar threats. When he had been in Rwanda for a year, the *Interahamwe* began targeting foreigners for execution. On returning from the gorillas one day, a villager stopped him to warn that armed Hutu militants were waiting for him farther up the road and had vowed to kill him. He took another way home.

Not long after, in January 1997, *Interahamwe* gunmen stormed into Ruhengeri, where Dave lived, looking for foreigners to kill. After a day with the gorillas, he was sitting on the verandah of his home in town and heard gunfire. "That was not unusual, because there were daily skirmishes between the rebels and the Tutsi army in and around the city," he says.

The following morning he learned that the *Interahamwe* had charged into a house, shooting dead three Spanish doctors and wounding an American aid worker. Dave escaped to the capital, Kigali, but to be with the gorillas each day he risked death by driving to the mountains along a road often targeted by Hutu rebels. At night he drove back to Kigali.

In June the next year, because of the extreme danger, Karisoke temporarily suspended the monitoring of the Fossey mountain gorillas. Dave moved across the continent to a research site, Mondika, a swampy jungle straddling the remote CAR–Congo border. The journey from Bangui took 15 hours by SUV, six hours by dugout canoe, and then an eight-mile trek through the jungle. There, he studied the unique behavior and diet of western lowland gorillas with a pair of habituated families.

He and fellow researchers made an astounding discovery. "Analyzing DNA from hair and dung found in gorilla nests," he tells me as we drive through the jungle to Bai Hokou, "we found tantalizing evidence that silverback dynasties—fathers, sons, brothers, and half brothers—like feudal lords, probably ruled over neighboring families. They supported each other."

That could explain why neighboring groups led by silverbacks are often seen "hanging out together in feeding places," and their intriguing lack of aggression toward each other. Dave believes this knowledge, as well as helping to understand human and ape evolution, is vital as the species is under serious threat from poaching, disease, and loss of habitat.

In 1995, Diane Doran-Sheehy, professor of anthropology at Stony Brook University in New York State, established the Mondika research center, which Dave came to direct. An important component of the research is to determine the difference between the terrestrial mountain gorillas and the arboreal western lowland gorillas who can spend part of the day high in the trees. Food is one of the biggest differences.

Doran-Sheehy and other researchers before her observed that the western lowland gorillas' diet is radically different from that of mountain gorillas: "Mountain gorillas eat mostly herbs—wild celery, nettles, bedstraw. Western lowland gorillas have a more diverse diet of fruit, leaves and herbs. They also consume termites, as well as the waxy green *ngombe* leaves and the bark of favorite trees. During certain times of the year, western [lowland] gorillas are practically frugivorous, seeking out jungle delicacies such as *bambu*, a seedy red fruit the size of a peach, or *mobei*, a larger yellow fruit that resembles a pineapple. At such times, fruit can make up 60 to 70 percent of their diet."

Dave was not deterred by the Central African Republic's violent reputation with bandits menacing roads and the constant threat of another bloody coup d'état. He says, "I was sad to leave the mountain gorillas, but not at all sad to leave Rwanda because I was becoming a callous person, seeing so many deaths and so many horrific things."

But violence is never far away in the Central African Republic, and a few weeks after Dave settled in at Mondika, just across the border, bandits attacked the camp in darkness at midnight. They shot over his head and stole his money and cameras. He shrugged off the danger: "I didn't feel the slightest bit fearful after Rwanda, because I didn't think there could be another place as hellish."

Dave soon became Mondika's director. Like Doran-Sheehy, he was intrigued by the many differences between the much-studied mountain gorillas and the shy western lowland gorillas. The easygoing mountain gorillas have a broader chest, longer and shaggier hair, and a more pronounced sagittal crest, the bony ridge on the top and back of the skull on males that anchors into place huge lower jaw muscles and tendons, That allows them to chew through the huge amount of vegetation they need to eat daily. Their hands and feet are broader, and their arms are longer.

The mountain gorillas, largely terrestrial, forage in family groups along lush alpine slopes, where their food at ground level is richer and denser than in rainforests. In contrast the lowland gorilla clans, led daily by a dominant male, seek the leaves, stalks, and sugary fruit of high jungle trees, shinnying up with astonishing agility to balance on boughs high above while they strip the branches. "The gorillas here are much shyer, much harder to habituate," Dave tells me as we drive along a track that cuts through the jungle. "They live in dense forests, and are hard to find because they're hunted for food and so fear humans."

For an as yet unknown reason, he tells me, female western lowland gorillas sometimes migrate to smaller groups than the mountain gorilla females. It could be that they believe this tactic gives them more chances of mating with the silverback. Also unknown is why western lowland silverbacks are generally more aggressive toward each other, despite the related silverbacks at Mondika living in peace in overlapping ranges. At Mbeli Bai, where gorillas are frequently spotted, the silverbacks commonly carry scars and wounds from fights with other silverbacks. Researchers observed six silverback deaths in 67 "silverback years," the addition of the number of years they observed each silverback. That is considerably more than among mountain gorilla silverbacks.

Dave tells me that while visiting the Mondika project's headquarters at Bayanga a couple of years ago, he met a young Italian gorilla researcher who has been running Bai Hokou, the other gorilla study site at Dzanga-Sangha since 1998. Chloe Cipolletta is slim, vivacious, and whip-smart, and some years later they were married in the jungle at her camp in a pygmy wedding ceremony.

The daughter of an influential Italian economist, Chloe could live in a fancy Roman villa, drive the latest Alfa Romeo, and be courted by sleek young men clad in Armani and Versace. Instead, she has chosen to live in the Bai Hokou wilderness. Like Dave, she is dedicated to the go-rillas' well-being. In mid–2000, Dave moved to Bai Hokou to co-direct the study site with her, habituating and studying two gorilla families.

An hour after leaving Bayanga in Dave's SUV we reach Bai Hokou, a handful of thatched huts clustered on a hill in thick forest and sur-rounded by a wire barrier hung with rattling tin cans to frighten off marauding forest elephants. They come close to camp and could flatten it if they wished, but Dave tells me that these behemoths of the jungle are terrified by the clatter they cause when they stumble into the cans and flee back into the jungle.

As we arrive, Chloe is separating gorilla dung with twigs to discover what fruits the apes had been eating. She has counted 326 different plants they use as food. The gorillas adapt their diet to the forests they live in, and in neighboring Gabon, primatologists discovered that the gorillas there were eating shoots and stems of a hallucinogenic shrub called *iboga*. The locals there use it in tribal rituals. Witnesses have seen gorillas deliberately seek out and eat the shrub, and then go "into a wild frenzy, jumping around and fleeing as though seeing frightening images."

The Gabon western lowland gorillas also eat the seeds and fruit of cola species that grow naturally in the jungle there. Gabon truck drivers chew the same seeds to stay awake on long journeys, and the seeds are used in manufacturing local cola drinks.

When Dave and I arrive, Chloe drops her work to leap into his arms and wrap him in a passionate hug. "He's my Tarzan," she replies when I ask what attracted her to Dave. "He climbs trees, and is the first to try anything."

"Does that mean you're his Jane?"

"No." She grins. "I'm his Cheeta."

Based at the camp with Chloe are 40 CAR nationals including pygmy trackers and three research assistants, a pair of them American women in their 20s. Californian Jessica Zerr found it hard going at first, following the gorillas through the heavy jungle, and she has had malaria four times. But she never despairs, saying, "To be with the gorillas was my life's dream."

Over coffee Dave tells me what I have seen already, that opposites do attract, the tough but soft-spoken American and the razzamatazz Italian. "Chloe's very fiery and outspoken, she'll tell you exactly what's on her mind immediately. I'm more reserved and quiet, but still firm in my opinion and beliefs."

Dinner is outdoors Roman-style, pasta cooked al dente by Chloe in a tasty sauce and accompanied by red wine under the flickering light of kerosene lamps. Our talk centers on Chloe's eight-year-long effort, in addition to pursuing gorilla research, to establish gorilla tourism here to provide funds for the locals so they will not be tempted to kill the gorillas. She runs the Dzanga-Sangha Primate Habituation Program and has habituated two groups of gorillas to human presence. Two teams of trackers and researchers observe each group from 6:00 a.m. to 6:00 p.m. every day.

"It's very difficult compared with habituating mountain gorillas," Dave says, because of their nature and the dense forest. "It's a long, arduous process and we've had limited success, and that's why we know much less about their behavior than the mountain gorillas, but we're persevering. We make daily contact to accustom the gorillas to our presence, to gain their trust so that they begin to understand that we're not going to harm them. Eventually, they come to see us as a neutral element in their environment, and that's when they begin to ignore us. Then, we can follow them without influencing their behavior."

Without the pygmy trackers, Dave says the habituation process would be impossible. "They pick up what to us are invisible traces of the paths the gorillas have taken, but you'd expect that, because they've shared these forests with the gorillas for thousands of years."

It is still tough work, because the western lowland gorillas forage over greater areas each day than mountain gorillas, with their home range averaging 10 square miles. They are also four times more likely to encounter other groups, though researchers at Mondika and Bai Hokou found these meetings, in contrast to those of their mountain cousins and the silverbacks at Mbeli Bai, to be "calm rather than aggressive." Dave says that the most frequent response he observed of a silverback to a strange male was to ignore him. But not always.

One of the groups followed by Chloe and her team consists of a silverback named Mlima and his young son. Once, many thousands of gorillas ranged through the CAR's forests, but not much more than 5,000 remain today, and Chloe's study of their habits will help efforts to conserve them. For several years, Chloe and her team of pygmies have tracked Mlima and his clan every day through the tangled foliage in the humid forest, slogging through the mud in the rainy season. As with the chimpanzee researchers, their starting point is ideally in the morning at the gorillas' night nests. The pygmy trackers follow the knuckle prints, and the mayhem of torn and shredded plants left behind by the gorillas from their early morning feeding.

In the beginning, Mlima and his family kept their distance from these strange creatures, but after several months accepted their presence so that the humans could approach to within 10 yards. If they went closer, they might put the gorillas at risk from any disease they were then carrying without their knowledge.

A year earlier, Mlima lorded over a clan of 11 adult females and young gorillas. Life was especially good for his children. Much of the youngsters' first years of life are spent in play—tumbling, chasing each other in circles and around the trees, pulling faces at each other, mock-wrestling, mock-biting, and, yes, banging each other on the head, just like the little mountain gorillas.

Play for young gorillas begins the bonding of their social relationships and also allows them to practice important motor skills, though social play is more frequent. Male youngsters play more often than females, and the little males prefer to play with other males. The young females also prefer to play with young males. Their mothers have little or no influence on their choice of playmates and rarcly encourage their youngsters to play.

But life is always tenuous for the infants. It is believed that western lowland silverbacks, like the mountain gorillas, do commit infanticide. A male who takes over a group kills all the infants to bring the nursing females back into estrus. At Mbeli Bai, no infanticide has been seen by researchers, though they suspect it occurred on two occasions when an infant disappeared after the death of a silverback and the mother transferred to another group.

Japanese researchers in the DRC witnessed two pregnant eastern lowland females transferring to another group and then giving birth three months later. The silverback killed the two infants several days after they were born. In the same month, the silverback spared the life of a baby born to a female with whom he had been observed living for at least a year.

A year before my arrival, Mlima himself had been attacked by another silverback, who gave him a severe beating. "We found him one day with scratches across his chest, [and] bleeding from a cut above his eye," says Chloe. "All around him the trees had been smashed from a massive fight. I was scared that the wounds would get infected, but there was nothing I could do about it—what happens in nature is not for me to interfere with." She has rules for intervention that she says are generally standard for great ape researchers. "If the gorilla is harmed by a human, I'll intervene, but if the injury is natural, I won't get involved."

What Chloe seems to have overlooked is that by habituating Mlima, a wild gorilla, she had already intervened in his life and may have influenced his behavior.

Mlima fought so fiercely not only to protect his status as the family's patriarch, but also to safeguard his infants. As Mlima licked his wounds, his females joined the new silverback even though it was most likely that he would kill any of Mlima's infants still on their mothers' breasts. "That's typical of gorillas," says Dave. "The females are looking for the strongest males to protect them. Just one young male, his four-year-old son, stayed with Mlima."

But a week before my arrival a wandering young female, what primatologists call a disperser, had settled in with the silverback and the youngster. "We've backed off from Mlima so as not to frighten away the new female, who's not habituated, and so give him a chance with her," Chloe tells me.

"It's best then that I don't go to see him," I respond, "because a strange face might spook his new female and frighten her away."

A few months earlier a friend of mine from Melbourne, Australia, Danny Korman, had journeyed to Bai Hokou as a gorilla tourist. Mlima had not yet met the female, and so Danny went into the jungle with

Chloe and the pygmies to spend a day with the silverback and his son. Danny wrote:

> *We watch Mlima sitting cross-legged on the forest floor, lazily munching on succulent termites from a nest he had broken into with his huge fists. I unwittingly break a fundamental rule; do not stare at gorillas. Mlima stops and glares menacingly back at me. I lower my eyes to demonstrate my submission, but it is too late. Propelled by massive arms he charges at a lightning speed. With two yards to spare he veers to the left. Casually, he returns to the half-eaten termite mound and resumes his insect meal.*
>
> *He feeds and sleeps through the day and then when the evening light filters through the canopy, Mlima finds a thicket cozy enough to spend the night, and stretches out comfortably on the floor. His son plays in a nearby tree until summoned by his father. The forest becomes a world of blue grey shadows and silently we withdraw from the thicket to leave the gorillas sleeping.*

To the south of Bai Hokou, in Gabon's Moukalaba-Doudou National Park, 500 miles from the capital, Libreville, three Japanese researchers from Kyoto University had encountered somewhat different reactions when they attempted to habituate western lowland gorillas. They were attacked by the adult females when they tried to get a group used to their presence. Unlike at Bai Hokou, it had been four decades since poachers had hunted bushmeat in the forest there, and the area contained a higher density of gorillas. The gorillas initially fled on contact, and it took the Japanese 21 months to get them to accept human presence, as against 24 months at Bai Hokou. As long as the silverback was in sight, the juveniles approached the researchers, curious about them, and the males accepted them fairly quickly.

However, the adult females reacted aggressively and attacked the Japanese team in one of every four encounters. They screamed an alarm vocalization, a "*wu, wu, wu,*" and charged the researchers, grabbing their feet and trying to bite them.

In 70 percent of these encounters, the group's silverback roared a "*wraagh*" warning and charged down the females, pulling them to the ground and chasing them away, allowing the researchers to retreat. It was more of a bluff. Silverbacks sometimes beat females, often to break up fights, but they rarely bite hard enough to cause wounds. Their most common threats take the form of displays, chest beating and charges, with the male ending the threat by thumping the female. The females usually move away or appease the silverback by embracing him and making humming vocalizations.

At midday on the next day Dave and I leave with a pygmy tracker, Ngbanda, to find the other habituated family. Dave works day and night, seven days a week, with the anti-poaching patrols and grabs the rare chance to visit his beloved gorillas. Like Ngbanda, he is barefoot. His feet have been toughened by years of trekking through African jungles without shoes.

The camp overlooks one of Dzanga-Sangha's many *bais*. They are marshy salt licks, small clearings threaded with streams and puddles laced with sodium, potassium, phosphorous, and manganese, minerals that the forest animals instinctively know is good for their health.

Thousands of white butterflies rise fluttering into the air from their nesting places on the ground as we walk by, giving the jungle a surreal, fairy-tale look. As Ngbanda leads us along a path carved by the giant feet of generations of forest elephants, the rain-speckled jungle presses against us, exuding a pungent odor of dank earth and foliage. We leap over fallen logs, paddle through streams, and duck away from *djeli*, vines studded with thorns, that festoon the tracks. Ngbanda and Dave, long practiced, are untouched, but sometimes I get entangled in vines that slash my bare arms and face, streaking them with blood.

Now and then we have to race through a patch of jungle teeming with armies of big ants, and we're constantly swarmed by thousands of tiny stingless sweat bees that swathe our clothes and hover around our ears, mouths, and eyeballs. I am more worried about the cobras, and especially the mambas camouflaged in the forest-floor mulch.

Suddenly Ngbanda halts. "Elephant," Dave murmurs.

In the shadowy foliage I spy a trunk and tusks. "Run like hell if he charges because they hate humans, with good reason," Dave whispers. I nod, knowing that already from my earlier encounter. Thankfully, the elephant does not see us as a threat.

Ngbanda knows this forest like I know the streets of downtown

Sydney, and he unfailingly leads us to the nests used by the gorillas the previous night. He has been alerted to the spot by the pygmies who had followed the gorillas yesterday. Unlike the sophisticated tree-nests of the bonobos, and even the chimpanzees, these look as if the gorillas flopped down on the forest floor with their massive bodies flattening the vegetation around them. At Bai Hokou, Chloe found that at 331 sleeping sites she documented, the gorillas slept on the bare ground 44 percent of the time. But a closer inspection on this day shows that the apes had bent the stems of large plants to provide comfortable beds. They look much like the night nests of the mountain gorillas.

Two hours later, as we push through a bamboo thicket where visibility is down to about five yards, Ngbanda halts us. He makes the same clicking noise as Wasse, a sign to the gorillas here that we come in peace, that we are not a hunting party intent on killing the patriarch, his five wives, and several children. Trackers back in Rwanda and the Congo used a stifled coughing sound to pass on the same message.

"*Ebobo*," Ngbanda mouths. "Gorilla." Dave and I see nothing, but he has learned to trust the pygmy trackers, who call on a heritage of thousands of years as forest dwellers. "They seem to have x-ray vision; they see and hear things in the jungle that we can't," he whispers. "But when they can't see anything, they fall back on their intuitive sense of the jungle."

Ngbanda points to a thicket a few yards ahead. It takes me a few minutes to make out the eyes, nose patch, and forehead of a female gorilla staring at us through the bushes. Then, Ngbanda points at a huge tree. Through binoculars I see, about 20 yards above us, a potbellied female feasting on fruit, while below her an infant nestles on another branch chewing leaves.

The females, like those of the other gorilla subspecies, give birth for the first time at about age 10, and thereafter every four years. Birth occurs eight and a half months after conception, with the newborn weighing on average four pounds. It clings to its mother's fur, and by five months old happily rides on her back. The mothers suckle each infant for three or four years, and it grows at about twice the rate of a human infant.

The females leave their natal groups when they become sexually mature at about eight years, and like the mountain gorillas go in search of a new family. Males also become sexually mature at between 10 and

12 years old, and leave home three or four years later to wander the tropical forest until they are big enough to attract females.

Moments later we hear a silverback pounding his barrel chest in warning somewhere in the thicket. The pygmy leads us toward the sound, deeper into the bamboo labyrinth where I can barely see beyond three or four yards. Leafy trees tower above us, their many branches spread wide like an umbrella, jammed together and blocking the sky so that the tree canopy's dense weave throws a shadowy gloom across the forest.

Without warning Ngbanda drops to the ground, immediately followed by Dave and then me. "Makumba," Dave whispers.

Ahead, I see the silverback eating bamboo stalks. A few moments later he disappears. Listening to gorilla sounds that Dave and I cannot hear, Ngbanda plots the path of Makumba, and we follow through the undergrowth and down an elephant track. Now and then we hear the silverback beating his chest, warning us away.

Suddenly, the silverback jumps from behind a spray of foliage, about five yards ahead, his huge face scowling. With a forearm as big as a man's thigh, he slams a bunch of striplings repeatedly against the ground. He screams a warning as we drop to our knees, and once again he disappears into the thicket. "He's displaying his authority over us, warning us not to come closer," Dave says.

Over the next hour we catch glimpses of Makumba's several females and their young up the trees and feeding. The silverback and his family barely tolerate our presence and will not let us get close, so encountering these giant apes as they go about their jungle lives is a precious experience.

Back at Bai Hokou I use the camp's shower, an ice-cold waterfall tumbling over a rocky outcrop, to strip away the dirt, sweat, and blood, and then rejoin Dave and Chloe. It was at this camp 12 months earlier that Dave made the fateful decision to pick up a gun for the first time in a decade. He and Chloe had heard the poachers' gunshots most days and knew from the pygmies that they were slaughtering gorillas, elephants, and other forest animals. Dave also saw that Bayanga's market was overflowing with bushmeat. There are about 4,000 gorillas and 3,000 elephants in the park, and he realized that if the slaughter continued they could be largely wiped out.

Before I went to the CAR, Jane Goodall had told me in an e-mail that you can buy gorilla, chimpanzee, and elephant meat in most cities in Central Africa—in Gabon, Cameroon, both Congos, and the Central African Republic. She wrote: "The trade is daily bringing many endangered species closer to extinction. Nothing large enough to eat is spared the hunter's gun. The trade is absolutely unsustainable. It's not about feeding hungry people, but about pandering to the cultural preference of the urban elite for wild animal flesh."

I also talked to Heather Eves, director of the Bushmeat Crisis Task Force, a Washington, D.C.–based consortium of 30 major international conservation groups. "Studies have shown that up to three and a half million metric tons of bushmeat, from insects to elephants, are traded each year in the Congo Basin alone," she told me. "It's six times the sustainable rate of the region's creatures, and it has become the most serious threat to Africa's wildlife. The best way to combat it is by law enforcement like Dave's anti-poaching patrols."

Both Dave and Chloe nod when I tell them this. "Now you know why I've given up gorilla research for a time to help combat the poachers here," he tells me.

One Bayanga family in particular, the Sangha, were ruthless elephant and gorilla killers, and Dave says he vowed to break their control to save the forest animals. Without a leader, the anti-poaching patrols in the reserve had fallen into sullen idleness. "The guards were de-motivated, and had a sense of helplessness with the barrage of poaching," Dave says, our talk almost drowned out by the hum and screech of nocturnal forest insects.

Though it meant seeing his wife only once a week, when Chloe came into Bayanga for supplies and a meeting with Dave, he volunteered with her encouragement to lead the anti-poaching patrols. It was a difficult sacrifice, he says. The two were obviously deeply in love and cherished their gorilla research at Bai Hokou, but they decided Dave's move was necessary.

As we eat fruit for dessert, he tells me that he began a year earlier, and one of his first decisions was to boost the Bayaka pygmy trackers' morale, essential to the patrols' success. "The Bayaka didn't want to work with the guards who treated them as the other villagers did, as *babinga*, or subhuman," he said. Dave stopped the guards from demeaning them. "In working with me, I'm trying to motivate the pygmies to reclaim their forests."

He next targeted information about the poachers, building an intelligence network, aided by his Sango fluency. During the following days in Bayanga I see Bantu and pygmies daily approach Dave at soccer games, while he is playing basketball, and in the dark night at his house, offering information about the poachers, usually for a small price. After one encounter I witness at the village basketball court, that same night Dave stakes out a track leading from the forest and collars a poacher with an elephant gun, which he confiscates and destroys.

With funding from the World Wildlife Fund, he has bumped up the patrols and hired new paramilitary guards, now totaling 48, maintaining an emergency team on alert 24 hours a day, with another team devoted to finding and destroying snares. His guards man roadblocks to catch bushmeat traders and constantly patrol areas in the reserve where the animal life is especially rich, up to 10 days in the field at a time. It is dangerous work. On one snare patrol, guards and poachers stumbled on each other in the night and in the confusion a poacher shot at the guards, but missed and killed one of his porters.

The guards have become devoted to Dave, who spends at least three or four days and nights a week with them slogging through the jungle, armed with his trusty 9-millimeter pistol. But, because the project is short on funds, the guards are heavily outgunned by the poachers, having only four AK-47s and five old Russian bolt action rifles among them. That means many guards go on patrol with only their bare fists as a weapon. "My wish is to have at least one AK-47 for every man," Dave tells me. "It would give them more confidence."

The most dangerous patrols are the ambushes, and one overcast night I join Dave and two guards crouching in the darkness by the side of a forest track known to be used by poachers. Dave's adrenalin is pumping as he remembers other ambushes that he says were very intense, "especially when you approach an angry poacher with a gun."

Fireflies flit through the black void, causing my heart to thump against my ribs, thinking they are poachers' torches. After two hours, Dave calls off the ambush. He assures me, "It's not a waste of time, because the poachers know we're out in the forest somewhere every night waiting for them."

His relentless tactics have worked, leading to the arrest and imprisonment of many poachers. In his bare office at Bayanga, the Dzanga-Sangha's project director, Jean-Bernard Yarissem, tells me, "The numbers of animals being poached here has dropped considerably since Dave took control of the anti-poaching patrols. It's now hard to find illegal bushmeat in the market."

One morning on a porch overlooking the Sangha River, I meet Jean-Rene Sangha, once the reserve's most notorious poacher and elephant killer. As tiny tsetse flies buzz about us, the wiry 26-year-old stares at me with hard, cold, cocky eyes as he tells how he joined the family's bloody trade at age 10. He learned from his older brothers as they stalked forest animals. In the decade and a half since, he admits with a callous laugh to killing "more than a hundred elephants," a toll that actually numbers many hundreds according to his brother who sits beside him and translates. He has also slaughtered dozens of gorillas, including silverbacks, females, and their young, for bushmeat. "Before, when I was a poacher, I didn't like gorillas and elephants; I hated them because they're dangerous," he says.

Flaunting a devil's courage, Sangha shot the elephants at point-blank range, mostly for their tusks, which he smuggled across the border into Cameroon. "Before, the price of one kilo of tusk was eight thousand CFA [$20], but now one kilo costs twelve thousand CFA," he tells me, the market reacting to the growing scarcity of ivory. With a pair of big tusks weighing about 130 pounds, the 700,000 CFA would keep several families in Bayanga for more than a year. The sale of the elephant meat was a bonus.

The poachers constantly plot to kill Dave. "They want Dave dead because he keeps on acting to stop them, but he is well guarded by his rangers," Jean-Rene tells me. Although his friends continue to poach the reserve's animals, Jean-Rene had a sudden conversion after guards caught him in the forest with his elephant gun a few months earlier. On his second day in jail, Dave came to talk to him, persuading the legendary poacher to stop killing the forests' animals. "Dave told me to stop poaching, telling me that by poaching I am destroying animal life."

Jean-Rene laughs when I refuse to believe his sudden conversion from animal hater and killer to animal lover. It seems to me his crossover was caused more by fear and a taste of jail, knowing Dave would not stop hounding him. Keen to use his know-how, Dave offered to make him a forest guard earning 80,000 CFA a month, or more than twice the wage of a laborer. Jean-Rene accepted, giving Dave one of his most significant victories. "I'll help the other guards because the forest is very large, and I know how the poachers work," he says. "I was a poacher; I know how to struggle against poachers."

With the poachers on the run, Dave's other major concern is Ebola, the highly infectious hemorrhagic-fever disease that suddenly appeared in Central Africa a few years ago. It is passed on by direct contact with body fluids or organs of victims, and kills up to 90 percent of people infected. No cure exists and sufferers die horribly, with massive internal and external bleeding. As primates, gorillas are also victims and vectors, and a recent Ebola outbreak in the neighboring Republic of Congo caused a devastating death toll among gorillas, killing up to 90 percent in some areas.

"At Lossi Gorilla Sanctuary in the Congo, 136 out of 143 gorillas died from Ebola," Dave tells me in the sitting room of his bungalow as we tuck into a bachelor's lunch—slices of bread and processed cheese washed down with bottled water. That left Dzanga-Sangha with Africa's only habituated western lowland gorillas. "Ebola has now been reported at the Congo's Odzala National Park, which has one of Africa's highest densities of gorillas. That's just a few hundred miles away in a contiguous forest with some barriers, but nothing too extreme to be able to block it."

He and Chloe brought together local health officials and village chiefs, and urged them to warn their people not to eat dead monkeys, gorillas, or chimpanzees. "That's how it spread in the Congo," Dave cautioned them. He admits that there is very little more he can do to stop Ebola's spread if it heads toward Dzanga-Sangha. "All we can do is keep our fingers crossed."

The bushmeat trade and its link with Ebola is even causing concern at high levels in the U.S. government. "Bushmeat is brought here and sold by Africans in markets in Washington, New York, Atlanta, and other U.S. cities, and health authorities are worried that it could cause an ebola outbreak," Heather Eves told me.

The battle to crush the bushmeat trade is far from won. Villagers at Bayanga and across sub-Saharan Africa remain unconvinced. Pascal

Dangbino, one of Dave's best guards, admits to me that "because they are very poor and have a difficult time finding jobs," the locals "feel they should have the right to utilize the forest. Conservation is a difficult concept for them to understand."

A few days later, Dave takes me with him on a routine patrol to seek out poachers in the nature reserve. The thick treetop foliage plunges day into night as we enter a forbidding, alien world. The jungle shakes with the haunting shriek of countless insects as we clamber over fallen logs and step around giant trees strung with thorn-studded vines. The forest floor is dappled with rotting leaves and toadstools and speckled with black-water puddles. Lurking in the mulch are deadly mambas and cobras.

Leading us on this dangerous patrol is Bambangu, a pygmy tracker, followed by two tall Bantu paramilitary guards clad in camouflage uniforms and toting AK-47s. Dave is also in camouflage and armed with his 9-millimeter pistol. The smell of damp soil, rotting fruit, and heavy-scented blossoms soaks the air as I follow in their footsteps. The day before, Bambangu had approached Dave and whispered that he had discovered poachers' snares in the jungle, brutal wire circlets that can maim a gorilla for life. Now, he leads us to them.

Twelve-gauge shotgun cartridges litter the animal track we are silently treading. "They're fresh; the poachers have probably been here today," Dave whispers. My throat tightens with fear, knowing that the poachers carry AK-47s and large-bore shotguns to kill elephants. The pygmy points to a snare, the wire hooked to a bent sapling with its deadly circle hidden in the mulch. With a grim smile, Dave rips it apart. "You've got to be careful, because you could lose an eye if it springs back," he murmurs.

The pygmy leads us to a duiker, a poodle-sized antelope, caught in a snare's death trap. "This could have been a gorilla, a chimp, or a leopard," Dave whispers as we squat by the stiffened body. His eyes harden and his lips tighten, because the duiker had almost torn off its hind leg desperately trying to free itself from the snare. "Poor thing," he murmurs.

We leave the duiker to creep through the jungle searching for more snares, finding and destroying another 171 hidden along the animal tracks. Dave loops the wire around his forearm to take back to Bayanga. Since he took charge of the anti-poaching patrols, in one year he and his men have destroyed more than 70,000 snares, putting an immense dent in the poachers' haul of forest animals. When we return, we find the duiker gone. Dave grimaces in frustration. "The poachers must have been nearby and knew we were here," he whispers. "If we'd stumbled on them, there could have been shooting."

On most days I have been meeting Wasse at the Bayanga market or at his village, and we have become friends. As my visit nears its end he invites me to join him on a hunt for the pygmies' favorite prey, blue duiker, the miniature antelope. "There's a good chance we'll see gorillas as we move through the forest," he says. A dozen men and women carrying nets rush down Mossapola's slope to cram into and on top of the SUV. Jacob, the driver from Bangui, takes us deep into the rainforest.

Leaving the SUV, Wasse leads us in skirting the jungle's edge for a few miles, and then we plunge into its depths, a carpet of treetop leaves turning day into night. The reek of damp earth and rotting fruit grips my nostrils, and the bell-like voices of the Bayaka tinkle with laughter as we penetrate the gloom. Wasse is silent. He focuses on the hunt.

After 30 minutes, Wasse halts us. "*Ebobo*," he murmurs. In the gloom about 20 yards ahead, a female adult gorilla crouches as if she were caught in headlights. I train the binoculars on her and see that her eyes pulse with fear on seeing the pygmies. We gaze back at her, me caught in the wonder of another rare encounter with an unhabituated gorilla, though perhaps the gleam in Wasse's eyes is prompted by a more basic desire—food. He has given up gorilla hunting, but the enthusiasm in his voice as he tells me of years past when he ate gorilla meat seems to signal that he has never lost the taste for it.

The gorilla suddenly breaks out of the spell and scampers over the hill. "Her family is nearby and she's gone to sound the warning," he says.

Wasse is the clan's greatest hunter, and while he tells me the pygmy tribes have no traditional chiefs, the Bayaka defer to his wisdom on the hunt. They halt at his sign and spread the nets quickly. Made from a vine known as *nkusa,* they are knee-high and each of the six nets stretches for about 10 yards, the pygmies curling them into a semicircle across the hillside. Wooden toggles hooked onto saplings hold the nets firm.

He explains that when the tiny antelope hear us coming they go to ground, hiding beneath the foliage. "We yell to frighten them and send them running into the nets."

Most of the Bayaka disappear up the hill and a few minutes later the jungle erupts with whoops, cries, and yodels. Small, stocky silhouettes flit in and out of the trees as they charge down the slope from all directions. Many of the men carry bundles of leaves, which they shake vigorously. A wild-eyed, poodle-sized duiker escapes past me. Wasse gives me a baffled smile. He asks the women if any are pregnant. "Women at the start of pregnancy bring bad luck on the hunt," he tells me.

Once again we set the nets, and once again a cornered duiker escapes. On the third try, a fleeing duiker crashes into the net. Jandu, Wasse's wife, grabs it. The duiker's eyes are terrified and its heart thumps against its tiny chest.

She is about to whack it on the head with the blunt edge of a machete when I beg her not to kill it. She looks at me, puzzled. My emotions are mixed. I want to see a pygmy hunt but feel responsible for this duiker's death. I offer her money, but she shakes her head. I double and then triple the offer, and she still refuses.

Wasse steps in and accepts my offer. "Just this one," he says. Jandu hands the duiker to me. As it struggles to get free, I let it loose and it scampers away.

The next duiker caught is not so lucky. Wasse stabs it with a pygmy-sized spear. Two hours later, the pygmies emerge from the forest with three duiker and a porcupine. Inspired by their success, his wife, Jandu, holds up a carcass and bursts into excited song, immediately joined by the other women. It is the most extraordinary sound I have ever heard—like angels in song. The song is a high-pitched melange of warbles and yodels, each woman drifting in and out of the melody. The music lulls to a murmur and then surges to wild chanting—a musical reflection of the pygmies' way of life in the rainforest.

Pygmy life is communal, and food from a hunt is always shared. Family ties are very strong, and rules of clan cooperation are rarely flouted. "Sometimes we fight among ourselves in the village when angry," Wasse says. "But we don't fight other clans. The jungle is large and there is enough room for everybody."

That night I return to Mossapola to meet the pygmies' most powerful rainforest spirit, Ejengi. The Bayaka worship a supreme being but also revere *mokoondi,* forest spirits, good and evil, who visit from time to time, and the clan gathers in a meadow at the jungle's edge to greet this great forest spirit. As I arrive, the pygmies are singing at full pitch, accompanied by drums.

The tribe suddenly goes quiet, and all eyes turn to the jungle. Emerging from the shadows, a handful of pygmy men accompany a creature swathed from top to bottom with hundreds of long strips of raffia. It has no features, no limbs.

"It's Ejengi," says Wasse, his voice trembling with awe.

I watch as Ejengi glides across the clearing. Without warning he charges, and the Bayaka scatter in terror because they believe that if any of the raffia strips touch their bodies, it will strip off the skin. Then he glides toward me. The forest spirit cannot speak, and his wishes are communicated by a squat man who tilts his head at an odd angle as he quizzes me. "Ejengi wants to know why you have come here."

Content that I am no threat, Ejengi begins dancing and charging again. The drums beat faster, the chants grow more frenzied. Drawn into the ritual, I, too, spin to the music. The hours pass as so many minutes. It is long after midnight when I realize it is time to go. Wasse grips my arm in farewell and our eyes meet. Then he turns back into the grip of Ejengi. As I leave, the chanting drifts back into the trees until it melts into the night sounds of the jungle.

On my final night at Bayanga, nearing dusk on a day when the moon will be full, Dave strides barefoot and bare-chested through the jungle about 20 miles from Bayanga heading for Dzanga Bai, the reserve's most spectacular forest clearing. He and I carry packs of red wine, cheese, and baguettes for his and Chloe's monthly ritual. At the *bai,* a sandy salt lick the size of three football fields, more than 70 forest elephants, usually hidden by the jungle, have gathered in the open. It was just two miles from here that Jean-Rene Sangha shot his last elephant, and until Dave put muscle back into the anti-poaching patrols the numbers of elephants visiting the *bai* daily had fallen drastically. But now they have returned to forest elephant paradise, and a researcher has identified more than 2,500 individual elephants using the *bai*.

Chloe is already there, and we settle down in safety on a tree-high wooden platform with stairs to watch as solitary bulls plod into the clearing, followed by calves scurrying next to their mothers and juveniles escorted by bulky matriarchs. Two young bulls jostle for dominance by locking tusks and shoving hard. Ignoring the ruckus, elephants kneel in the mineral-rich mud and use their trunks to suck up the water laden with sodium, calcium, magnesium, potassium, and clay. This lets the large herbivores overcome mineral deficiences. Others employ their tusks to dig fresh deep holes up to four yards wide in the substrata, or stand in line waiting patiently for their turn. The calves wallow merrily in the mud until they look like little chocolate elephants. We have brought mosquito nets and settle in for the night, smiling at each other as the elephants' growling, rumbling, screaming, and trumpeting echo around the *bai*.

As darkness arrives and the full moon turns the *bai* a ghostly silver, the elephants still come and go. "Every time I'm at the *bai*, gosh, it's a magnificent feeling, and it gives me a sense of optimism that there is the chance to have some long-term success," Dave whispers. "It's like a huge movie theater just watching all this unfold before your eyes at night, because you never get the chance to see it anywhere else. If everyone in the world knew about the place, I don't think we'd have a problem saving Dzanga-Sangha."

Back in Sydney a few weeks later, I get an e-mail from Dave describing a tragedy that has just shocked him and Chloe. A stranger silverback passing by had challenged Mlima, the habituated silverback, to a fight over the female who was with him and his young son. Mlima suffered severe bite wounds. The female scurried away with the victor. Mlima lingered for a few days, and then died. His son disappeared and has not been seen since. Five years of habituation had been destroyed in a few hours. Chloe, true to her word, declined to intervene and save Mlima with veterinary care, believing that to do so would interfere with nature.

Not long after, Dave and Chloe left Dzangha-Sangha. Dave became director of the Jane Goodall Institute's research on chimpanzees in the Congo/Brazzaville, and Chloe studied chimpanzees, her favorite great ape, in the Ivory Coast. They lasted there just over a year, with Dave dispirited by constant fighting with corrupt senior government officials and politicians. He was also in charge of a chimpanzee sanctuary near Pointe Noire on the Congo coastline and had more than 140 chimpanzees there, mostly orphans confiscated from poachers.

He told me, "The government kept putting pressure on me to quickly reintroduce them back into the wild, but they would not help facilitate this difficult process. These chimpanzees are huge, and they've been habituated to humans. Put them back into the wild and they'll raid villages nearby for food. Unlike unhabituated wild chimpanzees, they're not frightened of humans. If we do not carefully consider every step throughout the reintroduction process, I've warned the officials, the chimpanzees may end up killing villagers. The officials didn't seem to care."

There is another danger in the reintroduction of orphaned or captive-born chimpanzees to their native wilderness habitats. Primatologist Elizabeth Lonsdorf notes that behavioral studies show "the territorial nature of chimpanzees may make it risky to reintroduce animals into areas where wild chimpanzees reside."

Dave has since moved back to the WWF as the coordinator of its African Great Apes programs. That puts him in the front line of the urgent efforts to save the African great apes from extinction in the wild.

Since Dave's departure from Bayanga, regretfully antipoaching patrols there have fallen back into disarray. Jean-Rene Sangha returned to his old trade of elephant and gorilla killer, but he was arrested and sent to jail.

To avoid a repeat of Mlima's killing, which would threaten gorilla tourism there, in mid–2009 Chloe's successor at Bai Hokou, primatologist Angelique Todd, informed me that she and the trackers try to frighten away any silverbacks who come near the habituated silverback Makumba. But the threat of a murderous fight remains. She told me, "If there's a male around [we] normally make sure our presence is known to increase the chances he flees [because they are unhabituated]. But sometimes these males are so full of testosterone, it makes no difference or they leave it until later in the evening when we've gone."

4

Despite the evidence of tool-making by Leah at Mbeli Bai, many zoologists regard gorillas as the dullards of the great ape world, plodders in mental ability when compared with the bonobos, chimpanzees, and orangutans. Koko, a western lowland gorilla, was used in pioneering studies measuring the ability of great apes to comprehend human language, but her achievements pale when compared with Kanzi's linguistic brilliance. And so I was surprised when visiting Chicago's Lincoln Park Zoo, in the middle of the city, to find that the star there in mathematical ability tests is Rollie, a 12-year-old female western lowland gorilla. As *The Economist* put it when reporting on a paper given at the International Primatological Society Conference in Edinburgh in 2008, "She was very good at learning to put discrete items in a list, a skill that is used by people to memorise phone numbers, interpret calendars, and most importantly of all, to acquire and process language."

Making lists is a cultural skill learned early in life by humans, but the animals in experiments conducted by Steve Ross, a primatologist at the zoo, had never been exposed to lists before. Rollie was shown a computer screen with a line of random numbers on it and had to repeat each numeral in its correct order. If she succeeded she got a jellybean, but if she didn't, she got nothing. She started with the numbers 1 and 2, and after a year and a half has worked up to listing seven different numbers in sequence. "She gets them right 35 percent of the time," Steve tells me. "Rollie advanced to this level faster than any other animal I tested, including a bright young chimp. The chimp took twice as long as Rollie to learn the same sequences."

Four decades ago, Penny Patterson, a young American graduate student in psychology, first came across Koko at the San Francisco Zoo when the gorilla was a tiny undernourished baby. With the permission of the zoo, she took Koko in hand, and when she was still an infant began to teach the little gorilla American Sign Language (ASL), used by deaf people. This culminated in an ongoing partnership, the longest-ever continuous experiment to teach language to another species. Although she is in the shadow of Kanzi, Koko now has a working vocabulary of about 1,000 signed words.

Koko went to live with Patterson in Woodside, California, daily studying how to use signs. The initial ambition was modest, three words: *up*, *drink*, and *eat*. But Koko proved to be a whiz kid and began to string words together, nectarine yogurt becoming *orange—flower—sauce*, and a ring, *finger—bracelet*. She can now put together eight signed words in a row.

Koko even used ASL to summon a dentist. When she repeatedly signed that she was suffering pain in the mouth, Patterson drew up a pain chart on a scale from one to 10. Koko signed that her pain was extreme, number nine. She was put under anesthetic for a dentist to examine her. He found an infected tooth and extracted it.

Koko once used sign language when shown how her niece, Binti Jua, the daughter of her sister, rescued a three-year-old child who had fallen into the gorilla exhibit at Brookfield Zoo in Chicago in 1996. The little boy had climbed over a protective barrier and plummeted 20 feet into the enclosure. With her own daughter riding on her back, the eight-year-old Binti Jua picked up the unconscious child and shielded him from the seven other gorillas in the exhibit who excitedly scampered around the intruder. Then, gently holding the boy, she carried him to a gate where a rescue team was waiting. The boy recovered.

Patterson showed Koko a videotape of the incident and asked what she thought about Binti Jua's rescue. Koko signed *girl* and *good*. "Although Binti's maternal instinct might lead her to pick up the child, her intelligence was demonstrated when she moved the child to the human access door," Patterson said.

A decade later, another gorilla came to the rescue of a child at the Jersey Wildlife Preservation Trust in Great Britain, only this time it was Jambo, a male gorilla. Rather than picking up the child, Jambo stood over him and guarded him from the other gorillas until zoo officials arrived.

Silverback gorillas are among the most stressed zoo animals. Daily, they are confronted by crowds of visitors eager to see King Kong and his family. Staring at a silverback from a few yards is a challenge, and it happens hundreds of times a day for each captive dominant male. In a safari-style zoo, the gorillas would be far enough away not to be bothered by such rudeness.

That may not be realistic, because great apes in zoos are crowd pullers and open plains zoos are by necessity usually located on the outskirts of a city or even farther away. If they must be kept in zoos within a city's precincts, with their restrictions on space, Dian Fossey in *Gorillas in the Mist* set down humane instructions for their upkeep:

> *For captive gorillas, trees should be available to climb and material such as straw, branches and bamboo supplied for nest building. Food could be allotted in small portions throughout the day and require some degree of preparation such as peeling and stripping of stalks or even searching for randomly distributed items supplied at various locations within the enclosure. . . . Of prime importance to the reclusive gorilla are obscured niches where captive animals may withdraw, as desired, not only from the presence of people but also from one another, as is the species habit.*

However, few zoos anywhere have given their gorillas such privacy because it would spoil the spectacle for the paying visitors. Some zoos have put up curtains so that visitors view the gorillas through peepholes—see but not be seen. But most keep their great apes on full display, justifying the need to draw and please more visitors by the large costs of keeping the gorillas.

At Sydney's famed Taronga Zoo in Australia, I arrive at the western lowland gorilla enclosure at opening time, 9:00 a.m. I am the only person present, and half a dozen gorillas have already left their cages and are in the enclosure, which measures about 150 yards long by 30 yards wide. It looks attractive with a high, rocky back wall, a small waterfall, and logs to clamber over. But even captive-born gorillas, impelled by hundreds of thousands of years of evolving into semi-arboreal creatures, must feel the need to climb trees, and here they are denied the pleasure. The huge silverback ignores me as he bends over the tumbling water and scoops some into his mouth with a cupped hand.

Within an hour more than 100 people have crowded the viewing area in front of the enclosure, and many of the children are yelling at the gorillas in piercing voices. The silverback retreats to the rear of the enclosure and turns his back on the crowd, avoiding their stares. Other gorillas crowd into a small "cave" near their cages, perhaps to escape the clamor, and turn their backs on visitors staring at them through a toughened glass panel.

Courtney Keane and Nicola Marples, from the Department of Zoology at Trinity College in Dublin, recently conducted a study to assess the effects of zoo visitors on the behavior of western lowland gorillas at the Dublin Zoo. They noted that early research on this topic claimed that visitors were "a source of enrichment or stimulation for the animals."

That was wishful thinking. In contrast, they found that zoo visitors may affect individual gorillas differently, with the male gorillas showing significant changes in aggression as the crowds increased. The males sat with their backs to the visitors, "suggesting that they are more irritated by the visitors and had developed this avoidance behaviour as a coping mechanism." The adult gorillas also "spent less time foraging and more time inactive when visitors were present."

At one of Africa's most rugged and remote places, the highland border region that separates Cameroon and Nigeria, on steep mountain slopes in dense forest live the planet's rarest great apes, the Cross River gorillas. Cameroon is the CAR's next-door neighbor, and there I hope to witness the valiant efforts of a pair of primatologists to save these last of their kind from extinction. I also hope to meet the gorilla people, a secretive cult of villagers who claim they can transform themselves into gorillas. They roam the jungle slopes for a few days at a time with the true gorillas before morphing back into human form and returning to their villages.

The Cross River Gorillas

RAREST OF THE RARE

1

In the cloudy highlands along the border between Cameroon and Nigeria in western Africa live the world's rarest great apes, the Cross River gorillas. Wildlife Conservation Society primatologist Aaron Nicholas has undertaken an arduous and underfunded effort to save these unique primates. "There are only about three hundred Cross River gorillas left spread over about 770 square miles," Aaron tells me when I phone him from Sydney, "and the increasing destruction of their forest habitat for farmland and grazing is threatening their existence. They are surrounded by some of the densest human settlements in Africa."

There are 11 small family groups of gorillas that have been fragmented by farmland and settlements. "We're urgently attempting to get the government to protect them with rangers, and establish safe-haven channels through the forests so they can migrate between groups and so maintain a varied gene pool," he says.

Aaron is not exaggerating. The International Union for Conservation of Nature, which monitors endangered species worldwide, has designated the Cross River gorilla a separate subspecies and has placed it on its red list, which warns that the ape is critically endangered and threatened with extinction.

John Oates, professor emeritus of anthropology at the City University of New York, was one of the referees who studied the anatomical differences. "The Cross River gorilla has a smaller head, shorter jaw, shorter palate, and shorter hands and feet than the western lowland gorilla," he told me by phone.

The skull difference has probably evolved as a result of their diets. Whereas soft fruit makes up the larger part of the western lowland gorillas' diet, the Cross River gorillas' habitat has a longer and more intense dry season. So while they eat fruit when available, they feed more heavily on tree bark and liana.

For decades, most people including myself have believed erroneously that the mountain gorillas, with about 700 individuals, were the rarest great apes on the planet. Hundreds of journalists, writers, and filmmakers have reported from their native habitat on their plight over the decades, but not one has ever been to the Cross River gorillas' habitat. That is probably because hardly anyone knows about them.

Aaron is based in Limbe, on Cameroon's western coastline, a few hours' drive from the gorillas' habitat. They could be gone forever soon, he tells me: "They are now facing imminent extinction." Only seven outsiders have ever seen them. If I can get to Limbe, he says, he will take me to their habitat in a remote forest highland straddling the border between Cameroon and Nigeria. It is a biodiversity hotspot of global significance that supports a high variety of animal and plant species, many of them threatened.

I board a jet at Nairobi in the late afternoon and shoot across the continent, flying high above the Congo Basin in the heart of the African continent. I am doing it easy; European explorers took months to battle through jungles like those far below the plane.

This journey to the land of the Cross River gorillas will also let me visit pygmies in Cameroon, which has 40,000 of the world's smallest people, about one fifth of the African pygmy population according to London-based Survival International. The pygmies, gorillas, and chimpanzees have lived side by side for thousands of years in the rainforests, and so I plan a detour to visit a clan that lives largely in the traditional way to see whether they still hunt gorillas and chimpanzees for food.

In Uganda, Nzito and his people are by necessity now more focused on growing their marijuana plants than hunting great apes for food. At Mossapola, Wasse told me his clan had stopped killing chimpanzees and gorillas because they knew their numbers were rapidly diminishing and had to be protected. I aim to find out whether knowledge of the need to conserve the great apes has reached the pygmies in Cameroon.

Four hours later we land at the Cameroon capital, Yaoundé. It's "That's Africa" time again, as one-third of the baggage has not arrived with the plane. The empty carousel trundles around and around as if taunting the sad, angry, or resigned passengers. My suitcase is here, but my sleeping bag is missing. One of the many touts you find at most developing-country airports, like fleas on a raddled dog, sidles up to me. "It's okay," he says. "Happens all the time. The next flight from

Nairobi is in two days' time. Your sleeping bag will be on it. Give me five dollars and I'll make sure you get it."

The road into the capital is like a roller coaster as Yaoundé spreads across waves of hills. The street life here in West Africa looks much like that in East Africa, shantytowns that mottle the lush green landscape. Cameroon is resources rich, with offshore oil fields, logging, and ocean fisheries, but the BBC says it is one of the most corrupt countries on Earth. The president, Paul Biya, yet another ruthless Big Man, has been in power for almost three decades, and dissent is choked by a brutal secret police.

In *Tyrants: The World's 20 Worst Living Dictators,* David Wallechinsky lists Biya alongside the genocidal Robert Mugabe of Zimbabwe and the murderous Teodoro Obiang Nguema Mbasogo, president for life of Equatorial Guinea.

Cameroon is "one of the best-endowed primary commodities economies in sub-Saharan Africa," notes the CIA's *World Factbook*. The most recent per capita GDP is $2,200. Line that up against the dollar or two that many people here earn each day, as in most of Africa, and you get an idea of the billions of dollars plundered from the national income by the political elite. They splurge it on mansions swathed in marble and sparkling with gold fittings, squads of mistresses, fleets of expensive cars, hordes of servants, and a luxurious lifestyle that includes frequent shopping trips flying by private jet to Paris, London, and New York.

No one knows the approximate amount of this stolen treasure trove in Cameroon, but in next-door Gabon, an investigative report examined the known financial records of President Omar Bongo, the deeply corrupt leader of that oil-rich country for four decades. It found that he and his extended family spent on average around $100 million a year on themselves. His salary is less than one-thousandth of that. A French police report in 2007 noted that Bongo and his family were the owners of not only 39 properties in the country, mostly in the highly expensive 16th district of Paris, but also posh villas on the French Riviera. The Bongo family also had 70 separate bank accounts in France.

In another country that abuts Cameroon, oil-rich Nigeria, a federal government investigation two years earlier found that since independence in 1962, corrupt politicians and military leaders had stolen $600 billion. Almost all of it was untraceable.

Rock stars Bono and Bob Geldof pressured the G8 leaders at their 2005 summit in Gleneagles, Scotland, to grant multi-billion-dollar debt relief to nations across sub-Saharan Africa, claiming that the benefits would flow on to the people. Cameroon's debt was reduced by $1.26 billion, but there was hardly any cheering in the shantytowns. The people there knew they would see little of these savings.

Amid all of this ostentatious wealth, as usual, the great apes are largely forgotten. Cameroon, along with the neighboring Republic of Congo and Gabon, is one of the largest and most significant homelands of great apes living in the wild, but, typically for Africa, each government mostly depends on Western conservation organizations to fund the sanctuaries and habitat protection for their chimpanzees and gorillas.

It is now mid-evening and the shantytowns along the road to the capital roll on and on, fronted by rickety wooden roadside shops. Outdoor butchers, with their fat-streaked sides of beef hung on hooks and goat heads staring dumbly into space from blood-splattered benches, are pushed against mom-and-dad grocery stores offering not much more than packets of biscuits, cans of tinned meat, toothpaste, washing powder, and the ubiquitous cans of Coca-Cola. Peddlers sit by open-air stalls selling faded hand-me-down clothes. The most colorful are the numerous barbers, with just one or two dilapidated wooden chairs and cracked mirrors in their small shacks. At their entrance they boast vibrant paintings of the oiled and greased high-top styles you can have fashioned inside.

Most of the signs are in French. Cameroon is a curious hybrid state split between French speakers and English speakers. It is one of the many deformed children spawned in the 1960s in the ferment of independence from European colonialists. The English speakers of British Cameroon wanted stand-alone independence, but a heavy-handed United Nations in 1961 gave them two choices: join an independent English-speaking Nigeria (which abuts an English-speaking province of Cameroon) or throw their lot in with the French speakers of what was formerly French Cameroon. They had to pick what they believed to be the lesser of two evils.

Cameroon is slightly bigger than California. The English speakers are the majority in the two provinces with some of the most lucrative natural resources, off-shore oil, and fisheries. The French speakers have much of the agricultural land and forests, which turned out to be lucrative with logging.

Most of the English speakers, making up about one-third of Cameroon's population, want to break away and become an independent state, one that would be resource rich. Just as important, they could cast off the cultural imperialism imposed on them by the French speakers. The president in this hybrid state, the deeply corrupt Paul Biya, can speak only French. That goes for many of the elite who send their children to Paris for their education, but the imperatives of the commercial and bureaucratic worlds make it a must that English speakers are fluent in French.

The English speakers are forever restive, and so the French national military, stationed here with the approval of the Cameroonian government, brandishes a mailed fist at them, marshaling displays of power such as French warships visiting the waters of unhappy provinces. French troops are also stationed there. They are intended to cow any opposition to the French-speakers' rule, supporting Biya to the hilt. As the Americans used to say of the string of despots they supported, "Sure, he's a dictator, but he's our dictator." West African states such as Cameroon are the remnants of a worldwide French colonial empire, and the French government is determined to keep the few French-speaking states left within the Francophile fold.

My hotel in the middle of Yaoundé sits across the road from a marketplace. It is Saturday night and well into the early hours, very loud music booming from several bars keeps me awake. It does not trouble me. West African popular music is one of the world's glories, bouncy and zesty, spurred along by an irresistible beat; the singers' voices melt into smooth harmonies. At 2:00 a.m., unable to sleep, I go to an outdoor bar packed with young couples, old couples, and loners. They smile at me as I order a beer and a freshly grilled fish coated with *pili pili*, the explosive African chili sauce, and sit back to enjoy the music.

The next morning a taxi takes me to Yaoundé's small zoo to see if they have any great apes on display. The animals are packed into aging concrete enclosures fronted by rusting bars. Like animals in most zoos, they seem bored with their two-dimensional world. Many Western zoos have devised what they call "enrichment programs," especially for the great apes in their care. Keepers hide food in their enclosures, and offer puzzles with ropes or concrete termite mounds with tasty jam inside. It is all designed to get them off their rear ends, moving and thinking throughout each day, but there is no substitute for the thrills, spills, and challenges of living in the wild. Here in Yaoundé, the animals have little to stimulate their minds except anticipation of the next meal.

I spy the most handsome lion I have ever seen. He is a picture of noble dignity with his thick mane, sleek body, tawny fur, and imperious stare. He sits dreamily by a female, who nudges him in the manner of cats big and small. There are plenty of monkeys, but no great apes. A young attendant approaches. "All the great apes have been moved to the Mefou sanctuary, about thirty miles out of town," he tells me. "You should go there."

Sanctuaries are expected to be one of the saviors of the great apes, and so the following day I head out in a taxi on the road to Mefou. We are stopped by three military roadblocks, each about five miles apart. At the barrier, soldiers clad in riot gear gruffly order me to get out of the car, show my passport, and be on my way. At the second roadblock, a girl in her late teens is not so lucky. When she reaches the head of a line of passengers from a bus, a soldier demands she show her ID. She holds up her arms helplessly. He barks at her, she shakes her head, and then he hustles her into a Jeep that takes off in the direction of Yaoundé.

"She's probably an illegal immigrant," says Benjamin, my driver. "They come from the Central African Republic looking for work, and she probably hopes to be a maid. The soldiers will probably take it in turns to rape her at their barracks, and then let her go. If the girl resists she faces years in jail."

We turn off the main road and the car bounces along a rutted dirt track, passing a couple of villages and a small wooden church, and enters a national park. Tall trees close over us. As we near Mefou, I recall the scorn for such places voiced by a grumpy old Danish professor who told me he had once been the head of conservation for the World Wildlife Fund and was looking for rare frogs in the Congo where I met him. He claimed to be the world's top expert on frogs. "We should kill these orphaned great apes in sanctuaries because they're no use to the species," he told me. "They waste valuable resources in caring for them, millions of dollars that could be spent protecting the species in their habitats."

This is a view I have heard expressed by many conservationists over the years, and it seems in isolation cruel yet to them logical. But my heart rules over my head here. About 50 yards from the sanctuary's entrance is the nursery for western lowland gorilla orphans, and they are already out of their cage and romping around a fenced-off grass yard. They have that wonderfully solemn look of all the gorilla subspecies, a contrast to the impish and mischievous features of young chimpanzees.

One little male gorilla, perhaps on the way to becoming a dominant silverback a decade hence, keeps beating his already stocky chest and charging the other youngsters. They ignore him. Such behavior among young chimpanzees would spark angry fits of screaming, bashing, and fleeing, but there is an ethereal calmness about the little gorillas.

I wonder how anyone could condemn these little gorillas to death merely because they are orphans, and because it is expensive to run their sanctuary. But the frog professor had a steely gaze and the unemotional, disciplined mind of someone who had killed and pickled thousands of frogs in the name of research. I have no doubt that he would have condemned these young gorillas to be killed, humanely of course, in the name of efficient conservation.

The manager, Rachel Hogan, a 34-year-old Englishwoman, greets me. "The 2,500-acre sanctuary was set up in 1998 when all the chimpanzees and gorillas at the Yaoundé zoo were transferred here," she tells me. "They were living in terrible conditions there, but unfortunately the government is now putting pressure on us to transfer some of the great apes here back to the zoo. They say people who visit the zoo are disappointed they can't see any chimpanzees or gorillas."

We walk up a muddy slope with the forest pressing in on both sides, the dank, earthy smell of the soil, bark, and leaves fresh from overnight rainfall. A compound fenced in with strands of electrified wire contains a night cage holding eight chimpanzees. "We keep the chimpanzees and gorillas locked in cages at night to prevent poachers from coming in and killing them for bushmeat," Rachel explains.

The chimpanzees stare at me with curiosity as a keeper unlocks the barred doors. Then all interest in me is forgotten as they rush out to enjoy a spread of fresh fruit and vegetables bought from the local market, the first of three daily meals. They gulp down their food, much like children at a birthday party determined to get more than their share of cake and ice cream. I think of the difference of the chimps at Kibale calmly making their way soon after dawn to the fig tree, climbing up into the branches and choosing their own spot from which to enjoy a leisurely breakfast. After the chimps at Mefou gobble down the food, a pair settles down to groom each other, ruffling intently through their partner's fur. The others lope into thick undergrowth or scramble up the trees that dot their enclosure.

"We've divided the chimps into families, like in the wild," Rachel says. "They soon sort themselves out with a dominant male and a dominant female. It's natural behavior, even among chimps who were taken from their families as infants and didn't have the chance to learn how to behave in the chimp world from their mothers and fathers. It's wired into their genes. When each new chimp or gorilla arrives, we keep them in quarantine for three months and observe their behavior.

"This gives us the clue as to which family we'll put them in. Each family has three acres of enclosed forest as close as we can get it to their natural habitat. We have to think about the politics of each group to make sure they settle in without trouble. The gorillas are okay, but with the chimpanzees there's a lot of fighting if we get it wrong. Even at the best of times, the chimpanzees squabble every day in their tussles for dominance."

There are two sets of electric fences barring the chimpanzees from the outside world, made necessary by the chimps' intelligence. Rachel tells me, "When we had just one fence, the chimps quickly learned that if they placed a small tree against it that had fallen down overnight during the rainy season, they could then scamper up it over the fence and escape. We had a few get away into the forest, but eventually got them back because they were used to getting tastier food than they could forage themselves. Now, there is an inner and outer electric fence, so that even if they get over the inner fence using a fallen tree there's no way over the outer fence."

As at Ngamba in Uganda, the new arrivals are mostly orphans confiscated from poachers or from people who kept them as pets. "They suffered severe trauma when captured, and are usually in a pitiful state, psychologically and physically," Rachel says. But they must be happy at

Mefou, because the sanctuary is expanding, as several chimpanzee and gorilla females have recently given birth.

As we walk along the outer fence of the next three-acre compound, a big male named Max with the typical chimpanzee long-arm knuckle walk keeps pace with us a yard away on the other side. He has the look of a scarred and grizzled old boxer who has seen it all, done it all, and I notice through the dense black fur on his broad head that his ears are missing. About 20 yards back from the fence a dozen little chimpanzees are high in the trees, swinging effortlessly from branch to branch with breathtaking skill. When we stop to talk, Max plops down on the ground and stares through the wire at us.

"When we got Max he had no ears," Rachel says. "His previous owner had chopped them off for whatever reason. He'd spent years living in a small cage, and so he was frightened of the open space when he came here. In all the years he's been here he's never climbed a tree. But he's terrific with the young chimps, he really looks after them, and so we've put them in with him once they leave the nursery. He loves it, and so do they."

Angry screams suddenly tear the air. After Kibale, I would know the sound anywhere. A pair of male chimps in a nearby leafy enclosure are hurtling after another male. He seems to be fleeing for his life. They disappear into the undergrowth, and the chimp being chased scrambles up a tree and perches on a branch, safe for the moment. Rachel smiles. "As in the wild, the male chimps are always testing each other in displays of dominance, and there's always a lot of fighting. The male chimp who was chased wants to be the dominant male of the group, but he's got no allies, and so the dominant male with his allies has just been putting him in his place."

Farther along the track, a male chimpanzee with a lopsided, scarred face jumps up and down as we approach. Alone in a small enclosure, he stands upright, claps his hands, and does a jig with the expertise of a veteran street performer. There is something about his beat-up appearance that draws my heart to him. "That's Tommy," Rachel says. "His previous owners probably taught him to perform like that. He was confiscated from a family who had him for many years." He arrived with an open wound on his back, perhaps from a machete blow, as well as poor teeth and an emaciated body. "Despite Tommy's suffering at the hands of humans, he has a sweet nature. We had one of the volunteers, an Irishwoman, be with him twenty-four hours a day at first, and he's recovering well. In a few months we'll introduce him to a suitable family."

The sanctuary also has two western lowland gorilla families, all orphans brought here years ago. Despite the achievements of Koko in California and Rollie at the Chicago Lincoln Park Zoo, gorilla intelligence is often downplayed, unfavorably compared to that of the quickwitted chimpanzees. But the gorillas here also figured out how to scale the electric fence with fallen trees and Rachel had to put a second fence in place, a few feet out from the first. The keepers have yet to arrive with their food, and the six gorillas that form one of the families are still in their night cage.

The dominant male, 10 years old, is up to four years away from becoming a silverback, but he has already fathered a baby, a tiny bundle of fur and large, calm eyes, which the mother cradles possessively against her chest. The male chases the other adolescent male around the cage in a slow-motion mock battle, the gorillas taking it in turn to slap each other harmlessly. Beyond the cage is their lush enclosure, planted with dense shrubs and trees. Though it is not the wild, it is far better than any other gorilla compound in the many zoos I have visited around the world.

All six male and female volunteers at the sanctuary are from Europe, the United States, and Australia. I ask why there are no African volunteers. "It's too expensive," says Rachel. "Each volunteer pays three thousand dollars to spend three months here helping to look after the chimpanzees and gorillas; it helps fund the running costs. Not many Africans can afford that."

The volunteers live on site in huts, eat local food, and toil seven days a week, a labor of love. But the workers who know the individual great apes best are the sanctuary's African keepers. They come from a nearby village, and some have been here since the sanctuary opened a decade ago.

When Rachel came to Cameroon as a volunteer she planned to stay just three months, but that changed when she was given a two-week-old male gorilla named Nkan to hand rear, even though his prospects for surviving were slim. The tiny wizened gorilla was so small that he fit into the hands of the sanctuary's veterinarian. His eyes were unfocused, and he could not control his arms and legs. Rachel stayed with him 24 hours a day, sleeping by his side and feeding him each hour like a devoted mother caring for her sick baby.

Nkan was struck down by pneumonia soon after arriving, and had to take antibiotics for four months. He also contracted shigella, a disease that attacks the immune system. He struggled to regain his health and eventually succeeded, with Rachel by his side. "This bonded me closer to him," she tells me. "That's when I decided to stay in Cameroon and manage the sanctuary."

As so often happens with infant great apes reared by humans, Nkan forgot he was a gorilla. Rachel says, "He didn't know how to act like a gorilla, or to play and communicate. We solved the problem about a year after he arrived, when Shai came here. He was the same age as Nkan. Hunters had killed his family for bushmeat, but he was too small to eat and so they planned to sell him on the black market. We put the two together, and Shai taught Nkan how to act as a gorilla. You saw them together just now at the enclosure. One day we hope to reintroduce them both back to the forest."

The desire to one day return their orphan chimpanzees and gorillas to their native habitats is a worthy ambition I have heard at almost all the sanctuaries I visited. You even see it noted on the display boards outside their cages and enclosures. The sanctuary managers know it is for most of the orphans an unrealistic ambition, and I believe they use such "good news" to deflect any uncomfortable questions from potential donors and visitors. They might ask why the chimpanzees and gorillas, especially those that had been plucked from the wild, are kept behind bars once they have grown up and not then returned to their native habitat.

Most great ape sanctuary managers I met agreed with me when I pressed them. Like humans, great ape infants learn for years from their mothers, fathers, uncles, and aunts how survive in their habitat—how to make night nests, seek food, hunt, find water, and avoid danger. Some of this behavior is wired into the great apes' brains, but most of it must be learned firsthand by imitating the adults in the group. Orphans raised in a sanctuary can never have this long learning experience, and are therefore unlikely to survive if they are released back into the wild. It is a sure bet that Nkan and Shai will grow old together at Mefou.

I ask Rachel the inevitable question that all these sanctuaries face, the one the Dane posed. Why spend millions of dollars each year caring for a few hundred orphans in the great ape sanctuaries in Africa when the same amount could be better spent in providing armed guards to protect the great apes in their native habitats?

"It's a lot easier to persuade little old ladies in the U.S. or Europe to donate money to help save an orphan chimpanzee or gorilla than it is for them to give money to buy assault rifles and ammunition for rangers. And keeping the apes in sanctuaries does provide genetic insurance in case there is a disaster in the wild that wipes out a species there," Rachel says.

On the way back to Yaoundé, Benjamin, my driver, complains about the spiraling price of essentials, food, and gas. The price of bread, a staple here, had doubled in the past few months. I explain that the world's major suppliers of wheat, Australia and China, suffered severe droughts in the previous year, and that has pushed up the price because of the supply-and-demand factor governing international wheat markets. Add to that the increased cost of transport caused by the dizzy rise in the price of oil, and you can see why the cost of wheat this year has jumped on the world market by up to 40 percent.

Benjamin shakes his head, unconvinced by my reasoning. "I know all about the rise in the petrol price, because every time it goes up I suffer as a driver. The rich can afford the rises, it's nothing to them, but we ordinary people can't. Already we've had to cut back on the bread we buy, and also meat and vegetables, and my family is suffering."

I would usually expect this to prompt a plea that I give the driver a big tip, but this time it does not happen. Benjamin's face darkens. "Unless the price of bread and petrol drop back to what they were, there'll be trouble here." I do not think even Benjamin could have predicted how serious the trouble would be in the following days.

2

That night I visit Yaoundé's best nightclub, owned by one of world soccer's superstars, Cameroonian Samuel Eto'o, who plays for FC Barcelona, the champion Spanish team. Eto'o has clearly tipped some of his millions into the nightclub, because it is as good as any I have seen in New York, Paris, or London.

On the dance floor, under flashing multicolored lights, a swirl of sharply dressed young Cameroonian men and women gyrate as the music pounds from giant speakers. The only older people I see are about 20 middle-aged, expensively clad men and women seated in plush armchairs on a raised platform at the far end, I suppose so that they can be seen better. They obviously have no worries about rising food prices.

This is the first time I have encountered Cameroon's elite. I splash out on a single Coke at $10, but a stream of $200 bottles of Johnny Walker whiskey is borne up to the elite on the platform by a parade of beautiful young women. I wonder what kind of country this is where the elite can spend on a single bottle of whiskey the same amount a laborer might earn in 200 days.

The next morning, I take a car to the office of one of the many impoverished but good-hearted nonprofit organizations you find in all African capital cities. I have come to Cameroon primarily to go to the habitat of the rare Cross River gorillas, but I have decided also to visit a pygmy clan to see whether they are still hunting the great apes. The Rainforest Foundation in London, a lobby group for Africa's pygmies, has put me in touch with an organization they sponsor, the Centre for the Environment and Development.

The CED office is in a dilapidated two-story concrete building on a hill near my hotel. It looks like a relic from the colonial period, but the way the tropical heat and torrential rainfall lay waste to everything here, it could be just a few years old. The furnishings appear to have come from a down-at-the-heels secondhand shop. Here, I am told a familiar story. Samuel Nnah, in charge of projects aiding the Baka, says he struggles against a government that willfully ignores the pygmies' rights.

A ruggedly handsome bearded man, Samuel resembles Joe Frazier, the champion American boxer. He says, "The government legalizes timber companies to hack down Cameroon's rainforests at an alarming rate, and then exports the valuable hardwood. To get at the trees they've forced the pygmies from their forests, making them live in settlements by the roadside. The pygmies have to beg land from the Bantu owners, who then claim they own the pygmies—men, women, and children—as their personal property. It also means that the numbers of chimpanzees and gorillas in the wild are falling drastically."

"Do the pygmies still hunt gorillas and chimpanzees?" I ask. Samuel ignores the question.

The next day, my sleeping bag turns up at the airport on a flight from Nairobi. On the road from Yaoundé to Djoum, a rundown town near Cameroon's southern border, I witness the environmental vandalism mentioned by Samuel, passing more than 100 timber trucks, each bearing four or five huge hardwood tree trunks to the port of Douala to be shipped abroad. The Cameroon government seems proud of this lucrative destruction, placing on its 10,000-franc note (US$25) a forklift bearing a rainforest tree trunk toward a lumber truck. The trucks and logs are coated in red dust, just like the trees and plants that line the roadside.

At Djoum, the CED's provincial director, Joseph Mougou, tells me he is battling the government to protect the human rights of 3,000 Baka who live in 64 villages in the province. He says, "Starting in 1994, the government has forced the Baka from their abodes in the primary forest, designating them as national parks. They allow them to hunt in the secondary forest, mostly rat moles, bush pigs, and duiker. But that's where the government also allows the timber companies free rein to log, and that's destroying the forests."

Another problem is ID cards, essential for daily living in Cameroon. "The law requires mothers to register their babies within thirty days of birth," Joseph explains. "But pygmy mothers about to give birth, obeying tradition, must go into the forest for three months, and when they come out with their babies they must undergo a lengthy, costly court procedure to register them as citizens. The pygmies have little money, and we help them, but it's not easy, because our funds are limited."

I ask him whether the pygmies still go into the forest to hunt chimpanzees and gorillas. Joseph pauses. "You should ask them yourself."

* * *

Forty miles beyond Djoum along a dirt track, after passing scores of fully loaded timber trucks, I reach Nkondu, a pygmy village set up in 1998 after the government forced the clan from the nearby forest. It has 15 huts made from mud and wood in the cubelike fashion of the Bantu. Awi, the clan's chief, a diminutive man who barely comes up to my chest, welcomes me. Joseph has alerted the villagers to my arrival and the pygmies are readying to leave for the forest, men and women tying empty cane packs onto their backs for hunting and foraging.

With me is Manfred Mesumbe, a tall Cameroonian anthropologist, an expert on pygmy culture who lived with this clan for two months the previous year. "The Bantu government has forced them to stop living in the forests, their culture's bedrock," he tells me. "Within a generation much of their unique traditional ways will be gone forever." He nods to my question: "Yes, they still hunt chimpanzees and gorillas."

He tells me, "The older children attend a boarding school up the road during weekdays, and that's why they're not here, but the infants go to the pre-school in the village. They'll join us later today."

"Goni! Goni! Goni Bule!" Awi shouts. "Let's go to the forest!"

Pygmies, more than most people, are independent minded, and it takes another hour for Awi to round up the villagers, but at mid-afternoon we leave. For two hours we trek single-file through the rainforest near the Congo border. Balancing like tightrope walkers, we cross tree trunks the pygmies have thrown over streams and hack through heavy under-growth, using machetes to cut away bamboo stands and lianas hanging like drapes in our path. The sweltering heat swaddles us like a sauna. We reach a small clearing surrounded by enormous hardwood trees whose dense canopy almost blots out the sky, and here the pygmies put their cane packs on the ground. We have reached our home for the next few days.

For thousands of years the pygmies have lived in harmomy with their environment, equatorial Africa's magnificent jungles. They inhabit a narrow band of tropical rainforest approximately 4 degrees above and 4 degrees below the equator, stretching from the Atlantic coastline to Lake Victoria in Uganda. With about 200,000 still remaining, the pyg-mies are the largest group of hunter-gatherers left on Earth. But now their ancient way of life is under serious threat of destruction by the Bantu, the taller Africans.

The pygmies suffer severe racial discrimination, with many Bantu believing them to be subhuman. Many own pygmy families, forcing thousands to work as virtual slaves in their fields and gardens for their lifetimes with little or no pay. The pygmies do not know how to fight back.

As their ancestors have done for millennia, at mid-afternoon the Baka clan swiftly construct beehive-shaped huts in the clearing. The 30 women and men chop striplings from the trees and thrust their sharpened ends into the ground, bending them in arcs to form the skeleton of each chest-high hut. They then weave bundles of green leaves stripped from nearby trees into the latticework to form the huts' rainproof skin.

Like Awi, none of the pygmy men stands higher than my shoulder. The women are much smaller. As they carry firewood to the camp, Manfred and I put up our small collapsible tent. Not long after, about 20 unaccompanied children between the ages of three and five stream into the forest clearing where their parents have constructed the beehive huts. "Pygmies know the forest from a young age, and the children came here themselves along the same jungle paths we used," Manfred says.

Suddenly the pygmies stir, their faces taut with tension. They hear something that I cannot. Without warning, three scowling Bantu brandishing machetes stride into the clearing. They are bandits, I fear, common in this lawless place. I am carrying my money in a handibag strung around my neck, and news of strangers travels fast among the Bantu here. Manfred points to the most fearsome, a stocky man with angry features. "It's Joseph Bikono, chief of the Bantu village near where the pygmies live by the roadside."

Bikono glares at me and then scowls at the pygmies. "Who gave you permission to leave your village?" he roars in French, which Manfred translates for me. "You pygmies belong to me, you know that, and you must always do what I say, not what you want. I own you. Don't ever forget it."

Most of the pygmies bow their heads in submission, but a young man steps forward. "It's Jeantie Mutulu," Manfred murmurs. "He's one of the few pygmies who've finished high school."

"We've always obeyed you, Bikono, always left the forest and gone back to the village when you demanded it," Mutulu shouts in French. "But not now, not ever again. From now on we'll do what we want."

About half the pygmies begin shouting defiantly at Bikono, but the other half, fearful of his power, prepare to trek back to their village. Bikono swivels and glowers at me. "You, *le blanc*," he yells. "Get out of the forest now."

"This is a very dangerous situation," Manfred murmurs. "Bikono is a violent man."

When the villagers defy Bikono, he demands 100,000 francs (about $250) from me as a bribe to remain with the pygmies in the forest. I ask him first for a receipt. Then, with one eye on his machete, I refuse to give him the money and say he has committed a crime. I threaten to return to Djoum and report him to the police chief with the receipt as evidence. Bikono's face falls at this threat, and the three Bantu shuffle away.

The pygmies greet this victory with singing and dancing. Growing more frenzied by the minute, they continue until just before midnight, when they drop to sleep. American primatologist Dave Greer, who spent a decade with the CAR's Bayaka pygmies, once told me, "The pygmies are the world's most enthusiastic party-goers. I've seen them sing and dance for days on end, only stopping for food and sleep."

The following morning, Jeantie leaves for Djoum to report Bikono's threat to Joseph at the CED. Over the next three days I accompany Awi and his clan deeper into the forest to hunt, fish, and gather edible plants. The Baka here seem midway between the Bayaka in the CAR and the Batwa in Uganda. They have abandoned net hunting, and put out snares like the Bantu to trap small prey. They even hunt with guns. "Bikono sometimes gives us a gun and orders us to shoot an elephant," Awi tells me. "We keep some of the meat, while he takes most of it as well as the tusks."

"Hunting elephants is illegal in Cameroon and having a gun is very rare," Manfred tells me. "But highly placed policemen and politicians work through village chiefs like Bikono, giving guns to the pygmies to kill forest elephants. They get high prices for the tusks, which are smuggled out in ships from Douala to Japan and China."

"Do you shoot chimpanzees and gorillas?" I ask Awi.

Manfred and Awi talk for a few minutes, because the pygmies do not seem to know the French words for chimpanzees and gorillas. They discuss the way the two species of great apes look and behave. "They know them well; they call them 'the monkeys that are like people,'" Manfred tells me. "They say one kind is smaller, chimpanzees, and the other is bigger, the gorillas."

"Of course we shoot them," says Awi. "Our people have hunted and eaten the big and small monkeys forever. It's much easier now with a gun. Before we used a bow and poisoned arrows. They're easy to find because they stay in the same area, so we know where to go. With the smaller monkeys, we make a noise like them [he imitates a pant hoot] and hit the trees with our feet. They come looking for us because they are very curious."

He imitates a silverback's charge, causing the other pygmies to laugh. "We follow the trails of the bigger monkeys. It was dangerous when we only had bow and arrows, and our hunters were sometimes killed by them. But now Bikono gives us the gun and bullets, and we get an arm or a leg. He sells the rest in the market."

I ask why Bikono wants them to shoot the bigger monkeys. Awi shrugs. "We chop off their heads and paws, and give them to Bikono, as well as all the meat except for an arm or leg, which he allows us to keep."

This suggests that there is still a trade in gorilla heads and paws as grisly souvenirs, even though the United States and European nations have banned their import.

"What happens to the infant if you kill its mother?" I ask Awi.

"We never kill a mother with a baby or even a young one. Usually we kill males, because that means the females will continue to have the same number of babies."

"How many of the smaller and larger monkeys would you kill in a year?"

He holds up two tiny hands and counts off six. "It's how many Bikono asks us to shoot. He's more interested in elephants."

"The numbers of smaller and bigger monkeys are falling, and if you keep shooting them, one day there'll be none left."

Awi looks at me as if I am a simpleton. "There are many of these monkeys in the forest, I've seen them with my own eyes, and our god gave them to us to hunt and eat."

There seems no point in debating Awi. I have a Westerner's conservation view, which has developed only over the past few decades, while Awi holds the traditional view. If most of the 500 or more pygmy clans in Cameroon shoot chimpanzees and gorillas for their Bantu masters, then the slaughter of the great apes here must be on a terrifying scale.

I am interested to see how the pygmies here compare culturally with Wasse and his Bayaka clan, who live virtually next door in the CAR. With Awi acknowledged as village chief, the Baka here clearly have begun accepting Bantu customs, because traditional pygmy tribes never have chiefs. But they cling to the pygmy tradition of revering Ejengi, the greatest of their rainforest spirits. He had come out of the forest to meet me at Mossapola, and I am thrilled when Awi tells me that perhaps Ejengi will visit the clan on this, my final night with them.

As the light slips from the sky, the women gathered in the clearing begin swaying as they chant a hymn of welcome to the greatest rainforest spirit. In a tight circle, the men dance wildly to the thud of drums. When I join them they smile.

"Kwa! Kwa!"

The high-pitched scream and thumping on trees reverberates through the jungle gloom. "Ejengi is near," Awi says, his voice trembling.

"Does he come often?"

"Only rarely. This is the first time in many months. He is coming to meet you."

As the sky grows dark, just as at Mossapola, Ejengi emerges from the gloom accompanied by four clansmen. His head-to-toe raffia strips are ghostly-white and he has no head, no limbs, no features, giving him a spooky look. Ejengi dances gracefully with the men for about an hour, the raffia strips swirling about him, and then four little boys are brought before him. Shoulder to shoulder, they sway to the drum's thump as Ejengi solemnly and slowly dances along their line, letting his rippling raffia strips brush their bodies. "Ejengi's touch fills them with power to brave the forest's many dangers," Awi tells me.

Through the night Ejengi dances with the men and boys as the women chant ancient forest hymns. The effect is quite different from Mossapola's Ejengi, several hundred miles to the north. Where that ceremony had all the exuberance and sheen of a nonstop all-night dance party, this ritual is more dignified, more sacred, with the tree canopy soaring overhead like the arches of a rainforest cathedral.

Nearing dawn, as darkness drains from the sky, the women's chanting grows more frenzied. "Ejengi will leave us soon," Awi murmurs, an ache in his voice.

Six teenage boys step forward and form a line, shoulder to shoulder, before the rainforest spirit. The drums beat louder and the women's piercing singing shakes the leaves on the trees as Ejengi pushes against each of the boys in turn, attempting to tumble them off their feet. As one, they brace their shoulders and stand firm. Again and again in a powerful metaphor of the pygmies' life, Ejengi moves along their line, shoving each boy, trying to knock them off their feet. Steeling every muscle in their bodies, they do not budge.

"Ejengi is testing their power to withstand adversity," Awi tells me as my throat chokes, aware of the charged emotion among a people being buffeted daily by the powerful Bantu. "We Baka face hard times and our youngsters need all that power to survive as pygmies."

And then Ejengi is gone, uttering his birdlike calls and thumping trees as his presence fades away into the fastness of the rainforest. Quietly, the pygmies gather their meager belongings and lead me through the jungle back to their roadside village to say good-bye.

Later in the day at Djoum I meet the province's administrator, a Bantu named Frederic Makene Tchalle. "The pygmies are impossible to understand," he says, more smug than puzzled. "How can they leave their village and tramp into the forest, as they did with you, leaving all their possessions for anyone to steal? They're not like you and I; they're not like any other people."

He shows no sign of wanting to accommodate the pygmies' unique nature, no sign that his government has given any thought to how to help them retain as many of their rainforest customs as possible while integrating them into Cameroon's wider society. Although, with the right message, and funds to support it, conservation groups might be able to persuade the pygmy clans here to protect chimpanzees and gorillas, like at Mossapola, rather than kill them at the direction of the Bantu village chiefs.

3

On the way back to Yaoundé, once again a procession of timber trucks passes us, though this time they are returning from the Douala port empty, on the way to the forests to load up again. Manfred tells me that President Biya is nearing the end of his second seven-year term in power, in 2011, and even though he will be 78 years old then, he is manipulating his puppets, who sit in the parliament to vote out the constitutional provision that a person can serve only two seven-year terms as president. He clearly wants to be president for life. Biya controls the Cameroonian media with a brutal grip, and newspapers daily give him fawning front-page coverage.

In any true democracy, apart from all Biya's other sins against the legitimacy of governance, this Big Man would have been cast out of office in disgrace for what the CIA *Factbook*'s entry describes as widespread "trafficking in persons." It outlines the current situation:

> *Cameroon is a source, transit and destination country for women and children trafficked for the purposes of forced labor and commercial exploitation: most victims are children trafficked within the country, with girls primarily trafficked for domestic servitude and sexual exploitation. Cameroon is a transit country for children trafficked between Gabon and Nigeria and from Nigeria to Saudi Arabia: it is a source country for women transported by sex-trafficking rings to Europe.*

So, what is the estimable President Biya, received with honor by his peers at the African Union's meetings of heads of state, and on the world stage at the United Nations, doing to crush this evil trade? Not much. The CIA says: "Cameroon reported some arrests of traffickers, none of them were prosecuted or punished."

Once more there is trouble for Cameroon's chimpanzees and gorillas implicit in this report. A government that refuses to protect its own children and women from widespread sexual slavery, and refuses to prosecute those who trade in sexual slavery, is hardly likely to give a moment's thought to the fate of its great apes.

The next morning I leave Yaoundé, taking a four-hour bus ride to Douala, the country's biggest city and major port. My destination is Limbe, on the west coast, where I hope to meet the only Cross River gorilla in captivity, and then go on to the rare gorillas' habitat. The bus hurtles along a winding road, driven at such a furious pace that the driver must think he is in a Formula One race. The villages strung along the road are even more impoverished than the shantytowns in the capital.

The bus passes an uncountable number of timber trucks, each stained with the red dust from the forests from where I had just come. Each truck is carrying four or five enormous tree trunks, stripped of their branches, to the ships waiting at the port. The trucks make the journey day and night, and the eco-vandalism is so immense that it makes me wonder what hope there is for the forests of Cameroon and for the great apes that call them home.

We cross a bridge over a wide river, and its calm, dark waters remind me that the first Europeans to navigate it were Portuguese mariners in sailing ships three centuries before. They came in search of gold. They were amazed at the abundance of prawns in the salty river waters, and so they named this place Camãrao, the Portuguese word for prawn. The French colonialists later turned it in to Cameroon or Prawn Land.

Douala is a huge, sprawling city of 2 million people settled by a wide, deep river that washes into the nearby Atlantic Ocean. Apart from one precinct graced by stately colonial-era buildings, it is a dirty, crowded place, and it seems at trigger point, with a brooding sense of tension in the air. I transfer to a taxi, and we drive past the port where ships that have come to take the hardwood logs to Japan, China, and even the United States are at anchor at the many docks.

On the outskirts of Douala, we pass over a bridge and drive along what looks like an enormous rubbish dump. It is a marketplace, more than a mile long and one shop deep. Thousands of plastic bags litter the roadside and piles of malodorous trash rot in the hot sun. Most shops are wooden shacks with colorful depictions of their goods painted on their sides. The road is narrow, potholed, and crammed with cars, bumper to bumper.

I sigh with relief when we hit the open road, free of the traffic snarl, but my relief is short-lived. The driver must have decided to commit suicide this day, because he hurtles along the road to Limbe, overtaking cars with no thought of what is looming on the other side of the road. Several times we miss an oncoming car by inches, but while I shudder in fear, the driver turns each time to me and flashes a triumphant grin. In the weeks ahead, I learn that this madness is a national pastime of adult males in Cameroon. I rarely see a woman driving a car, but in the month I spend in the country I get a lifetime share of homicidal maniac male drivers.

We pass by rubber tree plantations and Del Monte pineapple fields. An hour later, Limbe and the coastline are signaled by the blue, smudgy outline of Mount Cameroon, at 13,435 feet an active volcano that is West Africa's highest mountain. My hotel, a dozen crumbling bungalows with rusting tin roofs, is the best inn in town. It perches by the sea where an American warship sits at anchor about 300 yards out, here on a goodwill visit to this oil-rich country.

Across the sea looms the small island of Bioko. It is the site of Malabo, capital of Equatorial Guinea, a tiny country with Africa's third most extensive oil fields. Equatorial Guinea is ruled by one of the worst of the worst, Teodoro Obiang Nguema Mbasogo. Along with his neighbor President Biya of Cameroon, he is high on the list in *Tyrants: The World's 20 Worst Dictators*. The BBC says that Mbasogo "rules through fear and repression," and that political rivals brave enough to challenge him have been assassinated.

Mbasogo has such a ruthless grip on the media that in 2006 the state radio described him as "The God of Equatorial Guinea" and added that he had "direct contact with the Almighty." Perhaps that is why the state radio also declared that he had "the power to kill anyone without having to give the reason why."

Equatorial Guinea is another transit point for the trafficking of children across borders as sexual slaves, but neither the CIA nor the U.S. State Department in their fact sheets makes mention of Mbasogo's murderous rule. Secretary of State Condoleezza Rice even welcomed him to Washington in 2004 with a warm handshake and a smile. She described him as "a good friend." Whatever diplomatic groveling an American secretary of state has to do before a ruler of an oil-rich state, that praise went far too far, and put an indelible stain on Rice's reputation.

I want to hire a small boat to take me across to Malabo for a few days to see what such a place is like, but when I check at the port no one is willing to go. "We don't go over there," the owner of a small fishing boat tells me, "because it's an evil place. Mbasogo is the Devil on Earth."

The next morning I walk up the hill to the Limbe Wildlife Centre, that dreadful place where I had seen dozens of chimpanzees kept in bare, stony enclosures and going mad. I have come to see Nyango, the only Cross River gorilla in captivity. The gorilla enclosure is much better than those for the chimpanzees, with a large open space and an enormous tree to climb, but the 12 gorillas seem listless. Nyango sits forlornly by the front wire fence, away from the other gorillas, all western lowlanders. As I approach, she stares into space. She looks to me much like any other western lowland female, with a potbelly, flat leathery nose, and broad head topped with a fluff of orange hair. But the International Union for Conservation of Nature has designated the Cross River gorilla a separate subspecies after extensive study of comparative skeletons.

Jacqueline Sunderland-Groves, a Cross River gorilla researcher, was the manager of the Limbe sanctuary when Nyango arrived in 1994. She remembers the then young gorilla as a real character. She had been purchased as a pet at a market and raised as if she were a human by an American missionary family. "Nyango slept in the kids' bedroom, ate toast, and drank tea," she told me. "Little wonder that when she came to Limbe [at the age of four] she didn't want to be stuck in an enclosure. We had a dedicated keeper walking around the grounds with her all day and staying with her overnight."

The Limbe sanctuary's manager, English veterinarian Felix Lankester, tells me, "I'm sure Nyango still believes she is a human."

Now, kept behind the fence day and night with the other gorillas, most of them brought here as orphans, victims of the bushmeat trade, she must wonder why she has to live with a bunch of hairy brutes instead of with humans like her on the other side of the fence. I sit by Nyango at the wire fence and she slouches against it, looking at me with sad, soulful, beseeching eyes, as if to plead . . . *help me, help me, please get me out of this nightmare.*

I ask Felix about the lack of greenery and the shocking state of the chimpanzee enclosures. He says it is caused by the huge amount of rain in the wet season. It supposedly washes away the plants and shrubs, but that excuse falls flat when the rest of Limbe, including the botanic park across the road, is lush and green from the abundant tropical rain. A much better, leafy chimpanzee enclosure was under construction.

The Limbe sanctuary is home to some of the most famous gorillas in the world, nicknamed the Taiping Four, and their fate is meant to be a salutary lesson for the illegal traders in wild-born great apes. The four adolescent western lowland gorillas are in quarantine at the sanctuary for another month, having arrived two months earlier, but I am allowed to get close to their cage in a fenced-off area behind Felix's office.

The three females sit on a bare floor behind bars and study their navels, but Izan, the young male, struts up and down the cage, displaying to the ladies his admirably bellicose nature. He stops now and then to take up a male gorilla's impressive stance, his broad arms thrust forward and planted on the ground, causing a dip in his back, giving the appearance of more bulk to his already powerful shoulders and massive buttocks. It is a seduction pose to impress the females, and a challenging pose to shake the hearts of rival males. The ladies glance now and then at Izan as he runs up and down the cage, thumping his chest.

A female named Oyin comes to the bars to share intensely primate gazes with me, and she is as curious about me as I am about her. As our eyes meet, I feel the same thrill as always when I look into the glinting eyes of a great ape, deep wells of intelligence and understanding. Oyin and I sit together on either side of the bars for about 10 minutes, looking at each other for a few moments, then glancing to the side in a polite manner, and then sharing once more our interest in each other.

Izan, seething with adolescent testosterone, perhaps grows jealous, seeing me as stealing Oyin's affections, because he charges up and down the cage, stamping his feet and banging his fists against the bars a couple of feet from Oyin's head. She gives him a look of indifference and then after a while rises and knuckle-walks to the back of the cage, maybe to calm him down.

I feel deeply sorry for all four gorillas. They should still be in some forest with their families, Izan one day to be a powerful silverback and the females to be mothers to a new generation of western lowland gorillas in the wild. Instead, they will live out their lives in this sterile place, just a shadow of what they might have been in the wild, all because of human greed.

According to the International Fund for Animal Welfare (IFAW), the Taiping Four were victims of the illegal trade in endangered species, grabbed as infants from a jungle in Cameroon and smuggled across the border to Nigeria. They next appeared at a zoo there. Documents were forged to show they were not wild but captive bred, and they were soon sold to another zoo in far-off Taiping, Malaysia. Trading in wild-born gorillas, whatever their age, is prohibited by international law, and when they were discovered at the Southeast Asian zoo and the forgery was detected, the Malaysian government confiscated them in 2004. It sent them to what supposedly was home—to South Africa, which has no gorillas in the wild—and they spent the next three-and-a-half years in the Pretoria National Zoo.

It was an odd decision, because under the regulations of the Convention on International Trade in Endangered Species (CITES) the four should have been returned to their homeland. But where? The IWAF reported that "two separate DNA tests have proven the country of origin of the gorillas to be Cameroon."

So home they came to great acclaim by the government in Yaoundé, which was and still is tight-fisted in funding habitat protection and sanctuaries for the country's great apes. The four were sent to the Limbe sanctuary, whose overseas donors are largely expected to fund their upkeep. It would have been better for them to go to the Mefou sanctuary, where they could live in a three-acre enclosure with plenty of bush to hide in when they tired of humans staring at them, and plenty of trees to climb.

The gorillas I saw at Mefou were far more active than the listless gorillas in the big enclosure at Limbe. Izan will also cause problems when he becomes a strapping silverback in a few years. Limbe already suffers from this common problem in great ape sanctuaries. The dominant male gorilla, Chella, is being challenged by a young male. Chella is nowhere in sight on the day I visit.

"We have to keep Chella locked up in the night cages on every alternate day, the same with the other male, so they can't fight," Felix tells me. Then what happens when another potential dominant male, Izan, is put into the compound? Felix frowns. "Maybe we'll have to build another enclosure just for the Taiping Four."

Four months later, Oyin, the beautiful young female western lowland gorilla being held at the Limbe Wildlife Centre, became ill and her white blood cell level plunged. Three veterinarians fought to save her but within a few days she died, possibly from malaria. It is endemic in the villages, towns, and cities of Cameroon. Not long after, Izan also died. The Limbe Wildlife Centre sent me the following report:

> *The results of the tests for Oyin and Izan never came up with a decisive conclusion about the cause of death, but we know it was a combination of stress caused by the introduction to the resident group and intestinal pathogens that they had not encountered before and had no immunity for (while our resident gorillas do).*

I understand the principle of returning captive wild gorillas to their native lands, to deter further illegal captures. But it was tragic that Izan, Oyin, and the other two were taken from Taiping and then Pretoria and then brought to Limbe in a blaze of self-satisfied publicity by wildlife authorities. The Limbe sanctuary already had an admitted problem with competing adult male gorillas and to introduce a new male gorilla into the group was asking for trouble. It would have been much better if the Taiping Four had been taken to Mefou near the capital, Yaounde, and given their own large lush enclosure like the other gorillas and chimpanzees there.

On the way back from the sanctuary, as I walk down the slope and through the botanical garden to my hotel, I see the U.S. warship still anchored offshore. The stars and stripes snaps in the stiff sea breeze. At the open-air restaurant by the sea, a waiter asks me where I have been. "Up to the sanctuary to see the gorillas."

A middle-aged white woman approaches me. Dressed in chinos and a T-shirt, she has mannish features topped off by a cropped haircut. "Are you interested in gorillas then?" she asks as she sits down at the table. Her fruity accent is pure Dublin. "I've just spent three months as a volunteer at the Mefou sanctuary near Yaoundé, and I've come here for a few days' holiday before goin' home."

"I was at Mefou a couple of days ago."

Her eyes light up. "Did yer see then me darlin' Tommy? I love that chimp. I took care of him when he arrived lookin' like he'd just lost a fight in Dublin's toughest bar."

"Yes, he's a charmer."

"Too right. I lost me heart to him. He was like me baby. I saved for a couple of years to go there as a volunteer, and if I had the money I'd spend a year there."

Her name is Ann Murphy, and she tells me she is a senior warden, a turnkey, at the men's maximum-security prison in Dublin. This must be one tough lady to daily stare down the notorious loyalist and Irish Republican Army killers and kidnappers there, and she looks it. I reckon one word out of place and she'd belt you with a baton.

"Why did you decide to become a volunteer, especially as it costs a lot?"

"Too right it did. But the apes are dyin' out, aren't they? We've all got to help them. I saw the sanctuary on a website and saved to go there."

"But your three thousand dollars, as well as your airfare and expenses, would do a lot more good for the great apes if you just donated it all to the sanctuary or a wildlife reserve."

"You're right, that it would, but I have me own feelin's and I wanted to make a difference by being there."

She seems a very nice person, and there seems no point hurting her feelings by carrying on the debate. At least she *is* doing something to help the great apes.

Although little is known about the day-to-day life of the Cross River gorillas, because they originated as lowland gorillas, primatologists assume that their social habits are broadly similar to those of the western lowland gorillas. But because their habitats are quite different, their diet has been adapted over the centuries to the highland forest ecology. In the tropical rainforests the diet of the western lowland gorillas is largely made up of soft fruits, but in the highlands the Cross River gorillas exist largely on leaves, vines, bark, and other roughage.

Jacqueline Sunderland-Groves was the first researcher to study the gorillas in the wild at length. From 2001 to 2006, she investigated their diet, range, nesting and group composition, and distribution on the Cameroon side of the border. In one rare encounter, she told me, all members of the group fled except the silverback, who barked and bluff charged her for 10 minutes, pounding his fists on the forest floor. Another time, she and a tracker followed a group up a very steep slope to a small ridge. Just as she reached the top, one of her trackers "charged around a rock along the ridge, brushing past me just in time for me to see a blackback male following a couple of meters behind at full speed. The blackback skidded to a halt a meter or two away from me, stared at me, and then turned and took off."

While there, Jacqueline saw the gorillas using "tools" on three occasions during display behavior. Chimpanzees and bonobos fashion tools to gain food, something not seen among the gorilla subspecies in the wild, though Leah, the western lowland gorilla at the Congo's Mbeli Bai, did use a stick as a tool to plumb the depth of a pool as she crossed it upright on two legs.

The Cross River gorillas' tool use is also unique. Because the gorillas were under severe threat from humans, Jacqueline told me that she decided not to habituate them, even though that would have made life a lot easier, because habituation would make the gorillas more vulnerable to hunters who share their habitat. The gorillas reacted angrily to the research team's presence, roaring threat vocalizations, barking, shaking trees, and beating their chests to scare them. The silverbacks and adolescents also charged them, normal behavior for unhabituated gorillas when confronted with human beings.

But on three separate occasions the gorillas took resistance a step further. The first time was in October 2004, when the researchers and a group of Cross River gorillas stumbled across each other in the grasslands. The silverback charged down a steep slope at the researchers, coming to within five yards. Three adults lined up behind the silverback and tore up fistfuls of grass with mud and the roots attached and threw them underarm at the researchers. The confrontation lasted three minutes, but none of the projectiles hit the researchers.

In the second attack, after bluff charging by the silverback, a young male picked up a branch about a yard long, eight yards from the researchers, and threw it underarm at them. He also missed.

The same year, a villager known to the researchers approached a group that was basking in the sun. Trying to frighten the gorillas, the man started shouting and banging his machete on the ground. He then threw rocks at them. The gorillas responded by tossing clumps of grass and rocks at him. The encounter lasted about an hour.

No one had reported such behavior before. Jacqueline believes the gorillas might have learned to throw objects at intruders after observing humans doing it to them to chase them away from farmland.

At mid-afternoon, I climb a dirt road overlooking the sea to the home of Aaron Nicholas, the primatologist who heads the Cross River Gorilla Project. The house is perched on a rocky outcrop that rises straight up out of the dark water. A sea mist swirls around the craggy hill and several small rocky islands near the shore, giving the scene the mystical, poetic look of a Chinese ink scroll painting. A Welshman, Aaron works for the Bronx Zoo's Wildlife Conservation Society.

When the door opens, I am surprised to find the highly experienced primatologist Ymke Warren, whom I had met in Rwanda a decade earlier when I went there to investigate the fate of the mountain gorillas soon after the genocide. At the time she worked for the Dian Fossey organization. Ymke was then a largely introspective young woman, but the years seem to have softened her, because she warmly welcomes me. She tells me she is Aaron's partner, and works with him.

Aaron's office, creatively cluttered with bound reports, computer screens, and gorilla bones, is perched directly over the sea, and frothy waves swirl about the bottom of the rocky cliff face as he shows me a dozen massive gorilla skulls, placing them on a table set on the balcony. "This is one of the natural world's most terrible tragedies, and yet hardly anyone knows about it," he says as he cradles one of the skulls distinguished by a prominent sagittal crest, the bony ridge that runs from the forehead to the back of the head. "Protecting the last remaining Cross River gorillas is one of the world's most important conservation efforts. They're on the brink of becoming extinct, the first of the great apes to do so."

I have heard similar stories around the world in regard to many species, and some are genuine, some are not. There is no better way for conservationists and zoologists to get funds than to claim the species they are studying or saving is on the edge. But there is no question about the serious threat to the Cross River gorillas, and yet the funds needed to protect them have been limited, far less than the $4 million required to implement a five-year protection plan.

Aaron's effort is supported by, among others, the Wildlife Conservation Society, the WWF, the U.S. Fish and Wildlife Service, the Great Ape Trust of Iowa, the Columbus Zoo, and the Disney Conservation Foundation. But, apart from those from the WCS, the donations are relatively meager, and that holds back ambitious plans Aaron and other primatologists in the International Union for Conservation of Nature's Cross River Gorilla specialist group have drawn up to protect them.

"There are only about three hundred Cross River gorillas left and hunters are still killing them in their habitats," Aaron says. Genetic studies show that there were many more Cross River gorillas up until about 100 years ago, when they began to suffer a marked decline. That was probably the result of guns being introduced, as well as growing human pressure for land for settlements and to graze cattle and grow crops. "The hunters now go after them not so much for their meat—duiker taste better and bring a better price at market—but for status. The hunters keep the skulls on display in their huts to show they're brave and strong." Shamans also use their bones for traditional medicine and for fetishes. Aaron got these skulls from hunters in the apes' native habitat, where he and Ymke are taking me in a couple of days.

"They're extremely shy, elusive creatures, driven into the mountains over the centuries by hunters and farmers, and only seven outsiders have ever seen them," he tells me. "Compare that to more than one hundred thousand outsiders who have seen mountain gorillas in their habitats. I'll bet hundreds of journalists have reported and filmed the mountain gorillas, but you're the first reporter who's ever taken the trouble to come here to tell the world about the Cross River gorillas and their plight."

I tell him, "I saw Nyango just now at the sanctuary and she looked much like the western lowland gorillas there. But Professor Oates had told me the difference is mostly found in the skulls."

"Yes, their skulls evolved differently over a long time from the other lowland gorillas. They were discovered in 1904, but outsiders hardly took any notice of them because they looked similar to other lowland gorillas, and by the 1980s we thought they were extinct. Then, they were rediscovered. There are just eleven family groups left, and they're scattered in forest habitats along the border between Cameroon and Nigeria. They evolved as lowland gorillas, but, because of extreme pressure from habitat destruction and hunting over the past few hundred years, they've retreated to terrain across the Cameroon and Nigerian border, high up in rugged mountains with steep slopes."

One of the gorillas' most serious problems is maintaining genetic diversity, especially with such a small, fragmented population. Like all other gorillas, the female and male adolescents leave the natal group, the females to seek out silverbacks with groups and the males to live alone until they are strong enough to attract females. Genetic analysis of the Cross River gorillas to establish DNA, from dung collected from the nests of the 11 known groups, shows that in some forests there has been successful migration. But young gorillas have found it difficult to disperse in areas where much of the forests have been razed and planted by farmers.

"That includes Kagwene, where we are taking you," says Aaron. That was where Jacqueline spent three years with the gorillas. "At Afi Mountain, on the Nigerian side of the border, the gorillas there are virtually isolated from the other gorilla areas by farms and a frequently used highway."

Aaron carefully places the skull back on the table. "Because they are a lowland gorilla, for a long time the mountain gorilla people led a campaign to deny they were a separate subspecies. After all, even though the Cross River gorillas are under dire threat, there are more than one hundred thousand western lowland gorillas in the wild. Researchers need a constant flow of funds, and one of the great appeals of the mountain gorillas is that there are just seven hundred of them, the rarest great apes on Earth, they claim. So, any knowledge that there is another subspecies with near enough to three hundred left, and far more endangered, might lessen the impact of the mountain gorilla researchers' appeal for funds and sympathy."

In 2002, the International Union for Conservation of Nature ended the dispute when it designated the Cross River gorillas a separate subspecies and placed them on the critical endangered list. It reported: "Critically endangered from habitat loss and bushmeat hunting, the remaining Cross River gorillas are surrounded by some of the densest human settlements in Africa. With their population splintered across the landscape, they are the most threatened of all gorillas and in urgent need of immediate aid and sustained conservation action."

"What kind of protection do they have?" I ask Aaron.

"None. We're trying to get the government to designate one of their habitats at a place called Kagwene as a gorilla sanctuary, but even that won't give them total protection, and it will only involve two of the eleven families. The forest corridors they need to link up so they maintain genetic diversity are under severe threat from habitat destruction."

"They're a world treasure, so why doesn't the Cameroon government give them total protection, and ban all hunting and logging in their habitats?"

"It's not so easy. There are plenty of villages in their habitats, and you can't deny the people there a living. But not all the villagers hunt them. In some villages there's a secret cult of gorilla people, villagers who claim they can magically transform themselves into gorillas and join the wild gorillas in the forests, and change back to humans at will. I'll introduce you to the cult's chief when we get there in a couple of days."

It would take another week before we could leave Limbe for the Cross River gorillas. This very night, Cameroon exploded into murderous violence.

4

After a couple of hours' sleep I go to dinner in the hotel's open-air restaurant and notice that most of the staff are clustered around the restaurant's TV, watching the nightly newscast in French. It is usually tuned into a soccer match; they are broadcast day and night in soccer-mad Cameroon. I am grateful to be in the southwest, which has a majority of English speakers, and as I settle into the seat for dinner, Monica, a young waitress, approaches with a look of fear. I had earlier told her I was going up the hill to the main road after dinner to an Internet café.

"You can't leave the hotel," Monica says. "The government has announced a national nighttime curfew. People all over Cameroon have rioted today over the jump in food and fuel prices, they've blockaded the roads, and the soldiers have warned they'll shoot on sight anyone breaking the curfew."

The suddenness of the uprising does not surprise me, because I have been caught in quite a few rebellions in Third World countries. At Douala, I sensed a deep and troubling anger among the people even in the short time I was there, and when I got to Limbe there were not many smiling faces on the crowded streets. Life is a hard enough struggle for families trying to exist on a few dollars a day, and when the wheat price doubles and the fuel price rises a third in just a few weeks, the social discipline that holds communities together can swiftly splinter and disintegrate under the stress.

Suddenly from somewhere up the hill we hear gunshots. "Fookin' hell!" Ann growls, and then downs a double shot of whiskey. It does not ease the frightened look in her eyes.

A two-man BBC crew who had been filming the historical botanical gardens sits with us, and they worry about missing their plane from Douala to London the following night. The TV news is reporting that the road from Douala to Limbe is the worst hit in the country, with many fiery barricades thrown up by rioters along the way. "We're going to see if we can hire a fishing boat from the port and sail up the coast to Nigeria if the road to Douala remains blockaded," says the producer. I have no intention of leaving yet.

"The soldiers are shooting at the rioters up the hill," Monica tells us. "We're all staying here tonight because of the curfew and the danger."

I wonder about the gorillas, chimpanzees, and other animals at the sanctuary, a few hundred yards from the gunshots. It will not be the first time many of the great apes there have heard gunfire, because their mothers were probably shot out of trees when they were captured. Would the soldiers or even the rioters break into the sanctuary and start shooting the animals there? "They're safe," Monica says. "They bring money here from tourists, and no one has any quarrel with them."

We cluster around the TV to hear the latest national news, given in French and English. Rioters have blockaded roads all over the country, but the worst trouble is here in the southwest, the English-speaking provinces. The soldiers have shot dead many rioters and have used tear gas and water cannons to break up a huge demonstration in Douala today. Young men are barricading the streets with burning tires across the roads, and countless shops have been looted and destroyed. The newscaster reports that the trouble began when taxi drivers in the cities went on a coordinated nationwide strike over the rapidly rising fuel prices. The riots spread across the country in a single day like rolling wildfires.

"The news says that it's the worst trouble in Cameroon in fifteen years," Monica tells me. She shudders when we hear more gunfire somewhere up the hill. "The soldiers are killing people," she says softly and crosses herself. Her lips barely move as she murmurs a prayer.

The waiters, manager, cleaners, washerwomen, and guests maneuver chairs around the TV and settle in. I decide to go up the hill to find out what is going on but do not tell Monica and the others, because they would probably try to stop me. The streetlights have all been turned off and I keep to the track's darkened fringes, spooked a little by shadows that seem to move. By the entrance to the botanical garden, I tread carefully across a small bridge, because I know that the rioters are very near by their shouting and screaming.

A hedge by the roadside shields me, and I peer around with the leaves brushing my face to see where the noise is coming from. It is like a scene from a twenty-first-century *Dante's Inferno.* Blasting over the street is the thudding rage of an American rap song from a CD player, turned at full volume but almost overwhelmed by the rioters' chilling yelling. This is true globalization.

Fifty yards away, about 30 rioters have piled tires across the road and set them afire to block vehicles. The flames crackle and spark as they shoot up into the air, and the rioters, bare-chested and in their 20s, jump up and down in a frenzied war dance while waving machetes and clubs. The flames flicker threateningly near their bare torsos. Gunfire closer to town freezes them for a few moments, and then they leap back into their celebration of street power. I wonder why the police and army have not arrived to clear the road and disperse the rioters, but hear the next day that there were so many barricades all over Limbe that the armed forces could deal with only some of them.

Having seen enough, I walk back down the hill to the hotel. The TV watchers are still staring intently at the screen. The next morning, Aaron arrives to share breakfast. "It looks as if the roads on the way to the mountains will be barricaded for a few days, and so we'll have to put off the trip to the gorillas until things settle down," he says.

I decide to use the time to visit a village about 10 miles down the coast that was once a major slave-buying center selling human cargo to sailing ships from America. The present-day villagers are said to be proud of their slave-trading ancestors. Aaron looks concerned but doesn't object. He has worked as a conservationist in many dangerous places in Africa and is probably used to slightly mad writers on the trail of an interesting story.

At mid-morning I walk along the seaside track, downhill from the city center. The street by the beach seems deserted—no rioters, no soldiers, no anyone. I pass by scraps of charred tires and burn marks stretching across the road like zebra's stripes, where the rioters had set up the flaming barricades the night before. There are no swimmers, no boatmen, and the waterfront restaurants are deserted. But about 200 yards ahead I spy a group of young men gripping machetes and clubs, and barring the way to a road through the mountains.

They wait for me to reach them, and then a boy in his late teens strides up and thrusts his face almost into mine. "You're a white man, you're a racist, you're no friend of ours," he snarls. He waves a machete in front of my face. "We kill white men."

I do not retreat, and my weapon is my smile. I have been in too many situations like this and know that I must show no fear and always smile, even if my heart is trembling. "Mate, of course I'm not your friend because I don't know you. If I get to know you, maybe we'll be friends if we like each other, and maybe not if we don't."

The others at the barricade chuckle. A tall young man with a smooth round face growls something in the local language to the teenager, who retreats into the mob. "Sorry about that," he says. "James is hot-headed."

His name is Robert, and we talk for about 15 minutes. He tells me that his father scraped together the money for his university education, he trained as an engineer, but when he graduated there were no jobs. "The jobs go to the French speakers," he says. "In the southwest we get very little funds from the government, even though we have the oil and the fisheries. We'd ditch the glorious Republic of Cameroon today if we English speakers were allowed to set up our own country."

The others murmur assent. Then Robert explains that they are demonstrating (his word) because oil and bread prices have increased steeply in the past month. The country is boiling mad, he says, and the only thing keeping President Biya in power is the implied threat of the French military forces. "All Cameroonians know that they'll move in and ruthlessly subdue any uprising to keep their puppet in power."

"Maybe Carla Bruni can persuade her husband to change French policy," I joke.

I explain how international trade works, adding that Cameroon buys its wheat on the open market, where the prices zoomed upward recently because of the tensions of supply and demand.

"Okay, I understand the wheat price jump, but we've got plenty of oil. It belongs to the people. Why doesn't the government sell it cheaply to us, and at the market price to foreigners? That's what they do in Saudi Arabia."

By now the other young men have surrounded Robert and me, and they listen intently. They all know the unspoken answer to this. The political elite are stripping the country's assets for their own use, and if they followed the Saudi example they would lose millions of dollars in rake-offs and under-the-counter payments.

"We also oppose Biya's attempt to stay in power for another term by manipulating the constitution. We're very angry, and that anger has now boiled over. We're not dreamy-eyed, we know the soldiers will crush us without mercy, but at least for once we've stood at the barricades and forced them to retreat."

"What if they shoot you?"

Robert reflects on this for a few moments. Then he says softly, as if moved by the enormous personal loss for himself and especially for his family, "Then I will have died a glorious death, and my family will be proud of me."

He asks where I am going, and I mention the coastal village of Bimbia. "It's not far; the road is closed but there are no barriers during the day. Up the hill I know a man who has a car—not reliable, but he can get you there." We shake hands. "Travel safe," he says.

About 300 yards up the winding hill, in a driveway, I spot a man working on the engine of a car that looks so decrepit that it would fall apart if I gave it a good kick. It is battered and rusted, and the windshield has a crack that runs from one corner to the other, held together by masking tape. Charley agrees to drive me to Bimbia and back for the Cameroonian equivalent of $10. The road is deserted as we head for the former slave village.

Bimbia is nestled against the sea on a slope, with about 30 bare-bone weatherboard huts scattered across the fields. A dirt track meanders down the hill toward the Atlantic Ocean. About 400 yards beyond the shore across the rippled dark water is a small uninhabited island covered with trees.

A boy leads me to the home of the village chief, Alfred Ejong, who is barefoot and clad in a weathered dark suit and a clean white shirt with no tie. One wall of his lounge room is covered with framed certificates from the government, awards for his excellent service. Like Joseph Bikono, the hostile village chief in pygmy land, the government in faraway Yaoundé depends on the loyalty of these men to control the villagers, make sure they vote for the government in elections, and identify troublemakers—political and criminal. Alfred knows about the riots in Limbe but tells me there has been no trouble in his village. "Our people till their own fields and fish in the sea, and don't depend on the market for their food."

Alfred is proud of his slave-trade ancestors, and seems surprised when I tell him that many people in far-off America might think their actions shameful. He says, "It was a good thing for our people. In the interior there were always wars, and the victorious kings and chiefs made slaves of their enemies—the men, women, and children. My ancestors would go to these rulers, buy the slaves, and march them back here for sale to the whites who came in big ships from across the sea."

The boy enters the room carrying a pair of rusted leg irons and a saber, its curved blade flecked with rust. "Our ancestors locked the slaves into chains and marched them here to the coast," he says. He grips the saber and with a grim smile brandishes it in the air. "Long ago a slave master from our village named Nafonde got this from a white sailor and used it to behead any slave causing trouble. He then licked the blood in front of all the other slaves to make them so frightened of him that none would dare try to escape. The slave masters who came after Nafonde all used this sword in the same way."

He takes me outside and points to the island offshore, which is called Niko. "Once the slaves arrived here, our ancestors took them by pirogue out onto the island. These inland people were terrified of the ocean and dared not try to escape. Our ancestors kept them there until the next slave ship arrived from America or Europe."

He walks me farther down the slope to where a large ship's cannon, the kind you see in pirate movies, is half buried in the damp earth. "The captain of an American slave ship gave this to the village chief, my ancestor, because pirates would attack to try and seize the slaves on the island. My ancestors drove them off with this cannon."

Back in his lounge, his wife serves us tea. I ask again whether he was ashamed of his ancestors' actions. "Why should I be? We didn't force people into slavery. My ancestors were the middlemen. The trade existed and if they didn't do it, others would."

He goes into a room and brings back a well-thumbed Bible. "I'm a Baptist lay pastor at our church and I know that in biblical times people had slaves," he says. "If the Bible doesn't condemn slavery, then why should I condemn my ancestors?"

On the way back to Limbe, the car splutters to a halt by an empty field. "It's run out of petrol," Charley says. I prepare myself for an eight-mile walk to town in the muggy heat, but he has an African solution. He disappears under the car with a spanner and soon emerges dragging the entire gas tank, which he had uncoupled. The dip at the bottom holds enough gas to get us back to Limbe. He pours it into a bottle and attaches this with an improvised fuel line to the engine, and then bolts the gas tank back onto the undercarriage.

"You've done this many times," I say with a smile. He laughs and nods as he climbs behind the wheel.

A mile down the road, I spy about 20 Cameroonian soldiers with assault rifles lying in the grass. Obeying hand signals, they rise to their feet and advance cautiously on what I suppose is an enemy position, maybe rioters. Charley does not seem worried. A French army officer clad in a camouflage uniform stands near the soldiers, watching. "He's a foreign legionnaire," Charley snarls. "The Man O' War army camp is nearby. We've got no hope of overthrowing those bastards in Yaoundé, Biya and his lot, because the French protect them."

I see what he means when I arrive back at the hotel. The American warship has gone, but in its place is a large French troop carrier anchored menacingly near the shore. The BBC crew have been busy while I was away. Essential to any film crew on a foreign assignment is a "fixer," a local who knows the best hotels, how to get the crew where they want to go, and how to grease palms. They are packing their gear into a hired car. One of the BBC men says, "We got the fixer to pay off the local army commander, and he's assigned a military pickup with soldiers to guard us on the drive to Douala. We're flying out from there tonight." Ann is going with them, and she gives me a farewell hug. I later learn that they arrived safely.

I decide to walk along the waterfront to find out who controls the streets. From about 100 yards away I see several soldiers standing over what looks like three bodies on the ground outside a government building by the seaside. A dozen more soldiers sit in the back of an army pickup cradling assault rifles. Up close, the soldiers are unsmiling but not unfriendly. "They're rioters; they got what they deserve," a corporal tells me as he points to the young men on the ground.

Two of the men are still alive, and their hands are tied behind their backs. Their shirts are bloodstained and they moan in pain, their dark eyes pulsating with fear as they stare up at me. The third man is still, the side of his face from the jaw to the temple smashed in and splattered with dried blood. It looks as if he has been hit hard with rifle butts. I cannot tell whether he is alive or dead. None of the three was at the barricade just down the road earlier in the morning.

That night at 8:00 p.m., President Biya addresses the nation on TV. The moment he blames the riots on his political opponents, claiming they organized the trouble to overthrow him, most of the watchers in the restaurant hurl abuse at him. Biya looks wooden and does not mention the actual events, not even a word about the barricades. "That looks like a speech he gave a couple of years ago," the restaurant manager tells me. "He's probably not in the country, and they've pulled this speech out to give us the sense that he's here and in command." That could be true. Biya spends months outside Cameroon every year, staying at a luxury hotel in Switzerland.

Biya gives his speech only in French. A few moments after he finishes we hear gunshots up the hill. Later I learn that soldiers killed two rioters up there.

The next morning, Monica tells me that hundreds of demonstrators were crossing the bridge in Douala when the army blocked them with soldiers at both ends. A helicopter gunship flew over the bridge and fired on them with bursts from machine guns. Many jumped off the bridge into the river to escape and were drowned. Others were shot dead on the bridge. People who went to the morgue to collect their relatives' bodies were turned away by armed soldiers on the government's orders.

The rioting continues for another two days and then suddenly stops. The taxi drivers ended their nationwide strike at midnight, and the steam seems to have gone out of the protesters. Survival for most Cameroonians is a daily struggle, and each day on the barricades means a day's wages lost and a day closer to starvation. The TV news reports that all the roads have been opened across the country. "Let's go to the gorillas tomorrow," Ymke says.

Aaron will follow us a day later. Along the road between Limbe and Douala, shops have been smashed and looted. The most serious destruction is at the hated gas stations with pumps torn out of the ground, illuminated gas price signs smashed into shards, and the offices torched, leaving charred remains. I later learn that at least 100 people were shot dead by the soldiers, with hundreds more injured and hundreds jailed.

The drive to the mountains on the border between Cameroon and Nigeria takes eight hours by SUV, half the way by a good road and the final half over a dirt track that winds along steep foothills. The track is littered with rocks that bounce the vehicle around as we pass many mud-hut villages. As we near our destination, Ymke points to another track high in the foothills running roughly parallel to us. It looks like a raw scar slashed across the landscape. "The Fon, the local hereditary king, built it," she explains. "He spent years as a colonel in the army, but the government finally got rid of him because he was incorruptible. When he left, instead of accepting his pension, he asked for and got from the army some cast-off earth-moving equipment, and built that private road leading to his palace."

About an hour later we reach our destination, the highland village of Njikwa. The blue sky is fluffed with clouds that hover close to or even below the many mountain peaks that surround us. Perched on top of a high hill is the Fon's palace, a modest bungalow made from mud bricks with a tin roof. He is the hereditary ruler of more than 25,000 people.

The Fon is a tall, genial man clad in a gray tracksuit and an embroidered cap. He looks about 50, but he is in his mid–70s. He is expecting us, and ushers Ymke and me into a small, dark reception room with seats pushed against the walls. The only other furniture is a side table containing two large bottles of Fanta. These must be for decoration, as the Fon does not offer us a drink. An antique curved sword hangs on a wall, a symbol of his family's power and martial spirit. We are now close to the Nigerian border, and for centuries the Fons, who form a patchwork of minor principalities across the mountains, fought each other until they were harnessed by the newly independent nations of Cameroon and Nigeria in the mid-twentieth century. Now, the Fon keeps the peace in his territory for the Biya government in far-off Yaoundé.

"I was the military attaché in our Washington embassy for a few years and enjoyed America very much," he tells me. "My wife and children still live in Washington, but I had to return to my people when my father died. While I was in the army, the best time was when I was a military attaché at our embassy in Kinshasa while Mobutu was in power. He'd take us on pleasure trips along the Congo on his yacht, with plenty of food and girls."

The Fon seems a lonely man with no family and no courtiers in sight. He has the power to judge minor matters of law, and can even order the flogging of what he calls a "miscreant." The Fon tells me he has never seen Cross River gorillas, whose habitat is just a few hours' trekking from here, much higher into the mountains. "The villagers where you're going have a secret cult of gorilla people," he says. "They can change themselves into gorillas."

"Aaron mentioned it. Do you believe they can really do that?"

"Of course. Our people possess magical powers that you as a Westerner must find difficult to believe, but they are real."

Maybe, but it sounds as if this is just like many remote village cults I have encountered elsewhere, such as that of the Amazon Indians, who told me they can transform themselves into jaguars at night and stalk their enemies. Such cults are usually so secret that no outsider is ever permitted to put them to the test, and can serve to instill fear in fellow villagers and enemies. After half an hour of polite talk, Ymke and I drive down another steep track carved out of the hillside by the Fon. It leads to his capital, Njikwa, of one the poorest villages I have encountered in Cameroon.

A straggly line of ramshackle wooden huts meanders with scant ambition down a muddy slope. Night is approaching, and though Ymke has brought a packet of instant noodles for herself, and a tiny portable stove, I had forgotten to go to the supermarket at Limbe and now find myself without food. I go from shack-shop to shack-shop, but am turned away by bulky women clad in cotton robes who claim on their doorsteps that they have nothing to sell. The village gives me the shivers. Men and women with blank faces sit on chairs outside their shacks or stroll with careful indifference along the muddy path that stitches the village from the top of the slope to the bottom. No one offers to help me; no one smiles or even looks at me.

Eventually, a barefoot boy in ragged shorts and T-shirt sidles up to me. He says there is no meat, no vegetables, not even eggs for sale in the village. He takes me to a dingy store that sells cotton cloth, bottles of warm beer, Cameroonian soda, and not much else. The storekeeper has just one item of food, packs of biscuits from far-off India that are two years beyond their use date. I buy a couple of packets, find they are rock hard, but eat them anyway to quiet my hunger.

Soon after daybreak the next morning, Ymke and I begin the daylong walk through the mountains to the huts where the project's trackers live while monitoring the Cross River gorillas. "There are two families of gorillas here, about eighteen gorillas in all, and over the centuries the subspecies has retreated from the intense pressure forced on them by humans to some of Africa's most rugged and remote country," she tells me as we get ready to trudge up a steep, grassy foothill scattered with boulders. Far away a jagged line of mountains forms the backdrop to the horizon. They seem impossibly high, impossibly distant. "That's where we're headed, right to the top," she says with a smile. My heart sinks.

By the village store an old man asks, "Are you going to the field station?" and when I nod, he raises his eyebrows in surprise. "Go majestically," he says, a kindly way of advising me to go slowly because it is a tough slog.

Within a couple of hours my thighs ache, my feet burn, my calves feel as if they have been pierced by knives, and blisters have blown up on my heels and burst—but we keep walking, walking, walking. As we climb higher and even higher, we pass scattered Fulani villages and pastureland. The Fulani are cattle herders who migrated here centuries ago from the far north, always in search of good grazing lands, and they have spread all over West Africa.

I have even encountered the Fulani in distant Timbuktu, and they are much the same here—the men tall and thin with long faces, clad in robes and colorful skull caps. The women mostly stay at home in these hills, but the men and boys grip long walking sticks as they herd their cattle across the slopes in search of pasture. The Fulani scorn the townsfolk and keep to themselves, speaking their own language and marrying among the tribe. As I pass, they look right through me.

On and on, up and up we trudge. The hours melt into each other. At midday we reach mountain forests, and the going becomes much tougher. The trees press against my shoulders, saplings slap my face, and I grip the branches to haul myself hand over fist up slopes that seem close to vertical. When we make a brief stop in a gloomy glade to drink water from flasks, Ymke points through gaps in the trees to a huddle of huts in the distance near the peak of a mountain that seems to float in the sky. It is still miles away. "That's our field station," she says.

I slump onto the forest floor, mumbling to myself that I cannot take another step. After a few moments of self-pity, I slowly rise to my feet with the aid of a stout forest stick and by sheer willpower start walking again.

At mid-afternoon, Ymke tells me we are now in gorilla territory. The forest is so dense here that the gorillas could be within 100 yards and we would not see them. We step cautiously along a mountain ridge not far from the field station. With sheer drops of thousands of feet, the slopes are so steep that looking at them makes me dizzy.

Ymke points to a big bush fire in the valley far below, burning not far from the forest's edge. Huge clouds of smoke almost obliterate the valley. "It's probably villagers burning off land," she says. "The fire encourages the growth of new shoots for their cattle. It happens all the time, especially in the dry season like now, and there's always a danger that the forest will catch fire and devastate the gorillas' habitat."

Beyond the valley is a line of blue shimmering mountains, their flanks streaked with cloud far below their peaks. They have the flimsy consistency of a watercolor painting. "That's the border with Nigeria," Ymke says. "The headwaters of the Cross River form a part of the border, and on both sides small groups of Cross River gorillas have their habitats."

We climb up a big rock perched on the brink of a steep slope, using cracks in the stone for footholds, and at the top, with intense relief I spy just a few hundred yards ahead our destination, four mud huts. The forest is so steep and dense that it takes another hour to reach them. At Limbe, the weather was very hot and extremely humid, but this high up, about 8,000 feet above sea level, I shiver from the wind and an icy mist that has settled over the huts. It is so thick that I can barely see my hand in front of me. The trackers are still out with the gorillas, and the only person here, the cook, greets us with boiling hot cups of tea and plates of omelets swimming in palm oil. Outside, I strip off my shirt and grit my teeth as I pour a bucket of very cold water over my head to wash away the sweat and dirt from the trek.

Ymke shows me to one of the small mud-floored mud huts with a pair of wooden beds. She and Aaron will sleep in the hut next door; the third is for the trackers and the fourth is the cookhouse. On one side of the small clearing that juts out from the mountain near its peak, the forest presses against the huts, while on the other side, a few yards away, the slope plunges to a valley thousands of feet below. The mountains are limestone, and strewn about us are giant stand-alone sugar-lump rocks, hundreds of feet high. They give the landscape a prehistoric look, as if a Tyrannosaurus Rex might come stomping out at any moment.

As dusk nears, four trackers arrive accompanied by Sama Billa, a young field biologist who is a gorilla monitor here at the field station. It is vastly different from any other great ape project I have visited, because the effort to protect the gorillas is considerably more important than the study of their behavior. "We try hard not to habituate the gorillas here so they don't become used to human presence," Sama tells me as we sit in the cook's hut, warmed by a log fire, our faces illuminated by the burning wood and the glow from a spluttering kerosene lamp. "Gorillas in the wild that have not been habituated to the presence of humans mostly flee at the first sign that humans are near, and that's the way we want to keep it."

Ymke had spent many years daily observing the habits of mountain gorillas at the Fossey site in Rwanda and has been a major force in form- ing this hands-off policy. "A researcher working with the Cross River gorillas a few years ago began to habituate them, but when one was shot soon after by a hunter, the habituation stopped," she tells me. "Now, our trackers deliberately stay one day behind the gorillas, tracking the group all day to their night nests from the previous night. The only time they come into contact with the gorillas is when they accidentally run into them. That's why only seven outsiders have ever seen them. I've seen them from a distance, but even though Aaron has been here many times he's yet to see them."

Her news makes me very happy, even though it limits the chances of an encounter with the rare gorillas. I tell Ymke about my time with Rugendo and his family in the Congo just before they were shot dead, and how the murderers were able to get so close because the gorillas were used to humans.

"You're the first reporter who's ever been out here, and so I'm glad you feel that way." That is probably because few people know about them.

Sama has had several encounters with the gorillas. He tells me, "It happens when we're tracking them from the day before and our paths accidentally cross. Usually, they'll get away before we get close, but sometimes the silverback will charge us, trying to frighten us away. Then, you must keep calm and not run away. I've had the silverback come close and roar into my face, but although I was scared I held my ground."

I tell him of my similar encounter with Kwakane in Rwanda, and he smiles.

One of the trackers, John Ndo, describes an even closer meeting he had with the gorillas: "When I saw them near, I tried to get away, but fell down. The silverback came up and pushed my shoulders against the ground. A blackback crouched behind him. The silverback shouted at me and I screamed because I was very scared. The silverback kept pushing and shouting at me for about five minutes, and then they went away."

The silverbacks had the strength to tear John and Sama limb from limb. No one has tested a silverback's strength, but captive gorillas in zoos have been observed bending steel bars, breaking welded joints, and tearing away metal plates in their enclosures.

There is something about mountain air that always makes me sleepy early at night. After dinner of an omelet washed down with tea, I slip into my goose-down sleeping bag to ward off the cold in the hut and immediately doze off. Just before midnight, I wake up and step outside. The sky is dark and a freezing wind whips through the mountains. Heavy clouds have settled at the edge of the clearing and spread back across the sky like a choppy sea. I walk to the edge and feel an almost overwhelming desire to dive head first into the clouds for a swim. My moment of mountain madness passes and I go back to bed.

5

An alarm clock wakes us at 6:00. The trackers emerge from their hut, blowing icy streams of air as they stamp their feet and clap their hands to get warm. The clouds have moved to settle on the clearing and I can see only a few yards ahead through the murk. A mug of hot tea jump-starts my frozen blood, and an omelet floating in palm oil puts the muscle back into my wobbly legs.

Soon after, Emmanuel Aseke, the head tracker and a former bushmeat hunter, leads us along the ridge and down a very steep slope crowded with leafy trees. Shards of faint light from the rising sun splinter the forest gloom. There is no path along the forest floor, just a thick carpet of rotting leaves. The only way I can get down is to grip a tree trunk and carefully edge toward the next tree within grasping distance down the slope. The trackers, mountain men, scurry down with the agility of goats, but it takes me more than an hour to reach what seems the bottom of the hill. "That was the easy part," says Emmanuel, smiling.

I am faced with the biggest test by a ravine, a vertical rock face we have to cross to get to the gorillas' night nests. We have to do it like rock climbers, using cracks an inch or two wide for foot and hand holds. I look below, see that the sheer drop is about 20 yards, and realize that if I slip and fall, that will be the end of me. I fearfully edge across the rock face. About halfway across the cracks disappear, but there is a small, stubby tree growing almost vertically out of the rock near my head. Following Emmanuel's lead, I grab a bough and swing across the rock face to where my feet and hands can grip more cracks.

With each dangerous step, my admiration grows for the gorillas who have retreated here over the centuries to escape the harm done to them by humans. The trackers walk and I stumble another hundred yards down yet another steep slope thick with undergrowth and emerge in a clearing enclosed by forest trees. With slopes surrounding it on all sides, it resembles a small amphitheater. "This is one of the gorillas' favorite nesting sites; they've come here many times to spend the night," says Sama. "We know from following them that the gorillas carry a complex map of their territory in their heads. They have favorite nesting sites, and know when the different kinds of fruit trees come into season."

Cross River gorillas live in small groups, commonly between four and seven individuals including a silverback, adult females, and young. Sama shows me the giant nest of the silverback who has snapped and fashioned reeds, plants, tall grass, and other vegetation into an oval bed. With the nest's back pressed against the slope, it looks very comfortable. Several smaller nests huddle around it. Next to one of the adult female's nests is a much smaller, rudimentary nest. "She's a nursing mother, and her infant is about two years old," Sama explains. "That's when infants begin making their own nests, copying their mother. This little gorilla has just started, and we know by the fur in the nests that he made his nest as practice because he spent the night sleeping with his mother in her nest."

The two trackers and Sama gather fur from the nests and dung scattered about them, placing each sample in marked plastic bags. Sama explains, "We get DNA samples from the hair, and other biological information, and by examining the dung we can tell what the gorillas are eating and if they are healthy."

The trek has exhausted me, and I lie back in the silverback's nest, which is so wide that when I stretch out my arms I still cannot reach the sides. I imagine the scene two nights before with the silverback grunting in contentment with his wives and children gathered around him, settling into their comfy nests for the night.

We must return the same hazardous way we came, and by the time we reach the ridge a couple of hours later I am exhausted and my big toes throb with pain. I take off my hiking boots and see that blood is seeping through the thick socks. Sama and the trackers are going to follow the trail of the family through the forest, where the great apes ate as they went, but my injured toes force me to return to the huts, guided by Ymke.

Aaron and a handful of porters have just arrived when we get there. They have hauled over the mountains several boxes containing solar cell panels and heavy batteries to provide electricity for the huts. With them are two electrical engineers from Yaoundé who will install the panels. "They cost twenty-four thousand dollars, but they'll make life much easier for the trackers living here, and also allow us to set up a hut for researchers who might want to come here with their computers," Aaron tells me. "Eventually, we should be able to connect to the Internet by satellite."

I take my boots and socks off and discover that both the big toenails have been torn loose and the fronts of my feet are soaked in blood. "My big toenails are pretty much gone," I tell Aaron.

"Whose haven't?" he says with an understanding smile. "The slopes are so steep that as you go down them, the front of your boots push up against the toenails and can tear them loose. The big toes take the brunt of it."

On the following morning after breakfast, Ymke leaves with the trackers, but I decide to stay at the field station for the day to help my toes heal. The nearest medic is back over the mountains, a day's trek for me with healthy feet, and I do not want to make them worse. Aaron and the electrical engineers spend the morning installing the solar panels, bolting them onto the roof of a hut.

At mid-afternoon, a man in his 20s walks into the field station. Banda is clad in torn pants and a torn shirt and has a machete strapped across his chest. The sheath is homemade, carved from forest wood. He is a deaf mute, talks to the cook in sign language, and smiles when the cook makes him an omelet.

"Banda's a poacher, one of the best hunters around here," Aaron tells me. "I'm trying to get him to stop poaching, but it's the only way he can make a living." The poacher has a wife and two children, and hunts small game in the forest such as porcupine, duiker, rat moles, bush pigs, and squirrels. After finishing the omelet, he waves good-bye and disappears into the trees.

"To the villagers here hunting is part of their culture, and there's the conflict between a traditional way of life and conservation of an endangered species," Aaron says. "We've got agreement in principle from the government to turn Kagwene into a gorilla sanctuary, and all it needs is the prime minister's signature on the document. It's on his desk right now. Then, the government will station rangers here to protect the gorillas while allowing hunting of non-endangered species."

"But the two gorilla families here make up only about six percent of the Cross River gorillas. What about the remainder?"

Aaron sighs. "It's a start. Hopefully, one day we'll have them all under protection in gorilla sanctuaries. If not, their chances of surviving are grim."

* * *

A torrential downpour hits the mountains an hour before dusk and gales of rain lash the huts. A howling wind, like a banshee on the loose, tears across the clearing. When the rain clears, in the gloom, wisps of cloud-like ghosts dance for us. Ymke and the trackers arrive back soon after, soaked to the skin. The cook has lit a log fire and we all settle in front of it on a pair of benches to eat dinner from tin plates and drink rainwater collected in a drum outside the hut.

After dinner we listen to the state radio from Yaoundé, with commentators still justifying the government's ruthless crushing of the riots.

The following day, I go with the trackers to the most recent gorillas' night nests even though I feel red-hot pain in my toes each time we descend a slope. The trackers lead Sama and me deep into the forest, and for about the six-hundredth time in my life I repeatedly mumble a vow that this will be the last time I ever go into a jungle or a desert. But, as always, the vow is thrust aside by my delight as we reach our destination. After two hours we find the nests, secure in a glade that looks like the abode of fairies in an enchanted forest.

Sunbeams break through the surrounding trees to paint the glade a golden hue. Massive trees tower above, their boughs twisted and bent this way and that so that with just a pinch of imagination they begin to look like giant woody creatures. The moss and lichen that clothe them glisten with dew. The silverback's nest resembles an oval throne, fashioned from branches he has snapped with ease to form the foundation, and with a layer of vegetation woven with striplings and leaves to make it soft and springy as a cushion. One difference between western lowland gorillas and the Cross River gorillas is that the latter more often build their night nests in trees.

The patriarch had built his nest in the crook of a big tree about five yards off the ground and overlooking the glade. Below, grouped around the tree, his wives and children have made their nests. The two-year-old is still practicing nest building and has snapped and bent some striplings together and added a few leaves for comfort. Once again, the trackers see by the fur in the commodious nest beside it that the little one cuddled up to his mother for the night.

While the trackers and Sama collect fur and dung for Aaron, I settle back in the mother's nest, moving around until I am comfortable. I begin to doze off but then suddenly, I am yanked back to the glade.

From somewhere not far away comes the *thock thock thock* sound I know well from the Congo and the Central African Republic. It is a silverback beating his chest. In silence, we peer in the direction of the sound. The silverback beats his chest once more and the hairs rise at the back of my neck. Sama whispers, "He's warning us to go away. We'd best leave now, because we don't want to provoke a confrontation."

As we climb the slope above the glade, I keep turning back to see whether the gorillas have gone back to their nests. But they can see and hear us far better than we can see and hear them, and they must be waiting until we are far away.

We trek back to the huts to find a sprightly old man waiting for me. He is with his son, one of the trackers. "It's Chief Zakaria Anji of Kinchi village; he's the gorilla people's leader," Aaron tells me. The village is about one hour's walk down the mountain from here, in the valley, and is relatively well off, growing coffee, yams, and ground nuts as cash crops.

The chief has on a colorful tribal tunic and a pair of pants. Like those of most mountain people I have met across the globe, his feet are much wider than you would expect from his height, perhaps from the daily effort of climbing and descending steep slopes. He remains silent as we sit by the fire and drink hot tea. I ask if he is hungry and when he nods, the cook makes him his standard dish—an omelet.

When he has finished eating, Zakaria warms his hands against the fire, bows his head for a few moments, and turns to look at me. "For much longer than our people can remember, some of our villagers have always been able to turn themselves into gorillas and live with the gorillas in the forest. After a few days, they turn themselves back into people and return to the village. That's why we've never hunted gorillas here and never will. If we kill a gorilla, then we might be killing one of our relatives. Not long ago, a woman I knew, Lydia Olou, had taken on her gorilla form and was in the jungle when Tanga Senseboy, a hunter from another village, shot her, believing her to be a gorilla. Lydia returned to the village and cried out, "I'm finished." She became very sick and died the next day, though she had no wound. A few days later, when Tanga went back into the forest the gorillas attacked and badly wounded him in revenge."

"Why do your people turn themselves into gorillas?" I ask.

He shakes his grizzled head flecked with curly gray hair. "I don't know."

"But aren't you chief of the gorilla people?"

Zakaria levels me with a hard stare. "Who told you that?"

He is clearly unhappy that I know, and I decide not to involve Aaron. Some of his trackers are from the gorilla people's village, and I do not want them to get angry with him.

"I thought you might have been because you're the chief here."

"We have many villages around here, and not all of them have gorilla people. As I've told you, some villagers even kill gorillas."

"Can children be gorilla people?"

He shakes his head.

"So, when do they become gorilla people?"

He stays silent for a few moments. "When they're in their mid-teens, both men and women can be gorilla people. They come from the same families, and not everyone in the village is a gorilla person."

"Do you have a ritual that welcomes them as new gorilla people?"

He slaps his hand together. "That's all I know. Please don't ask me any more questions."

Zakaria turns to talk to his son in the local language.

"That's more than he's ever told me about the gorilla people," Aaron says. "It's something we'll never understand, because that woman who was shot when she was in her gorilla form did die though there was no visible wound. When we first came here to monitor the gorillas, the villagers thought we were going to capture them and take them to zoos in the United States. I was told that some gorilla people transformed themselves and went into the forest to warn the gorillas to escape. Just after that, an entire gorilla family disappeared, and we've never seen them since. In 2003, there were up to thirty gorillas here, but now there are only eighteen."

What can the outside world do to help the gorillas? A UN-sponsored workshop held across the border in Nigeria in 2006 came up with a regional action plan for the Cross River gorillas' future. "With just under five million American dollars spread over five years we can save them, but so far only a third of that has been committed," Aaron explains. "The funds would provide for many more rangers in the gorillas' habitat, ensuring the forest corridors between the groups remain untouched so that young gorillas can disperse safely and maintain genetic diversity. We can also help villagers living around the habitat develop small businesses so they don't have to poach anymore."

The plan calls for the establishment of the Takamanda National Park on the Cameroon side of the border. The park would include four other localities where the gorillas are found. The experts also considered the development of eco-tourism with gorilla encounters the highlight, but foresaw many problems. "Habituation presents risks to gorillas including the introduction of human illnesses and loss of fear of humans, including hunters," the report noted. Then there is their habitat. "In addition to their wariness of humans because of a recent history of intense hunting, and the dense vegetation of their habitat, the small population is fragmented, and individual groups range over large areas in rugged and inaccessible terrain."

The next morning Ymke and I begin the trek back to Njikwa, the Fon's village, for an overnight stay before returning to Limbe. Aaron will come back a few says later. We know from the radio news that it is safe; the rioters have not barricaded the roads again. Because of my injured big toes, Aaron had arranged for a man with a pony to meet us halfway, at the edge of the forest with its steep rocky slopes. No horse ever goes into the rocky forest, because a fall could break its leg.

As we emerge from the trees, I am praying that the horse and its owner will be waiting for me, because my big toes are burning with pain. Although I have poured antiseptic onto the wounds several times a day, they seem to have been infected with some kind of super-microbe. Both toes are now stiff and red. My heart sinks when the hill comes into view. There is no man, no horse. I will have to walk all the way back to the Fon's village, where we left Ymke's SUV.

About 30 miles across the border in Nigeria in the Cross River National Park, the gorillas also have to contend with the destruction of their habitat by pastoralists whose settlements fragment their home ranges, as well as the anger of farmers when the gorillas raid their banana and plantain crops. In nine of the 10 villages visited by a researcher, the villagers reported raids by the gorillas during the dry season when their own food resources were limited.

In 2006, villagers with a history of hunting gorillas killed one of the rare apes. Rangers visiting the village at that time found that one man had killed another two Cross River gorillas in 2005. The rangers were told that a pregnant woman who had gone into the forest to collect edible leaves was chased by the two gorillas. A hunter nearby heard her screams and killed both gorillas. The gorillas were butchered in the forest and their body parts were carried back to the village to be shared among family members. The villagers were hostile to the rangers asking questions about the killings, and warned them to leave or they would be attacked.

A few weeks after I return home from Cameroon I get an e-mail with good news from Aaron. The prime minister has signed the document approving the establishment of a gorilla sanctuary at Kagwene. With headquarters at the Fon's village, its four rangers will monitor the gorillas daily and keep hunters away. So, for now, the chances of the two families of Cross River gorillas at Kagwene surviving are good. That leaves another 280 Cross River gorillas on their own.

Then, toward the end of the year, Aaron e-mailed more good news. The Cameroon government had just created a new national park to protect the Cross River gorillas. The 261-square-mile Takamanda National Park is expected to curb the destruction of forests and also hunting in the gorillas' habitat. Aaron told me that the park would make it possible for the gorillas to move freely between the park and Nigeria's Cross River National Park just across the border, helping to bring back together much of the fragmented habitat. That would allow greater opportunities of migration between the groups and thus help protect genetic diversity.

Aaron added that a survey of the rugged, steep slopes in the transborder area found that at least one group of 16 Cross River gorillas ranges across the border area, with possibly another two smaller groups living there as well. A separate search over 18 months during 2007 and 2008 in areas not before surveyed found single-night nests, probably those of solitary blackbacks who have dispersed from their natal groups. This showed, beyond all expectations, that the Cross River gorillas might yet survive in the wild.

*   *   *

Borneo calls. I plan to head into the Pacific to one of the world's largest islands. A mite bigger than Texas, Borneo is the home of the only great ape not to live in the wild in Africa, the orangutan. Whereas Africa's great apes are intensely social and live in groups, the "red ape" of Southeast Asia is mostly a peaceful loner.

Its habits and physiology are so distinct that one orangutan researcher joked to me that the only way we can begin to understand its evolution is by assuming that it is an alien creature that came from somewhere in outer space. The orangutans also face considerable threats in their homeland and within a few decades could be difficult to find in the wild.

The Enigmatic Orangutans

BORNEO'S MAN OF THE FOREST

1

Before heading to Borneo, I honor a commitment to go to Afghanistan on assignment for *Smithsonian* magazine. At a police base in the east, a suicide bomber attacks, killing 20 policemen and seriously wounding 32 people, including myself. Two policemen standing by my shoulder are cut down by shrapnel and killed. The bomber is a 12-year-old boy pretending to sell newspapers. But after three months of recuperation, with a dozen pieces of shrapnel still embedded in my brain, chest, and shoulder, I am ready to make the final journey to be among the great apes, this time traveling to the enigmatic and highly intelligent orangutans.

Mankind and orangutans have shared a long history in the steamy Southeast Asian jungles of Sumatra and Borneo. Long ago humans recognized that the red apes were very much like us when they gave them their name, *orang*, a Malay word meaning "man," and *utan*, meaning "forest." Our similarity did not stop prehistoric humans from eating the apes. Cooked orangutan bones dating back 350,000 years have been found in the Niah Caves near Kuching on Borneo's west coast.

I will never forget the time I entered the biggest Niah cave, its earthen floor reeking of ancient times, its giant limestone alcove big enough to hold a couple of 747s, its vast pitted ceiling home to thousands of swifts and bats. It takes little imagination to picture long-ago humans huddled around a fire in the cave at night, chomping on the meat and bones of an orangutan caught earlier in the day in the rainforest that surrounds the caves.

Our closest relative is the chimpanzee, but the great red ape, the tree-dwelling orangutan, is our first cousin. The orangutan split away from our common hominid ancestor up to 12 million years ago. Orangutans we see today date back at least 2 million years to the Pleistocene era, when fossil records show that they ranged from Borneo and Sumatra all the away to Vietnam and southern China.

They have not changed in all that time, perhaps because the rainforests where they live have remained much the same and so there has been little environmental impetus to evolve. The orangutans always had enough food and space. They differ from the African great apes in having coarser fur, whose hue ranges from red to orange; elongated hook-like hands and feet, to better cling to high branches; and more flexible, longer hip joints that allow them to rotate their hips so that they can use their legs up in the trees almost like a second pair of arms. Their big toes and fingers are opposable to their other toes and fingers, allowing a better grip.

They are the largest arboreal great apes, spending most of their lives high in the tropical trees, but, unlike the chimpanzees and bonobos, their great bulk prevents the adults from leaping and swinging through the branches.

Like Louis Leakey's other "angels," Dian Fossey and Jane Goodall, in 1971 a young German-born Canadian woman, Biruté Galdikas, decided to forgo personal comfort and live in the wild to study great apes. She was captivated by the orangutan's closeness to humans, seeing in a photo of a male orangutan his human-like appearance. In 1969 she met Leakey at UCLA, where he was lecturing, and said she wanted to study orangutans in their native habitat. He encouraged her, saying that he believed women to be better observers than men, that they were more patient and "did not excite aggressive tendencies in male primates the way men did."

Galdikas, who spent three decades with the Borneo orangutans, once observed a 250-pound male "briefly hang from a branch suspended by one and then two fingers, his legs dangling in the air as he ate." She found that orangutans choose from more than 400 types of jungle fruit and vegetation including pith, fungus, eggs, seeds, stems, honey, nuts, sap, flowers, and bamboo. Sixty percent of their time is devoted to finding and eating fruit.

Like the local humans, they love durian, a football-sized fruit that tastes like tangy custard but emits a putrid odor. Southeast Asians say durian smells like a toilet but tastes like heaven. The orangutans know the durian is in season between August and December, and will travel long distances to get to trees they are familiar with in their home range. They also know the ripening times of many other fruits, demonstrating that they construct complex mental maps of their home range making use of exceptional memories.

The orangutans favor termites as a tasty snack. Though meat is not a part of their daily diet, Galdikas once saw a big male eat a handful of baby squirrels in one gulp. One of her students witnessed a male consume an adult squirrel, including the skin and bones. Other researchers have seen orangutans eat slow lorises, small mammals that look like big-eyed squirrels, and on one occasion make a meal of a baby gibbon.

Dominant males can weigh up to 300 pounds, though they usually are closer to 200 pounds, and have an astonishing arm stretch of up to eight feet, fingertip to fingertip. A female weighs between 60 and 110 pounds. The massive dominant males, surging with testosterone, develop distinctive, wide cheek pads that look like big black blinkers when they are about 20 years old. Their booming "long calls" are magnified many times by a large air pouch that droops beneath the chin. The call can travel a mile through the jungle as they let any females within range know that the big boy is around and looking for them. The long call also warns any adult male within hearing to clear out.

A Bornean mating pair stays together for up to two weeks in what is called a consortship, having intercourse several times a day, while a dominant Sumatran male and a willing female can remain with each other for several months. A mother suckles her infant for six months, and then gradually introduces the little one to the hundreds of jungle foods including mineral-rich soil, which neutralizes the acidic fruit. She also teaches it to be wary of clouded leopards, mammoth pythons, and the dominant male orangutans.

Unlike the other great apes, orangutans are largely solitary and secretive creatures, with adult males living by themselves most of the time while adult females are usually accompanied by a single offspring, who stays with its mother from birth until it is about seven years of age. Orangutan females generally give birth for the first time when they are 15 or 16 and can lay claim to being the world's most devoted mothers. They keep their young at their side 100 percent of the time for the seven years. Orangutans have the slowest reproductive rate of any land mammal, and eight years can pass between births as the mother patiently protects and instructs her youngster in the myriad thrills and dangers of the wild.

Orangutans are like humans in that the first-time mother shows great curiosity toward her newborn. Researchers witnessed a first-time mother, Jessica, examine every part of her new baby's body. She seemed most amazed by its tiny fingers, gently holding and rubbing them. In turn, her infant licked her mouth and gently bit her face. The baby's reward was Jessica hugging, kissing, and cradling it.

There are two subspecies of orangutan, one found in Borneo and the other across the South China Sea in Sumatra. Both are under severe threat of extinction in the wild. There are only 6,000 Sumatran wild orangutans left, and they are not expected to survive there for too many more decades as new settlements and plantations destroy their habitat. The estimated 50,000 orangutans in Borneo are also listed by the International Union for Conservation of Nature as endangered.

Despite going our separate ways millions of years ago, orangutans and humans do share some similarities that are not found among chimpanzees and gorillas. A female has sex at any time during her menstrual cycle; females have no genital swelling during estrus; they do not knuckle-walk; their teeth and bone structures are more similar to ours, and we both have a particular vein in the arm absent in the chimpanzees and gorillas. But, alone among the great apes, orangutans are quiet creatures and make a noise only when it is necessary.

The flight to North Borneo, at the head of the great island, takes most of one day. As I settle in my seat, I recall one of my most thrilling encounters with the great apes. It was in 1985, in North Borneo, at a sanctuary for young orangutan orphans that had been confiscated by the police from people keeping them as pets. They were taken to Sepilok, a chunk of pristine jungle near the Sulu Sea, a government-funded school for infant and juvenile orphan orangutans, where attendants trained them to return to a life in the jungle. The Sepilok Orangutan Rehabilitation Centre was a 30-minute drive from the coastal town of Sandakan and was hardly known then, although in the past decade it has become a star on the Southeast Asian tourist circuit.

Back then, a small compound at the jungle's edge held about a dozen young orangutans, earnest-faced creatures, far more solemn than chimpanzee young and even small gorillas. Each day they climbed the rainforest trees by the center with a ponderous grace and, on the ground, wrestled each other, in slow motion when compared with the hyperactive and excitable young chimpanzees. The orangutans had deep, soulful brown eyes and their expression rarely changed, earnest and focused like the smartest kid in school. As with many of the young great apes in sanctuaries near their native habitats in Borneo and Africa, they had seen their mothers brutally killed up close, usually by gunshot to bring them down from the trees, the only way poachers can grab the baby.

At Sepilok, a cage in the compound was empty save for a very large python that lay almost comatose on the concrete floor behind bars. "We used to keep some of the little orangutans in this cage," a keeper told me, "until we came one morning a few months ago and found the python in the cage with a couple of bulges in its stomach. It had got in overnight, attracted by the chance for an easy meal, and swallowed two young orangutans. We put more bars in the cage and kept it here, and made new python-proof cages for the orangutans."

The little orangutans spent up to seven years living in the compound, learning how to climb trees in the nearby rainforest, select the proper jungle foods, and coexist with their own kind. "Since they were young, most have been living with human families, and they think they're a kind of human," the keeper said. "We have to get them used to the idea that they're orangutans, and sometimes that can be tough on them."

Sepilok's young males, in particular, had a mischievous streak and played tricks on visitors. One little terror named Datuk (a Malay honorific similar to the British title "Sir") reminded me of those three- or four-year-old boys who enjoy causing havoc and mayhem around them. Somehow, somewhere, Datuk had discovered that female humans hated having their skirts pulled up, and I watched with amusement as he stalked every female visitor to Sepilok wearing a skirt. He would creep up behind them, grab the skirt at the hem, and yank it up to expose their panties. As each victim screamed in horror, he escaped up a nearby tree.

Datuk's behavior was not unusual at Sepilok. In their book *The Orangutans* (2000), orangutan researchers Gisela Kaplan and Lesley J. Rogers tell of juvenile males at Sepilok acting like teenage boys: "They all tried to measure their strength, and were on the whole, more dangerous than the rehabilitant females. They loved playing tricks and setting ambushes for people." They grabbed handbags and pulled down tripods. The researchers imagined one young terror, Raja, "playing the role of an aspiring gangster boss in 1920's New York."

Such a feisty attitude is not possible in the wild because at that age, young males are still alone with their mother. The unanswered question is whether the disruptive behavior comes from the young males' experience of being dragged from the body of their mother when an infant, and losing forever her wise guidance, or whether it is wired into them. The young male bonobos at the Kinshasa sanctuary, deprived of their mother's care, were also disruptive, and yet I witnessed no such willful behavior among them in the wild.

Once the many years of training were up at Sepilok, the orangutans were taken to the forest near the compound and released into the wild. To make sure they did not starve, field assistants each day took buckets of bananas about a mile into the jungle to a small wooden platform, a kind of halfway house, where those who were hungry could come and eat. It was there that I had one of the most memorable of all my experiences among the great apes.

North Borneo is one of the muggiest places on Earth, and sweat streamed down my face and body as I trekked through the jungle with a field assistant. Waiting in ambush were armies of rubbery reddish leeches. Streamers of vines dangled from the moss-swathed giant trees, and as I brushed past them, leeches landed on me and dug into the flesh with their three-pronged fangs. Only when they began to fall off me, engorged into little bubbles with my blood, did I see them. The field assistant smiled helplessly as I shuddered each time a swollen leech dropped to the forest floor.

After a 30-minute trek through the rainforest, we reached a rickety, narrow wooden platform, about 15 feet high, and climbed the rungs to the top. The keeper placed a bucket of bananas he was carrying on the boards and uttered a high-pitched yell, a signal to any hungry orangutans nearby that food had arrived.

He called for about five minutes, and then across the jungle canopy I saw a bundle of reddish-brown fur at the top of the trees. The forest shook and shivered as if a heavy wind were buffeting it as the orangutan approached. Unlike chimpanzees, because of their weight orangutans do not leap from branch to branch, but they have evolved their own way of efficiently moving through the jungle trees. Near the canopy, an orangutan perches on the stout outer branch of a pole-like tree and uses its bulk to move the branch forward and backward in a rocking motion, again and again, in longer and longer arcs, until the branch gets near enough to a solid branch in the next tree for the orangutan to grasp it and clamber onto it.

They do not move very fast, but then there is no need for speed. It took the orangutan some time to reach us, but on the platform I saw it was an adult female. Dominant male orangutans are more than twice as heavy as females, and are swathed in dozens of reddish strands of hair that give them the look of bedraggled robed monks.

My camera bag was by my side, filled with about $5,000 worth of gear, and as the orangutan began to eat the bananas, I took pictures with a wide-angle lens from just three feet away. I became frustrated when she had eaten half the bananas and showed no sign of stopping because the field assistant had told me there were other orangutans not far away, including a big dominant male. I wanted pictures of more than one orangutan on the feeding platform. I was so focused on my aim that, without thinking of the consequences, I leaned over, took hold of the bucket, and placed it on my side of the platform to stop her from eating any more bananas.

A female orangutan is six times stronger than an adult male human, and she could have bashed me or thrown me off the platform. At the least, she could have grabbed my camera bag and sent it flying into the jungle. But this great ape was a diplomat. She captured my complete attention with a look from her deep brown eyes that I clearly remember to this day. Anyone who has been up close with a great ape knows that they have eyes that seem human. Or perhaps I should say that we humans have eyes that are eerily like those of the great apes, and of all the great apes, the most thoughtful I have seen belonged to orangutans.

On my most recent trip to Borneo, I would meet a French veterinarian, Marc Ancrenaz, who had spent the past decade directing a project that protects and studies several hundred wild orangutans in a North Borneo rainforest not far from Sepilok. He told me that because orangutans communicate much less with their voice than the other great apes, they have developed a much higher level of eye-to-eye communication. "It's so much more sophisticated than any other creatures', even humans', that they seem to be aliens," he told me. "I've seen them countless times in the jungle communicating with each other only through their eyes."

Moviegoers who saw Clint Eastwood's 1978 film *Every Which Way But Loose* would know what he means. Clyde the orangutan, Clint's sidekick, fixed the superstar with some of the most baleful, quizzical, humorous, and wry looks ever seen on the screen.

All those years ago, the orangutan on the feeding platform, with perfect timing, gazed at me for a few moments with her keenly understanding eyes, and then reached across the divide, took hold of my camera bag, and placed it by her side. She followed this with another meaningful look into my eyes. Even to a dumb human like me, it was obvious what she was signaling. I had taken away something she valued, and so she had taken away something I valued. Give mine back, she seemed to say with her eyes, and I'll give back yours.

I kept my eyes on hers, nodded, and put the bucket of bananas back beside her on the platform. She looked at me again for a few moments, almost as if she were prolonging the game for her own pleasure, and then placed the camera bag at my side and went back to eating the bananas. It was a singular encounter between two species of great apes, and I have never forgotten her uncanny ability to silently trade with me.

On the plane on my most recent visit, as we fly north over Borneo, I smile at the memory of that long-ago encounter and then look through my file for a printout of the world-famous San Diego Zoo's website entry on the red apes. It seems that the intelligence of the female who traded with me was not unusual. The website notes: "Orangutans have even been known to watch villagers use boats to cross the local waterways, and then untie a boat and ride it across the river on their own." Another orangutan who lives near a settlement sometimes unties a canoe by a stream, enters it, and propels it around for her own enjoyment, paddling with her hand.

The San Diego Zoo has its own comparison between the great apes, labeling orangutans relaxed problem-solvers: "If a chimp is given an oddly shaped peg and several different holes to fit it in, the chimp will immediately try shoving the peg into various holes until it finds the hole the peg fits in. An orangutan will approach the challenge quite differently. It will stare off into space, or even scratch itself with the peg. Then, after a while it will offhandedly stick the peg in the correct hole while looking at something else that has caught its interest."

This reminds me of what Carsten Knott, head keeper of the great apes at the Frankfurt Zoo, told me. If you give an orangutan a screwdriver, 30 minutes later you'll find that it has picked the lock on its cage and escaped. The San Diego Zoo's Ken Allen (a strange name for an orangutan), according to the zoo's website, would "unscrew bolts with his fingers, reach around to unlatch things, climb up a steep incline by the back of his enclosure to slip over the wall.

"Every time his keepers blocked his escape routes he would figure out new ones. It seemed more like a game to him, to see whether he could outwit his keepers because he never minded being led back to his enclosure. That gave him the chance to figure out a new escape route."

In mid–2009, Karta, a 27-year-old orangutan at the Adelaide Zoo in Australia, used a stick to deliberately short-circuit the electric wires surrounding her enclosure by twisting them together and cutting off the power supply. She then built a makeshift ladder from shrubs and sticks to escape over the wall. When she was discovered, keepers led her back to her enclosure. "We're looking at an animal that is extremely intelligent," said Peter Whitehead, the zoo's curator. "She's always trying to outsmart the keepers and today she showed a touch of genius."

In *The Octopus and the Orangutan*, Eugene Linden tells of the ingenuity of orangutans in Borneo who made insulating gloves out of straw so they could climb unharmed over electric fences. They can also use them for grasping spiny fruits and branches. Linden also describes the feat of a 12-year-old female orangutan named Unyil, which involved "tool-making, tool use and deception as well as great dexterity."

Unyil was at a sanctuary in Indonesian Borneo after she was rescued near death. She had been kept as a pet by a colonel. Her rescuers found that "Unyil was kept in quarters so small that she could not stand up and her arms stuck out from both sides." When she was released from her cage, the first thing she wanted was for someone to hug her.

Suspected of suffering from tuberculosis, she was kept in a quarantine compound where a pile of fruit was stored out of her reach. When the fruit began to disappear, workers were accused of the theft. But then the chief technician observed Unyil yank long hairs from her shoulder and knot them together into a rope five feet long: "She tied the rope around the end of a banana peel. With this weight now securely fixed she held one end of her rope and tossed the weight into a pile of the fruits. She repeated the process until she snagged an apple, and by jerking the line pulled the fruit within reach of her long arms."

Unyil knew she was taking something that belonged to someone else because she gobbled down the apple, seeds and all, when she was caught. A mango she stole by the same method presented a more difficult problem. "Her solution was to tip the lid of the septic tank and toss the large seed into a place where no one would look for it."

At the Leipzig Zoo in Germany, a researcher placed a banana beyond the bars just out of an orangutan's reach in an experiment to see whether orangutans plan ahead. The researcher gave the ape a hooked rod, which it immediately poked through the bars to snag the banana and draw it near enough to grab. The orangutan kept the rod with it all day and took it to bed that night. The next morning it carried the rod down to the bars and used it to hook a new banana.

Also at Leipzig, at the Max Planck Institute for Evolutionary Anthropology, researchers placed a peanut in a long, narrow, transparent tube bolted to a length of wood to hold it vertically in place. They filled one-quarter of the tube with water so that the peanut floated. A jug of water was placed close by on a table. A four-year-old boy was asked to get the peanut using anything in the room. The boy tried to reach it with his hand but the tube was too narrow. After four minutes he gave up.

The next boy, age six, after some thought tried to break open the tube. He failed. An eight-year-old girl puzzled over the problem for a minute and then used the water in the jug to raise the tube's water level to get the peanut.

The researchers took the tube to the Leipzig Zoo and tested a female orangutan, placing the tube near a water fountain from which she drank. The orangutan looked at the peanut in the tube for about five seconds, and then took water from the fountain into her mouth and spat it into the tube to raise the water level. The second mouthful brought the peanut to the top of the tube. In effect, the orangutan had used water as a tool.

In addition to being the most original creative thinkers among the great apes, orangutans are the greatest imitators. At a research camp in

Borneo, a male named Apollo Bob watched a guard play his guitar on a verandah. When the guard went to get a cup of coffee, Apollo Bob grabbed the guitar and disappeared. After searching the camp for the guitar, the guard went back to the house and heard the plunkety-plunk sound of a guitar. In a nearby bunkhouse he found Apollo Bob imitating him, gripping the guitar with one hand and using the other hand to play it.

Biruté Galdikas and another Canadian orangutan researcher, Anne Russon, documented over 30 separate cases at an orphan rehabilitation center where the orangutans imitated human behavior. They copied siphoning fuel from drums into cans, fanned the embers of a fire with a lid, imitated a gardener weeding, tried to paint a floor, and copied workers building a bridge over a stream with logs. The orangutans scrubbed themselves with soap, and one female tried brushing her teeth. She put the toothbrush in one side of the mouth and moved the bristles to and fro. When finished, she spat the used toothpaste onto the ground just like the humans she was copying. At the Great Ape Trust in Iowa, researchers were delighted to find that an orangutan there had learned to whistle by imitating them.

An orangutan named Ujian at the Heidelberg Zoo in Germany developed his own multiple-tone whistling technique, which proved so popular that a CD was released mingling his unique whistling with jungle sounds. The zoo's ape expert, Bernd Kowalsky, described Ujian as "a clever and talented fellow."

Orangutans are more creative in their sexual practices than chimpanzees or gorillas. In contrast to female chimpanzees, whose cycle comes around every 35 days or so, orangutan females, like humans and bonobos, are generally sexually receptive all month. It does not seem coincidental that only they among the great apes regularly have sex face to face, enjoy same-sex relations, and even indulge in oral sex.

One time at the San Diego Zoo, I watched through a clear glass panel, at a distance of about half a yard, a young female approach a huge male who was reclining with his back against the enclosure wall. A keeper had told me that she was besotted with him, but he preferred another female. Undeterred, she sat between his legs and bent over, and I clearly saw her take hold of his penis, put it in her mouth, and manipulate it. This prompted the male to have an erection, but he turned his face toward the glass panel, affected what seemed to me a bored look, and did not hide his disinterest in the female.

In the wild, males have also been observed practicing oral sex in foreplay before intercourse, hugging and kissing the female and licking her genitals. Alone among the great apes, the orangutan practice the face-to-face position exclusively. Sometimes the male is on top, and sometimes it is the female straddling him and making thrusting motions. And, like bonobos, orangutan males are sexually precocious, with infants under the age of one often masturbating.

Because of their solitary nature and remote jungle habitat, orangutans are among the hardest of the great apes to study, and that is why much less is known about their lifestyle than that of the gorillas, chimpanzees, and bonobos. British zoologist John MacKinnon wrote:

> *Steep terrain and thick vegetation made travel difficult. The dense canopy and the height of arboreal animals from the ground made observation conditions generally poor. As orangutans were shy, quiet and dispersed, they were hard to locate and easy to lose again. Heat, high humidity, rainstorms, floods, and gales added to the discomfort and hazard of fieldwork. Leeches, wasps, mosquitoes, horseflies, and ticks added further problems, and bears, wild pigs, snakes, crocodiles, elephants and, in Sumatra, tigers, also produced anxious moments.*

Even so, Western scientists have long known about the orangutan's extraordinary intelligence. Jacob Bontius, a Dutch physician stationed in Batavia (now Jakarta) in the Dutch East Indies, in 1658 described an encounter with orangutans, "this wonderful monster with the human face." His translator, James Burnett, wrote: "He relates that he saw several Orang Outangs, of both sexes, walking erect: and he particularly observed the female, that she shewed [*sic*] signs of modesty, by hiding herself from men whom she did not know. And, he adds, that she wept and groaned, and performed other human actions: So that nothing human seemed to be wanting in her, except speech. The [native people] say in truth, that they can talk, but do not wish to, lest they should be compelled to labor."

Burnett, who was a linguist, claimed in a book examining the origins of language that the orangutan was a "man," but one without language:

> *I still maintain, that his being possessed of the capacity of acquiring it, by having both the human intelligence and the organ of pronunciation, joined to the dispositions and affections of his mind, mild, gentle, and humane, is sufficient to denominate him a man. And it is very extraordinary to suppose that he is of another species, not because he wants any organs that we have, such as the organs of speech, but because he does not make the same use of them.*

He added his belief that orangutans did not speak because they were solitary creatures, and "nor do they practise the arts."

What Burnett did not know was that orangutans, and all the great apes, are not physically equipped to speak like we humans. Robert Shumaker, the American evolutionary biologist, noted: "Humans and the other great apes have all the same anatomical features related to speech production, but in a slightly altered formation. This relatively small difference has profound consequences related to vocal abilities. Human and nonhuman great ape infants have their larynx in a relatively high position in their throat. This limits the types of sounds that can be reproduced."

By the age of three months the human infant's larynx descends into a lower position in the throat, allowing the baby to reproduce a wide variety of sounds. However, as great ape infants grow, the larynx does not descend, and this limits the sounds they can make. They cannot reproduce consonants and complex vowels and are limited to basic vowel sounds.

In 1871, when Charles Darwin published *The Descent of Man*, he shocked people by suggesting that humans were descended from apelike ancestors. Thirty-three years earlier, he had visited an orangutan named Jenny, who was a star attraction at the London Zoo. To ward off the frosty London weather, she was being kept in the heated giraffe house.

In a letter to his sister Susan, Darwin wrote:

. . . the keeper showed her [Jenny] an apple, but would not give it to her, whereupon she threw herself on her back, kicked & cried, precisely like a naughty child.—She then looked very sulky & after two or three fits of pashion [sic], the keeper said, "Jenny if you will stop bawling & be a good girl, I will give you the apple."—She certainly understood every word of this, &, though like a child, she had great work to stop whining, she at last succeeded, & then got the apple, with which she jumped into an arm chair & began eating it, with the most contented countenance imaginable.

Darwin wrote in his diary: "Let man visit the orang-utan in domestication . . . see his intelligence . . . Man in his arrogance thinks himself a great work . . . More humble, and I believe true, to consider him created from animals."

Not everyone was captivated by our resemblance to the orangutan. When Queen Victoria visited the same Jenny at the zoo in 1843, she exclaimed that the great ape was "frightfully and painfully and disagreeably human."

2

In 1971, Biruté Galdikas journeyed by jet plane, speedboat, and dugout canoe to the jungle that was to become her home for three decades. The farther away from human settlements she went, the better she could find orangutans in the wild. Despite the harsh conditions of the tropical jungle, she set up a 20-square-mile research station in Tanjung Puting National Park and named it Camp Leakey.

Located on the peninsula on the south coast of Indonesian Borneo, Tanjung Puting National Park consists of 988,000 acres with huge tracts of tropical and mangrove forests containing enormous rookeries. It has more than 400 species of trees, whose leaves and bark are favored by orangutans as food. The camp was set up in a clearing at the edge of a swamp. At this time, knowledge of the reclusive red ape's lifestyle was scant, and that excited and motivated Galdikas. "I was born to study orangutans," she wrote.

Her first glimpse of wild orangutans was brief, a blur of orange fur, a mother and juvenile high in the canopy. They hooted, squeaked, and rained branches down on the intruder. Galdikas was in a canoe and could not get closer because of the shoulder-deep swamp water. It was no place to step out of the boat. She already knew about the jungle's "fifteen-foot-long human-eating crocodiles; poisonous cobras, vipers and kraits; pythons as thick as a person's thigh; sun bears with unpredictable tempers; bearded pigs whose razor-sharp tusks could shear a human being in half; and mysterious clouded leopards that pounced out of nowhere."

Galdikas refused to give up. For months she tried to get close to the orangutans but as soon as they saw her, they fled. Then, on Christmas Eve at Camp Leakey, came a breakthrough when she spied a female orangutan climbing a tree on dry ground. The orangutan squeaked, hooted, and dropped branches but did not flee. Galdikas named her Beth and her infant Bert. Beth peered at Galdikas for about a minute, and then began building a day nest from the branches and leaves for a siesta. Three hours later, when Galdikas next saw her, she was chewing on a chunk of bark.

Galdikas followed Beth and Bert for 10 hours that day, filling nearly 30 pages of her notebook with their behavior and the interactions of mother and son. She noted: "This is what I had come to Borneo to do. The cat's cradle, lining my work to Jane Goodall's study of chimpanzees, Dian Fossey's study of gorillas and Louis Leakey's vision of a living picture of human evolution, was coming into focus."

By the fifth day, Beth had become used to Galdikas following her, which seemed to the researcher a small miracle. Nothing of much account happened over the following days as Galdikas followed Beth and Burt, but that in itself was a revelation. Orangutans live in slow motion, a measured and mostly serene life, when compared with other great apes. Jane Goodall later told Galdikas that it took two years to witness as much orangutan behavior as Goodall saw in two hours with chimpanzees.

Not all orangutans were as benignly accepting as Beth. One female she named Cara led Galdikas into a trap in the swampy jungle. Suddenly, she heard a crack and looked up to see Cara trying to break a large branch. The orangutan stared Galdikas straight in the eyes as she aimed the branch at her. "She wants to kill me," Galdikas thought. Her body shook, her face flushed, and her legs quivered as she expected to die. But the branch was too wet; Cara was unable to snap it, and she gave up. Galdikas later saw male orangutans with their superior strength using branches as weapons. A male she named Howard once broke a branch without warning and sent it tumbling toward her.

During two decades in the jungle, Galdikas learned to treat the huge dominant males with extreme respect. "When faced with an angry three-hundred-pound wild adult male orangutan, with an arm span of perhaps eight feet, I try not to react," she noted. "I force myself to do nothing. Even if people with me turn and flee, I remain where I am, silent, immobile, expressionless. The orangutan must make a split-second decision; engage me in combat and risk injury, or withdraw into his meditative state and save himself for the real battle."

Galdikas saw dominant males challenging each other to a fight only a handful of times in all the decades she was in the jungles. The two behemoths will shake branches at each other and try to warn the other off with stares and booming long calls. Most of the time, one of the combatants will flee. Once, though, as she was wading through a waist-deep swamp in search of a female she had been following for days, Galdikas spied two big male orangutans face-to-face and glaring at each other. They were readying to fight over the female. Suddenly, they lunged, snarling and wrestling as they tried to bite each other on the head.

One of the orangutans broke off the fight and fled, leaving the other male peering at Galdikas. He swung down on some vines until he was just a yard above her head and glared into her eyes. He was so close she could smell his sweat. His message was clear, she says: "Leave me alone." She did. In a similar fight, a dominant male named Kakusi, who lived in the forest near Camp Leakey, almost tore off the face of another big male.

But death rarely results from such fights. "In the decade we've been observing wild orangutans, we've never seen one kill or even seriously injure another," Marc Ancrenaz, the French veterinarian, would later tell me. "There is a test of strength, and they slap each other about, but once it's clear that one is stronger than the other, the weaker male goes away."

The fight over a female that Galdikas witnessed was rare. She has found that an adult male's range is at least 30 square miles, and he can go for months without meeting another big male. Their relatively gentle nature is surprising, because orangutans evolved in tougher habitats than those of the other great apes in much of Africa.

When the nineteenth-century naturalist Alfred Russel Wallace asked a Dyak chief in Sarawak about the red apes' enemies, he replied that in a fight between an orangutan and a crocodile, the orangutan "gets upon him and beats him with his hands and feet, and tears him and kills him." Now, the orangutans are facing a far tougher enemy, modern man, who has put these forest giants in extreme peril.

Loggers have been plundering Borneo's rainforests since the nineteenth century, but orangutans in the wild there are now under attack from a relatively new threat—the wholesale destruction of their habitat for settlements, agriculture, and plantations. Slow and awkward on the ground, the orangutan moves easily along the branches just below the rainforest canopy; and when the trees are cut down as lumber or to clear the land for people and crops, such as the ubiquitous and invidious palm oil, there is no other place the orangutan can go. This clumsiness on the ground makes them vulnerable to poachers, who steal infants to sell for thousands of dollars on the pet-trade black market.

Borneo rainforests are one of the wonders of the natural world. They support at least 15,000 plant species, including more than 2,500 kinds of orchids, as well as many of the word's rarest fauna and flora. There are flowers as big as deck chairs, one of the world's largest butterflies, pygmy elephants, flying snakes, huge crocodiles, rhinoceros hornbills, a true rhinoceros so rare that there are just a few dozen left in the wild, and the orangutans.

The rainforests once covered 90 percent of the huge island, but in the 1970s, the Indonesian government shipped a million trans-migrants under the *transmigrasi* policy from overcrowded Indonesian islands such as Java to Indonesian Borneo to inhabit settlements hacked out of the jungles. Bulldozers and chainsaws tore down vast tracts of forest, causing thousands of orangutans to die.

The jungles are now being chopped down and burnt to ashes mostly because of a single plant—palm oil. This destruction ranks as one of the most scandalous acts of environmental vandalism ever on Earth. Indonesian and Malaysian farmers have planted millions of acres of palm oil. A producer can get up to 5,000 liters of oil per hectare each year, and the crop is 10 times more productive than soya beans.

Palm oil has become an essential in modern life, used in 10 percent of all the products found in an American supermarket. According to the conservationist group Friends of the Earth, these include staples such as soap, chocolate, toothpaste, biscuits, and cosmetics. Indonesia and Malaysia together provide 86 percent of the world's palm oil, with the former boasting more than 20 million acres under cultivation. In 2008, Indonesia produced 19.7 million metric tons of palm oil, while Malaysia was not far behind with 17.4 million metric tons. In the same year, the rest of the world manufactured only 5.8 million metric tons.

As a result, Borneo's rainforests are being hacked down at a startling rate, to be replaced with millions of rows of palm oil trees. The United Nations has estimated that based on current trends, "98 percent of Indonesian rainforest would be gone by 2022, and the rate of illegal intrusion by large logging companies into supposedly protected national parks would severely degrade those areas by 2012."

Over the past century, orangutans have lost over 90 percent of their population. Greenpeace USA claims that there are only 15,000 to 25,000 left, although it is more likely to be closer to 50,000. The world environment is also suffering because of this massive vandalism. Indonesia now ranks third in the world after the United States and China in carbon emissions, mostly because of the deliberate burning down of huge swaths of Borneo's rainforests to create palm oil plantations.

It is an enormous scandal, and involves high-level government corruption. Indonesians call it KKN, or *korupsi, kolsi*, and *nepotisme*—corruption, collusion, and nepotism. In just one of countless recent cases, the Indonesian forestry minister, Malam Sambat Kaban, intervened to drop criminal charges for illegal logging against Adelin Lis, a prominent businessman. Lis had been illegally tearing down rainforests and had paid bounty hunters to kill the orangutans inhabiting them.

Deprived of their habitat as "pests," the orangutans are forced to forage among the new palm seedlings to survive. One infant lost his hands when bounty hunters chopped off his mother's head with a machete. A bounty hunter explained, "Because orangutans are six times stronger than even a well-trained athlete, the only way to subdue them is with a series of violent blows to the head."

The Indonesian government has set up several national parks to protect the remaining orangutans. But while the politicians in Jakarta spout high-sounding plans for saving the red ape, at the ground level corrupt officials look the other way when the rainforests are illegally chopped down by logging companies, or when young orangutans are captured and spirited out of the country to be sold at high prices abroad. Four to five orangutans die for every baby reaching the pet-trade market, and a high proportion of these infants do not survive the harsh journey to market due to poor care, disease, injury, and psychological trauma.

From the time she arrived in Borneo, Biruté Galdikas was moved by the devastation caused to the orangutan population by logging and the destruction of their habitat. Although her primary purpose was research on their lifestyle in the wild, she decided to rescue wild-born orphaned orangutans, teach them forest skills, and then return them to the wild. She had been at Camp Leakey for only a month in 1971 when her assistant brought in a six-year-old female Galdikas named Akmad. Villagers had probably killed her mother and had been keeping her in a cage.

"How human she appeared, like an orange gnome, with her intelligent, quietly inquisitive face," Galdikas wrote in her book *Reflections of Eden* (1995). She let Akmad hang around Camp Leakey, even sharing her afternoon meal of rice, sardines, and sweetened tea. When Galdikas moved camp, Akmad disappeared. Six months later she turned up at the new camp starving with a gaunt face, all skin and bone. The allure of the jungle was too strong, though, and Akmad appeared less frequently, at one time disappearing for a year. When she was 15, Akmad became pregnant and gave birth to a baby male.

Akmad proved to Galdikas that she could rehabilitate orphan wild orangutans and return them to the jungle, and this became a major part of her work. In 1986 she co-founded the Orangutan Foundation International, whose headquarters are in Los Angeles. It funds research work at Camp Leakey and the rehabilitation of orphans.

Galdikas went against the wisdom of the times that claimed orphan orangutans could not be returned to the wild. Barbara Harrison, wife of the curator of the famed Sarawak Museum, was one of the first to try, in the 1960s, when she and her husband were given young orphan orangutans to raise, hoping to return them to the wild. She wrote in her memoir, *Orangutan*: "The babies that had come to us were in their helpless stage, when they are still entirely dependent on their mothers. Being adaptable and intelligent animals, they had accepted the human being as a suitable substitute mother and started to learn and live in a human way." Harrison did not believe she could teach them the skills necessary to live in the jungle: "Unless I learned to be an ape first, it would be impossible for me to teach an ape baby."

Galdikas set out to prove the skeptics wrong, and she established a flourishing rehabilitation center near Camp Leakey that has since cared for hundreds of orphan orangutans. However, in the late 1990s, she was plunged into controversy when articles claimed that she kept more than 100 little orangutans at her Borneo home and that at least 21 orphans had died from medical neglect. She was breaking the law, because Indonesia forbade anyone to own, sell, or buy an orangutan.

Research assistants also claimed she had largely given up her research work in the early 1990s, with her observations stacked in rotting piles that were being eaten by cockroaches. Earthwatch, an American-based eco-tourism organization that sends mostly Western volunteers to work on projects such as saving the Tsavo lions in Kenya and cheetahs in Namibia, cut its connection with Galdikas. It claimed that she had failed at the time to produce any reports on orangutan ecology.

Galdikas ignored the Earthwatch criticism but hit back in a letter to *Newsweek*, stating, "A handful of disgruntled individuals, jealous former volunteers, began to circulate rumors about my work that were outlandish." One report claimed that after the actor Julia Roberts had cuddled a three-month-old orphan at Galdikas's rehabilitation center near Camp Leakey and fed him a bottle of milk, Hughie died of an infection a few months later. Galdikas shrugged off the report, claiming that Hughie was still alive.

She could not so easily shrug off the criticism contained in a report in 1998 by an Indonesian government team sent to Camp Leakey to investigate Galdikas's care of orphan orangutans. Behind her house they found 89 orangutans, which were held in four "secret lodges," in conditions that did not meet health standards. Each isolation cage "measured 1.5 meters x 1 meter x 1 meter," and such a cage held three to five orangutans. Some of the cages had feces in them. A few small sheds contained a total of 10 isolation cages. Several baby orangutans were observed suffering from diarrhea and one baby was feverish. The report added that foreign tourists were being used to care for sick babies, including three young orangutans who were found in a hut without ventilation or light.

Indonesian government officials wrote to Galdikas warning that she was violating Indonesian law in keeping orangutans in her house, and that the care they got was substandard. Galdikas in reply ignored the damning accusations and countered that the investigation team rudely entered people's homes. She stated that she cared for the orangutans, "and they were able to live in the forest and former primary jungle."

Galdikas was allowed to keep Camp Leakey and her Orangutan Care Center operating, but her critics lined up to damn her rehabilitation program as outmoded and harmful. The Canadian primatologist Anne Russon, who had worked with Galdikas at Camp Leakey, believes that the "ex-captives," as she calls them, should be taught how to live in the forest rather than with each other, as happens there now. In her experience, orphans at rehabilitation centers become "moochers" rather than forest dwellers.

Russon also attacked Galidikas for keeping the orphans around the camp and allowing tourists to come close to them, stating that "the problems generated by Galdikas' style of rehabilitation outweigh the benefits." Russon believes that rehabilitated orphans "should be released only into areas free of wild orangutans," to minimize the chance of passing on diseases for which they have no immunity. She also advocates that the orphans' contact with and dependency on humans should be curtailed.

This seems sensible if you turn your back on the money-making opportunity of offering visitors, local and foreign, the experience of being close to young orangutans. One Internet tour site offering trips to Camp Leakey and the Orangutan Care Center states: "You can interact, hold hands, feed, carry and cuddle the 'great man of the forest.'" It adds that the orphans "often respond warmly to human interaction." This is the opposite of what orphan orangutans need if they are to be rehabilitated back into the wild, but if you take away the hands-on experience, the visitors' numbers will plummet.

Camp Leakey attracts more than 1,000 foreign tourists each year. It has an attractive Internet site with pictures of cute little orangutans, and offers tours to the remote site in Indonesia through Orangutan Foundation International. It also peddles posters, CDs, videos, and bumper stickers featuring the orphans. Donors can even become "foster parents" of individual orphans for $75. The site's latest entry gives few details of how the orphans are rehabilitated but says that 340 are being cared for there. This results in harmful overcrowding, which makes it difficult to train the orphans for life in the jungle, a return to their innate solitary nature.

While Camp Leakey continues to thrive as a money-making business, Galdikas now spends much of her time in her native Canada and in the United States, giving talks. Aside from the controversy, she and her research never captured the attention of the Western media as widely as Goodall and Fossey. Although popular in zoos, the serene loner red ape in Southeast Asia's remote rainforests never gripped the Western imagination the way chimpanzees and gorillas did in the jungles of Africa.

Africa, with its "wild tribes" and charismatic animals, was a far more seductive lure for Western explorers. Movie directors were attracted by the same bait and used the continent's great apes to both delight and terrify their audiences. In contrast, the orangutans, rarely seen in their jungle-top abodes and hard to transform into simplistic caricatures of angelic and demonic demi-humans, have never held the same appeal.

3

To be with orangutans in the wild, and witness efforts to rehabilitate orphans, I planned to go to Indonesian Borneo and then on to Sabah in Malaysian Borneo. But at the Indonesian consulate in Sydney, the press attaché warns me that the Foreign Ministry in Jakarta is refusing media visas for orangutan coverage. She says the Indonesian government believes there has been far too much bad foreign media publicity over the destruction of forests for palm oil plantations and the corruption linked to it. She is right—my visa request is refused.

Sabah is a welcome contrast to Indonesia. It offers some hope for its wild orangutans, because the state and federal governments are co-operating with international wildlife groups to protect its remaining red apes. Borneo is divided into Kalimantan (an Indonesian province), which takes up most of the southern and eastern parts of the island, the much smaller Malaysian states of Sarawak and Sabah as well as tiny Brunei. The first three boast huge stretches of palm oil plantations, replacing immense rainforests where hundreds of thousands of orangutans once lived but from which they are now gone forever.

As the plane approaches Kota Kinabalu, Sabah's capital, it is like coming home. I know Sabah well because I first came here four decades ago to meet the parents of my soon-to-be wife. I have been back many times, and these visits gave me an insight into the destruction of the state's hardwood rainforests. In the 1960s the politicians introduced legislation granting themselves the licenses to cut down much of Sabah's forests. The valuable logs were sold, principally to the United States and Japan.

It was legalized multibillion-dollar robbery, stealing from the public purse, with some senior politicians each banking hundreds of millions of dollars offshore in Hong Kong and Switzerland. To the south, in Sarawak, the leaders lusted for more, becoming billionaires from their hardwood licenses with the consequent disastrous destruction of much of their primary rainforests. No one knows how many orangutans lost their lives to fatten the bank accounts of the Bornean leaders, but the toll was enormous.

The leaders of the Malaysian federal government in Kuala Lumpur turned a blind eye to the plunder as long as the Sabahan politicians pledged allegiance. I have heard senior Malaysian federal politicians sneer at Sabah as "The Wild West" and scorn the local politicians as country bumpkins. But over the past decade a new generation of Western-educated politicians, while still profiting from the continued plunder of Sabah's forests for their own personal gain, have given some support to the preservation of the state's wildlife.

The jet swoops low over the glinting coral islands and aquamarine waters of the bay fronting Kota Kinabalu and lands at a modern terminal. The plane is packed with foreign tourists come to experience Sabah's abundant wildlife and some of the planet's best dive sites. About 700,000 foreign tourists now visit each year, and a major attraction is the young orangutan orphans at the Sepilok Orangutan Rehabilitation Centre. On the way to the immigration counter, we pass several tourist posters showing infant red apes staring at us with compelling eyes.

On my first visit, Kota Kinabalu was a sleepy city, snug in a tropical torpor, its city center mostly two- and three-story concrete blockhouses thrown up after the Japanese air force flattened the city during World War II. But today, with a sizzling economy fueled by palm oil, logging, and tourism, the city is crowded with new skyscrapers and shopping malls. The streets are clogged with cars. The local newspapers, as ever, are filled with racy accounts of the tangled alliances and power struggles of the political bosses.

The next morning, a pall of gray smoke blankets the city and visibility struggles to reach beyond 100 yards. Even the gathering heat of a shadowy sun cannot burn it off. My eyes sting from the smog. "The Indonesians down south are burning forest fires again," a longtime friend, Ahmad, tells me as we drive out of Kota Kinabalu. For several months a year planters in Indonesian Borneo burn off massive stretches of virgin forest to plant palm oil, and the heavy clouds of smoke sometimes stretch for more than 1,000 miles, fogging up even the streets of distant Kuala Lumpur across the South China Sea. Millions of Asians are adversely affected by the smoke from the fires.

Our destination is Sandakan, a seaside city not far from Sepilok and the Kinabatangan River, a stronghold of wild orangutans. On a smooth asphalt road the journey takes just under six hours, and for much of the way the landscape is swathed with palm oil fields on both sides of the road and stretching back into the hills. The last time I drove along this road a few years ago, it was bordered by pristine rainforest, marred by bare patches here and there where the loggers had hacked down the much valued towering tropical hardwood.

The change is troubling. Where once the great diversity of a tropical rainforest delighted the eye with hundreds of different kinds of trees, plants, and flowers, the hills for almost the entire journey now look as if an enormous army of stubby, spiky palm oil trees is swarming over them.

The roadside is dotted with *kampungs,* the Malay word for villages, and they have hardly changed for decades. In all the many developing countries I have visited, most of the nation's wealth is trapped inside the cities before it has a chance to seep out among the country folk. The villagers here live in simple weatherboard homes raised on stilts to keep the floor above the annual monsoonal rain that swamps the land. Most toil in the palm oil plantations, but the immense gush of wealth they generate barely trickles down to this level.

In each *kampung* there is at least one *kedai kopi,* an open-fronted wood-plank coffee shop with aging plastic chairs and tables, where the locals gather to swap rumors and whisper complaints about the politicians. The coffee shops are mostly owned by Chinese, the hard-working traders and merchants of Borneo. At night you often spy a sleepy-eyed child of 10 or 11 perched on a high stool operating the cash register at the family *kedai kopi,* taking payment and giving change, and turning aside in less busy moments to scribble homework.

On this visit, Sabahans are grumbling as usual in private about the neocolonial behavior of their "masters," the federal government leaders in Kuala Lumpur. Sabahans are roughly divided into three groups—the Chinese, the indigenous Christian Kadazans, and the Muslim Malays. But, as I have seen in Africa so many times, the dominant national race, the Malays across the sea, command the military and most departments of the federal government. They have used these to force all the other racial groups to accept the Malay language and Malay culture as the national language and culture.

It is dangerous to put your head up and protest in Malaysia. The politicians retain a much-hated and feared remnant of the British colonial days, the Internal Security Act. Under it a dissenter can be sent to jail indefinitely without trial, year after year, never knowing when he or she will be released.

Sandakan is signaled by the salty whiff of a sea breeze and coconut trees lining the road. We drive for about thirty minutes inland to the Sepilok Orangutan Rehabilitation Centre.

The rehabilitation center was established in 1964 by the Sabah

Forestry Department to take care of the many young orangutans orphaned by pet-trade poachers who killed their mothers, and by the destruction of their habitat by loggers. It is part of the Kabili-Sepilok
Forest Reserve and takes up 11,000 acres of dense equatorial forest.

The next morning, I stroll up to the center through a garden boasting many different types of tropical flowers. After a night of rain, the
heady odors of damp earth and flowers perfume the air. In the tropics,
as if to mirror the extravagant weather, the flowers are bold and colorful. I spy my favorite, the bird of paradise, its long stem bearing what
looks like a flamingo's beak splashed with patches of red, green, and
blue.

Because of its remoteness, Biruté Galdikas's rehabilitation center
could never rival this place as a magnet for visitors from across the globe.
At Sepilok's entrance, the signs of mass tourism greet me. A parking
lot is packed with tourist buses bearing misleading brand names such
as "Wild Adventures." A shop peddles souvenirs including postcards
of cute little orangutans, T-shirts adorned with the faces of cute little
orangutans, key rings bearing the faces of cute little orangutans, refrigerator magnets showing the faces of cute little orangutans, and baseball
caps featuring orangutan logos. All the orangutans here are orphans,
confiscated from people who were keeping them as pets, in zoos or
small circuses, and even as performers in nightclubs.

Streaming toward the "performing area" are about 150 tourists from
more than a dozen countries. Most are families, the adults with children in hand, lured here by advertising campaigns promising a unique
experience among the little orangutans. The Internet has thousands of
sites devoted to Sepilok, and one prominent site wrongly describes the
center as "home of the last wild orangutans in northern Borneo."

A sign at the entrance says that orangutans will come to the feeding
platform at 10:00 a.m. This puzzles me, because it took me half an hour
to trek through tough rainforest to get there two decades earlier, and
the many small children, as young as three and four, surely could not
walk out there and back.

All the great ape sanctuaries I have visited in Africa and Southeast
Asia inform visitors of the cruel treatment their little ones suffered
before they took them into their care, and Sepilok is no exception. A
one-year-old was rescued from a logging camp where she had been kept

as a pet. She was suffering wounds on both sides of her waist and the wrist, caused by ropes holding her captive. She had fallen 30 yards from a tree to the ground and cracked several ribs. After years of training at the sanctuary, she disappeared into the jungle but later came back to show her new baby to the Sepilok staff. There are more heartwarming rescue stories.

I join the visitors strolling along a wooden boardwalk raised a few yards over the rainforest floor. Amid the many orchid stands, huge dipterocarp trees—tall, solid giants of the jungle topped with crowns of branches and leaves—loom over us, their canopy casting a shadow on the boardwalk. From all around comes the sweet warbling of rainforest birds. A short walk from the entrance, we reach the place where orangutans are fed twice daily as an entertainment. We crowd a raised deck about 20 yards across the forest floor from a small wooden platform bolted up a big tree. An attendant is already seated there beside a bucket of bananas. The tourists press forward against a railing, eager for the best view of the orangutans.

Just before 10:00, there is a rustling among the leaves above the platform, and a pair of small orangutans, each about two years old, poke their heads out of the foliage and climb down to sit beside the bucket. For about five minutes the attendant feeds them bananas, and then they climb back up the tree and disappear into the leaves. "That's it," an attendant on the deck with us says. "They won't come back again until feeding time this afternoon."

The visitors shake their heads in disbelief and mutter among themselves. A woman from California, who holds her small son's hand, says to me, "This is a rip-off. We've come all this way, but we could have seen orangutans better at a zoo." Urged on by their children, eager to see more orangutans, most stay on the deck for about 20 minutes and then leave disappointed.

On my last visit, the platform was deep in the forest and was a genuine halfway house for orangutans attempting to adjust to a life in the jungle. But what I had just witnessed was a performance, because the little orangutans clearly faced several more years of "study" before they graduate from the Sepilok center and enter the nearby jungle.

However, the decision to keep visitors at a safe distance from the young orangutans is understandable. Humans can pass on diseases to great apes, and if a young orangutan contracted tuberculosis or even influenza from a visitor cuddling it, that disease could decimate the orangutans at Sepilok. At the Lake Victoria chimpanzee sanctuary, any visitor wishing to get close to the chimpanzees needed proof of several vaccinations.

By the entrance, I seek out the office of Sepilok's director, Sylvia Alsisto. She looks surprised when I tell her that someone we both know had arranged for me to speak to her. "I don't know anything about it," she snaps. "Get a letter of approval from the director of the Sabah Wildlife Department in Kota Kinabalu and I'll speak with you. That's the protocol." She turns back to her work, ignoring me.

I remain seated and politely ask again if I can speak with her about the sanctuary. She shakes her head. "I've told you that you've got to follow the protocol," she says. I apologize for the mix-up and she relents. "Okay, Paul, I'll answer your questions, but let's go up to the training area first because it's time for lessons in survival skills."

At the back of Sylvia's office, in an area closed to the general public, I find the old compound largely untouched. It is set against a flourishing jungle, with a tangled weave of vine and bush crouching like a mob of courtiers at the feet of the towering dipterocarps, long-living jungle trees whose canopy can be as high as 200 feet. One of the old cages contains four young orangutans, who cease their casual tumbling and come to the bars to peer at me. "They don't see so many strangers, and so they're fascinated by you," Sylvia explains.

Two decades have passed since I saw an orangutan up close, and I am instantly captured by the infants' precociously wise faces and unhurried manner. "They were confiscated as pets and brought here. It usually happens when the mother is killed, either to get the baby for the pet trade or because the plantation people are trying to frighten them away from the fields."

"So, the destruction of their forests is causing a crisis for orangutans."

She looks at me as if I were a leech. "You Europeans always put it that way. I prefer to call it land development," she says smugly.

"Sylvia, that's a term I'd expect from the developers rather than from someone who works to protect wildlife."

She bristles at the rebuke. "You've got your opinion and I've got mine. It's land development."

Nearby, a couple of little orangutans are swinging from branches high up a vine-covered tree. Their gymnastic display is breathtaking, both in their expertise and the casual way they seem to court disaster from a sudden drop of 10 to 15 yards.

"They adapt quickly here," Sylvia tells me.

A portly middle-aged Englishwoman is "supervising" their training, though I suspect she would have great trouble climbing beyond the lowest branch. I have a vision of her high in the trees, swinging from branch to branch to show the young orangutans how it is done. Orphaned young orangutans who have been raised as humans by their owners can find the prospect of climbing high trees terrifying when they arrive at Sepilok, but once they discover their natural aptitude they quickly learn.

"The older ones teach the younger ones," Sylvia says.

I ask her about the kind of food they eat in the jungle.

"We show the little ones what they should eat, and they quickly catch on."

"But an orangutan raised in the wild by its mother as it grows up learns to eat up to four hundred types of fruit, vegetation, and also termites. How is it possible for your little ones to learn that?" She shrugs. "And how is it possible for them to learn how to find fresh water, and avoid leopards and pythons?"

Ignoring the questions, Sylvia walks me back to her office. There she tells me that Sepilok has set aside a large chunk of jungle for the orangutans. "They start at the nursery and move up to the jungle gym. We feed some at Platform A, where the visitors come, and then when they're ready, at about age six to seven, we take them into the rainforest. If they can't find food at first, there is a daily feeding at Platform B, where you went in the 1980s. Soon enough, we find they begin to get enough to eat and disappear. Many we never see again, but some return now and then, especially females who bring back their babies to show us."

"How many rehabilitated orangutans have you returned to the wild?"

"Almost seven hundred."

Sylvia's answers raise my interest in the eventual fate of the "rehabilitated" orangutans. "Have you done research on the orangutans you've returned to the forest to see how many survive?"

She lowers her eyes. "No. Once they go into the rainforest, we don't check on their progress."

"But surely you have a duty to check whether your methods are successful in returning the orangutans to the wild. You make piles of money from the tourists coming here, and from local and international donors. You could be returning many to a more natural life, but you might also be putting back many unprepared for the rainforest's rigors, and so sending them to their deaths."

Sylvia abruptly stands up. "Thanks for coming, Paul. I've got plenty of work to do. And if you want to come back again, remember to observe the protocol."

I walk back to the hotel, not seeing the trees or flowers, focusing on a nagging thought. Sepilok sells itself in tourist markets around the world as a rehabilitation center. It is a seductive thought as you look at the pictures of Sepilok's orphans on the many websites, thinking that when they grow up they will be set free in the jungle and so help the survival of their species. And when you travel across the oceans with your children to get here, you can explain this wonderful rescue to them. But I am troubled by Sylvia's admission that no follow-up research is done on the orangutans once they are released into the jungle.

That night, the French veterinarian Marc Ancrenaz picks me up at the Sepilok Hotel. He is one of those wildlife heroes who have shown that individuals can make a difference in saving the last of the great apes in the wild. Marc is tall and handsome with a shock of salt-and-pepper hair, and his words tumble out as if they have trouble keeping up with his agile brain.

We drive through the darkened suburbs—dim or nonexistent street lighting seems mandatory in many developing countries—to an outdoor seaside restaurant. As we eat grilled sea prawns, steamed fish, and calamari and slap at mosquitoes, Marc tells me he abandoned his practice in Paris to travel to Africa and work with great apes at sanctuaries there. But on a holiday in Sabah with his wife a decade ago, he visited the Kinabatangan National Park when he heard it was home to hundreds of wild orangutans living in a secondary forest by a big river near Sandakan.

He says, "Until then it was believed that orangutans could not survive in a secondary forest, regrowth after the original forest had been logged out, because the trees would not be high enough for them, and there would not be enough food. But when we visited Kinabatangan, the *orang sungai*, the river people, took me to the orangutan habitat, and I saw that they were doing well. My wife and I stayed, and we've been here since then studying and protecting the orangutans. There are about eleven thousand orangutans living in the wild in Sabah, and about eleven hundred of them are in the forest where we have our project. They live in the Kinabatangan Wildlife Sanctuary, ten blocks of fragmented forest flood plain totaling twenty-six thousand hectares."

"When I was in Sabah two decades earlier, conservationists claimed that all the wildlife, scared away by the commotion of trucks, chainsaws and lumber workers, would never return to a logged-out forest," I tell him.

"You'll see for yourself when you go there tomorrow," he says. "I've even found recently for the first time anywhere in Borneo that orangutans pushed out of their forest are living in the palm oil plantations. I don't know how they survive there, but they do."

"I wonder if they are eating the palm oil fruit. It looks tasty enough, and probably the heart of the palm as well. Won't that anger the plantation owners, who might retaliate by killing the orangutans, as in Indonesian Borneo?"

"That's a big risk, but we can't stop the orangutans. The palm oil plantations have fragmented the Kinabatangan forest so that migrating orangutans often have to cross the plantations. Some decide to settle. Our big worry is a loss of genetic diversity. The eleven hundred orangutans living in our sanctuary need room to migrate so they don't become inbred."

"That number gives me the sense that a lot more orangutans are living in the wild, here and in Indonesian Borneo and Sarawak, than estimated," I say.

"I feel that also, but it can also work the other way. Remember that the estimates are not one-by-one head counts, but come from transects where researchers follow grid lines to survey a small area within a forest, maybe one percent. They then extrapolate those figures to cover the entire forest. That's not accurate, because in the same area you might have vastly different landscapes from swamp forest and woodland to savannah. That's not to say that we should believe there are many more great apes than estimated and so ease up on their protection."

Not surprisingly, orangutans are Marc's favorite great apes. "They are amazing creatures," he tells me. "They are the only known mammals on Earth that have two separate kinds of males. There is the flange male, the dominant male in a territory, who weighs up to three hundred pounds, up to three times as big as what we call the sub-adult males, who suffer from arrested development. The dominant male has enormous cheek pads, or flanges, made from subcutaneous fatty tissue, and an air sac dangling from the throat, which resonates his booming mating calls for up to a mile away. He controls the females in his territory, but we've found by DNA testing at the project that about forty percent of the babies are sired by the smaller sub-adult males."

"So, they're sneaking matings with the females when the big boy is not around."

Marc takes a swig of Tiger beer and laughs. "In the jungle vastness there's not much the dominant male can do about it. The sub-adult males have a covert lifestyle, creeping around the forest hoping to avoid the flange males. What is amazing, as we've seen in at Kinabatangan and in zoos, is that if a flange male dies or is taken away, one of the sub-adult males will grow into a flange male within months, more than doubling his size and growing huge cheek pads. He'll then take over the territory."

"Won't that mean he knows about the tricky ways of the smaller males?"

Marc laughs again. "Probably, but there's not much he can do about it."

Another puzzle is why the orangutans are the only great apes that evolved into a society of solitary creatures. Marc says, "It's my belief that long ago, while the other great apes evolved into societies based on family groups, the orangutans decided that there was too much hassle living in groups, and that life was much easier when you lived alone, or as a mother with an infant, with the occasional meeting to mate."

Biruté Galdikas had suggested much the same thing. She wrote that chimpanzees, being political animals, "devote much time, energy and intelligence to learning the social hierarchy and social etiquette of their community, monitoring and manipulating the emotions and loyalties of other chimps, forging alliances, staging coups, and resolving disputes."

But adult orangutans do not need each other, and therefore have no wish to manipulate or use each other.

This, says Galdikas, is why they seem so benign and innocent to us. "When chimpanzees befriend other chimpanzees, they invariably have ulterior motives; the more friends a chimpanzee has, the more powerful those friends, the better. But adult orangutans have nothing to gain from associating with one another."

Marc has close connections with the Sepilok sanctuary, and I tell him about Sylvia's admission that the sanctuary does not carry out any follow-up research on the rehabilitated orangutans they release back into the jungle. His brow furrows. "That's what she told you, but they know what's going on and they believe that more than forty percent of the orangutans they release back into the forest don't survive."

"More than forty percent?"

"Yes, and it's what you'd expect. As you know, orangutan mothers in the wild prepare their infants thoroughly for when they go their own way. They teach by daily example with the infants growing up by their mother's side, copying her in choosing what food to eat, how to find it, and following her around the rainforest. They learn how to react when they meet another orangutan. It's not possible to teach an orangutan orphan those same lessons at an orphan sanctuary like Sepilok, and so when they go back into the jungle many are not able to cope with life there and die."

"So, all those good-hearted people from around the world, visiting here and sending money to support the sanctuary's work, have been fooled."

"Maybe, but we should still keep the sanctuary operating because its most valuable benefit is in saving the virgin rainforests in Sepilok and at Tabin, a new orangutan sanctuary the Wildlife Department here has set up. It's about twice the size of Sepilok. The end justifies the means."

I find it tragic to send so many juvenile orangutans at age seven or eight to what is for many a cruel death in the rainforest. But Marc contends that many of them do survive, and that is a lot better than none at all.

The following day, at midday, an SUV takes me to Kinabatangan to see Marc's project. My companion is a young *orang sungai*, Malay for "river person," one of Marc's most trusted research officers at the orangutan site. He is round-faced and smooth-skinned. "Call me Mincho," he says cheerfully as the SUV bounces along a rough road that runs parallel to the broad, muddy Kinabatangan River. "Politicians!" he snorts. "For the past three elections all the politicians standing in this electorate have promised to make a proper road here, but when they win they do nothing. They're leeches, the lot of them."

Mincho parks me at a small lodge within a few yards of the riverbank. He says, "I'll come back at five tonight and take you out on the river. We'll go to the orangutan study site early tomorrow morning."

Just before 5:00, Mincho walks me down to the coffee-hued river, where a small, wooden flat-hulled boat waits us. The boatman is also an *orang sungai;* wiry and middle-aged, he has the unemotional expression of a man who has spent far too much of his life at the end of a fishing line. He guns the outboard motor, shooting us along the Kinabatangan River, hemmed in on both sides by scraggy rainforest.

"It was logged out in the nineteen-fifties, and this is secondary forest that has grown back," Mincho says. Imagine a man with a head of thick, dark hair, and then imagine another man with a balding scalp, and you have the difference between a virgin rainforest and one that has been logged out and regrown. Despite the shopworn look of the trees by the riverside, Mincho surprises me by revealing that most of the animal species, including the orangutans, came back after the loggers moved out, and the trees and plants began to grow once more.

"Our forest sanctuary is about 112,000 acres, and we've got about eleven hundred wild orangutans living there. We've tried to get the government to declare it a national park, but the loggers and palm oil growers wield money power, and have too much influence over the politicians. It's my dream that one day the plantations will all be gone and the wild creatures will have their forests to themselves."

The 350-mile Kinabatangan River is the country's second longest and tumbles down from its headwaters to the Sulu Sea near Sandakan. It holds one of the most diverse collections of wildlife on Earth including orangutans, the fabled pygmy elephants, and the extremely rare Sumatran rhinoceros, much shorter but stubby like its African cousin. In all of Borneo there are just 30 of these rhinoceroses left in the wild.

In centuries past, the Chinese emperors' huge junks sailed along the lower reaches of the Kinabatangan where we are, seeking the birds' nests of swifts formed from their solidified saliva, which the Chinese use in a highly desired soup, ivory from the elephants' tusks and hornbill beaks, and rhinoceros horns. Traditional Chinese medicine healers to this day still grind the horns into powder to use as an aphrodisiac. Mincho tells me the junks even came to trade for the giant tree trunks that once lined the riverbanks, taking them back to Beijing to hold up the ornate ceilings of the capital's soaring temples and palaces.

But the riverine forests are under attack by the ubiquitous land developers. "We're in a desperate situation. More than 300,000 hectares of forest over the past decade has been cut down for palm oil plantations," he says. "About eighty-five percent of the flood plain of the lower Kinabatangan, one of Asia's last forest flood plains, has been converted to land for agriculture, and most of the remainder is constantly under threat." In the 1980s, surveys in the Kinabatangan flood plains showed that the orangutan density was three to 39 square miles. But more recent surveys show that the population density is now between three and six orangutans crowded onto 217 acres, or up to eight times what is normal.

Mincho points to one of Borneo's rare sights, a group of long-limbed, slender potbellied monkeys with long tails perched on the spindly, bare branches of skeletal fig trees that arch over the riverbank. "They're proboscis monkeys," he says. Like so many animals in this crumbling island paradise, they are highly endangered, with only about 8,000 remaining in the wild. "Those trees are a proboscis monkey's five-star hotel. When night nears, you can see them edge out onto the branches over the river where they sleep. They prefer branches with no leaves so they can see if their most feared enemy, the clouded leopard, approaches. If a leopard attacks they drop into the river. They're good swimmers."

The proboscis dominant male, twice as big as the females, sprawls across a spread of branches, like a sultan on his throne, surrounded by his devoted harem of 10 females. Three of them have babies suckling at their breasts. He boasts a droopy, broad nose, at least six inches long, that settles below alert dark eyes. It looks more like a misplaced phallus at rest than a monkey's snout. "The females choose to live and mate with the males that have the longest noses, and just the one male can have up to twenty females in his family," Mincho says. "The dominant males have twenty-four-hour erections, and they're ready to mate at any time with any female."

"That sounds like hell to me."

Mincho gives me a wicked smile. "Yes, but they need to be ready for action all the time. I've seen plenty of times where the dominant male is mating with one of his females while the others are tugging at his legs demanding their turn."

"Perhaps it's something in their diet?"

"Maybe. They have a curious digestive system that only allows them to eat fruit that's unripe. They'd die in agony if they ate ripe fruit, though they never do of course. They're too smart for that."

Something large breaks the water's surface for a few moments a few yards to the right, causing a flurry of ripples, and then disappears. It may be one of the giant crocodiles that haunt the Kinabatangan's waters. "I once saw a crocodile on the riverbank that was a little longer than the boat, at least fifteen feet," Mincho says. He gifts me with his lovely smile again. "Maybe that's why the proboscis monkeys become such good swimmers, to escape from the crocodiles when they fall into the river."

Mincho, the descendant of countless generations of the river people, clearly loves the Kinabatangan and its creatures. Besides his work with the wild orangutans, he has spent thousands of hours since boyhood observing the animals and birds that live by the river and inland. He also draws on the knowledge passed on through the generations by his *orang sungai* ancestors. "The proboscis monkeys like to walk upright just like us when they travel through the jungle on the ground," he tells me as the boat idles in the river under the bare fig tree.

Moments later, I spy a helmeted hornbill swoop down to settle among the leafy branches of a riverside fig tree about 20 yards downriver from the sultan and his harem. It is a behemoth among birds, a flying battleship. Its spectacular yellow casque, a chunk of ivory mounted on its bill and curving out from its head, makes up to 10 percent of its weight. It is so large and heavy that you wonder how the hornbill can keep its head up from the weight of the casque. It must have the neck muscles of a champion wrestler.

The casque probably evolved to this size because the hornbills are monogamous and each male battles to secure and hold territory, the rivals bashing each other with their casques while in flight. The bigger the casque, the better chance you have to send a rival fleeing.

"Look, two more hornbill species," Mincho murmurs in excitement. On another branch perches a rhinoceros hornbill, its distinctive double-tiered ivory casque contrasting with its heavy black feathers. Nearby is a much smaller bird with a fluffy white head and shortened beak. It looks like hornbill junior, but Mincho says it is fully grown and a separate species. "You're lucky; I've rarely seen three hornbill species in one tree."

A stout-bearded pig pokes its head out from of the undergrowth, stares at us, and then, unafraid, ambles down the bank for a drink. The river people are Muslims and do not eat pork, and that might explain the pig's boldness. As we drift along the river I get the feeling that we are in one of those special but fast-disappearing places, a tropical rainforest that is host to a wide diversity of unique creatures and where time once had no time. But then the spell is broken as an ominous sight appears ahead at a bend in the river. Half the steeply sloped hillside is jungle, while the other half is a palm oil plantation, a clear line separating wilderness and "land development" running down to the riverbank. The contrast is brutal.

As if scripted, a tugboat chugs around the bend pulling a huge floating tank, its deck at water level. "There's three tons of palm oil in that tank," Mincho says. "Once a week the tugboat brings the production from one of the palm oil plantations to the Sandakan refinery."

Our boatman points to heavy dark clouds low in the sky over the jungle and swarming toward us. "It's monsoon season," Mincho explains, that annual month-long daily deluge of rain in the tropics. The boatman swings the boat around and shoots home, but the rain travels faster and hits us a few minutes before we reach the landing. A solid sheet of water crashes down from the skies, and it drenches us within a few moments. At the river landing, Mincho and I clamber ashore and head toward the wooden steps leading to the lodge, laughing like children at the tactile pleasure of the rain as it streams down our faces and soaks our clothes.

After changing clothes we eat dinner, grilled giant river prawns and boiled rice given fire by a helping of *sambal,* the Malay paste made from crushed shrimps and chili oil. The rain abruptly ceases, but the grass around the lodge is swamped by puddles. "They'll dry up by the morning," Mincho predicts.

I ask him how the orangutans here cope during the monsoon. I had seen photos of orangutans fashioning umbrellas from large leaves and holding them over their heads to ward off rain. "Yes, they're very clever," he says. "One of the females in our forest, Jenny, she's the smartest I've seen. We wear baseball caps when we go to the forest. I've seen Jenny, when it rains, looking at us and then making a cap out of leaves and bamboo and placing it on her head. If we're lucky we'll see Jenny tomorrow."

Other orangutans have been observed fashioning tools such as probes to extract insects and wild honey from tree hollows and manipulating short, thin twigs to extract deep-seated calorie-rich seed from fruit covered with stinging hair that can irritate their skin. They also tear off leafy branches and use them to swat away mosquitoes.

As with the chimpanzees, orangutans learn to use tools from the adults as they grow up, and the methods can differ in widely separated populations. As an example, only one group of orangutans has been observed using probes to get to the honey, insects, and fruit seeds. Sometime in the past, these techniques must have been nutted out by one or more super-smart individuals and then passed on from generation to generation.

Mincho has been a research officer at Marc's project for five years and has spent thousands of hours observing the red ape in the wild. "You know about the males raping the females?" he asks. I do, but only what I have read about it in research papers.

"A flange male has his own territory and constantly roams around it looking for females to mate with," Mincho tells me. "They have no choice. A female with an infant will try to hide when she hears him coming; she usually doesn't want to mate at that time but knows he'll force himself on her. But if she doesn't have an infant with her then she can be as enthusiastic as he is. I've seen it many times. They stay together for up to a couple of weeks mating, and they do it just like us."

He is amazed when I tell him that gorillas, chimps, and even the hedonistic bonobos mate for no longer than a minute at a time. He flashes that wicked smile again. "The orangutans do it for up to half an hour. Once I had the binoculars on a flange male we know well,

Simon, when he was mating, and when he had his climax it was just like a man." Mincho rolls back his eyes and affects a look of ecstatic joy imitating Simon, holds it for a few moments, and then laughs fondly at the memory.

"That's not exactly rape," I comment.

"We see that more with the sub-adult males, the orangutans who don't develop into flange males. The available females want to mate with the big males, and so the smaller males, when they find females, often have to force them to mate, and that looks to me like rape. The female screams as she resists and struggles with the male. Usually, he wins."

"It must happen a lot, because Marc told me that DNA testing here proved that about 40 percent of the infants have been sired by the smaller males."

"The jungle is very big, and so the smaller males have plenty of time to find and force females to mate with them because the big males can't be everywhere at once. But, if a flange male finds a sub-adult male mating with one of his females, then he'll sometimes try to catch and kill him." At the risk of their lives, sub-adult males will even force themselves on females who are carrying infants, though the little ones are never hurt.

Mincho slaps his hands together. "I'll come for you at seven in the morning, and we'll go meet the orangutans."

I remain on the verandah for the sheer pleasure of it. The dank, earthy smell of rain mingled with the forest mulch makes me dizzy with delight. There are other guests at the lodge, and we sit on the verandah and chat after dinner. Peter Riger, assistant director of conservation programs at the Houston Zoo, joins me. "It's a great job," he says. "I get to go around the world and give money to worthy projects like Marc's. His is one of the best conservation efforts for orangutans. We give him about twenty thousand dollars a year. It's not much, but every bit helps." Peter believes that from what he has seen, Marc's wild orangutans are thriving.

Our conversation turns to zoos and their place in modern times. After visiting zoos all over the world, I believe they are an anachronism, places where captured or zoo-bred animals are kept in cruel confinement for the daily hordes of visitors. I believe that even the best zoos treat their captives in a dreadful way. Chicago's Lincoln Park Zoo is one of America's best, and yet the gorillas and chimpanzees there are confined in small, artfully designed spaces that hardly give them room for a decent charge. It is the same for the great apes, including orangutans, at the famed San Diego Zoo.

I suggest to Peter that confining great apes in tiny compounds is like keeping a human in a room the size of a bathroom for the remaining days of his life with hundreds of strange creatures filing by the glass walls each day to peer at him.

He smiles indulgently at me. "We play a valuable role with captive breeding to ensure each endangered species does not die out, and in supporting efforts such as Marc's to make sure threatened species in the wild survive."

"That's true," I respond, "and we should commend you for it, but it still means you keep animals such as the great apes in conditions that are barbaric. I've heard primatologists call them sacrificial animals. They allow zoo visitors, especially children, to be educated about the need to preserve such wonderful creatures in the wild, but that doesn't lessen the mental stress we must cause them in keeping them in such unnatural conditions."

"That's the wrong approach. We call them ambassador animals. They are the ambassadors presenting the wild animals to zoo visitors."

"Ambassador? There's not a hint in that word of the cruelty you inflict on zoo animals."

Peter abruptly stands up. "Nice to talk to you. I've got to read some reports. Have a good night."

I look up the Houston Zoo's website and find its ambition is to provide "a fun, unique, and inspirational experience fostering appreciation, knowledge and care for the natural world."

Maybe, but I do not think it is much fun for most of the 4,500 animals kept captive there for the entertainment of the zoo's 1.5 million annual visitors. I am necessarily unfair in this, because I have been fortunate to witness wild animals in their native habitats all over the planet during the past three decades. Most people can see them only in documentaries or at zoos. But that still does not excuse the cruelty inflicted on the "sacrificial" zoo animals.

4

There is a hush over the river just before dawn. The sultry heat is soporific. Nothing moves, not even the insects, and there is no cackle or warble from the thousands of birds that perch shut-eyed in the trees along the riverbank. Then, silvery beams of light splinter the forest gloom. A silky white mist begins to lift over the water as a troop of pygmy elephants, about two-thirds the size of their African cousins, plod with slow-motion grace down the riverbank and slip into the muddy water up to their necks. They are practiced swimmers, even the toddlers, and hold their trunks above their heads as they swim to the other side. Led by a wise old matriarch, they climb up the riverbank and disappear into the jungle's shadows.

I have just finished a bowl of noodles laced with *sambal* when Mincho arrives on his motorbike. The boatman is waiting for us and we head upriver once more. We pass by the fractured hillside and about 15 minutes later, turn into a murky stream the width of a country lane that meanders through the lower reaches of the rainforest. The canopy soars over us, the trees on both sides of the bank meeting like joined fingers to form a leafy archway.

The boat pulls into the bank, and Mincho leads me along a much-trodden path through the forest to a wooden hut sheltering under an enormous hardwood tree. In the two rooms, sleeping bags and mosquito coils are laid out on the bare boards. It is home to the field assistants who aid Mincho and Marc in daily observing the orangutans' behavior. Mincho and I pull on gumboots on the porch and wait for one of the assistants to arrive. He is also an *orang sungai*. "All our workers are locals," Mincho says. He hands me a walking stick, a sturdy straight branch, about five feet long and stripped of its leaves.

I grit my teeth, not knowing how I will cope with a jungle for the first time in three months since the attack in Afghanistan. The blast messed up my inner-ear balance and affected my vision. Mincho knows about the attack and leads me through the forest at a stately pace. Even so, I have to jab the stick into the ground time and time again to stop from pitching forward into the mud or falling to the side.

But I am once again in a rainforest and buzz with pleasure, even though the air is so humid that it seems to sweat. I have trouble getting across the large tree roots that weave across the forest floor like giant arteries without stubbing my toes and jeopardizing my balance. The knee-high mud baths and moss-covered logs that have fallen across the path, which is carpeted with layers of slippery rotting leaves, would have been easy to cross a few months earlier but are now like an obstacle course. Mincho patiently waits as I struggle.

About two miles into the forest I spy a tent pitched below a high leafy tree. Three of Mincho's colleagues hail him in their native tongue. "You're in luck," he tells me, "because they're following Jenny and her infant son. They're up in the tree above us."

I stare up and see nothing, then use my binoculars but still cannot see any orangutans. "Jenny and the other females are very clever at hiding themselves in the leaves. They do it when a flange male or a sub-adult male is around and they don't want to mate. She's hiding because you're a stranger, she's never seen you before, but once she decides you're no threat, she'll show herself and her infant."

From high above I hear strange sounds, like someone planting juicy, noisy kisses and then uttering rasping raspberries. "That's Jenny; it's an orangutan's warning sounds," Mincho explains. "It says 'stay away from me.' They use it whenever they see stranger orangutans, or people, or dangerous animals. She rarely uses it when she sees us, but it's a lesson for her infant to do that when he sees a stranger like you."

The assistants all have notebooks and every 15 minutes they dutifully write down what Jenny and the infant are doing, even though they have yet to show themselves. One of the assistants digs up the roots of a sprouting seed coated with earth. The seed looks like a small yellow button. "We're replanting the riverbanks with these to set up a corridor of trees for the wildlife here," Mincho says. "These plants can only germinate from seeds that have passed through an orangutan's gut."

Moments later, 13-year-old Jenny peers at me from among the leaves, about 20 yards up the tree. She must have given the all-clear because her two-year old-son, Mallotus, cautiously creeps from the foliage and stares down at me. Jenny begins eating tree leaves—a signal for Mallotus, named after the species of tree in which he was born, that it is playtime again. He swings one-handed from a branch as he stares at the strange sight below—me.

Mallotus has a fluff of orange hair standing straight up, human-like ears, and dark, shining eyes. The skin around his eyes, with their ethereally calm expression, is heavily wrinkled, and his flat nose sits above a pair of thick lips that pucker from time to time. Is it my imagination that sees a smile on his face as he enjoys his jungle fun? His fingers resemble mine but are many times stronger. Jenny is a grown-up version of Mallotus, but with a potbelly and shaggier hair. Should he grow into a flange male, however, he will dwarf his mother.

After 10 minutes, Jenny gently pulls her little son onto her hip and climbs higher into the canopy. From there she rocks forward and backward on a stout branch, to and fro, for a few seconds until she can reach the branches of the tree next to it. Safely perched on a new branch, Jenny begins eating leaves once more while Mallotus enjoys his jungle play.

I am struck by Mallotus's lack of playmates. He and other infant and juvenile orangutans rarely get to meet others of their own age, and so they grow up accepting that being solitary is the natural way of life. Orangutans in the wild have years to develop the calm, contemplative nature that sets them apart from all other great apes.

The morning becomes a slow march through the rainforest following the two orangutans, although little happens besides foraging through the canopy and taking naps in day nests. At about 11:00 a.m. Jenny and her son disappear from sight high up a tall tree, hidden by leaves. "She's either breastfeeding Mallotus or she's built a day nest and they're taking a nap," Mincho says. I take this chance to rest, because the walk through the rainforest has exhausted me. Mincho, seeing me stagger, spreads a tarpaulin over the mulch, and I lay down and fall asleep.

A squelching sound deep in my right ear wakes me. There is liquid moving about down near the eardrum. I tear a scrap of paper from my notebook, screw it into a point, and carefully lower it into the ear. My knees buckle when I remove it, seeing that the top half is soaked in blood. This could be very serious. Perhaps my inner ear is still damaged from the bomb blast. If so, I need to get to the hospital in Sandakan urgently, and it is about an hour back along the river by boat and then another two hours in an SUV.

I feel something moving inside the ear. It could be a poisonous spider or some other dangerous insect. Mincho shines a flashlight in my ear. "I can't see anything," he says. "It must be very deep."

Whatever it is moves again. I sit on the forest floor to think it out. It only causes more trouble to panic in a situation like this. I remember the story about John Speke, the nineteenth-century English explorer, who was the first European to discover the source of the White Nile at Lake Victoria in Uganda. One night on an expedition into the African interior, he woke to find a beetle moving about deep in his ear. It nearly drove him mad, and he used a knife to stab it. He went temporarily deaf in the ear.

So, I instead decide to try and drown the tiny creature. I tilt the ear to the sky and pour water into it, hold it for a minute or so, and then pour it out. The creature must be tough because it moves once more, and so I try again. After the third dousing I feel it moving up the ear canal. Mincho shines his flashlight down it. "It's a leech," he says. He sounds shocked. "I've never seen or heard of this before. It's still too deep for me to hook it out." He pulls a can of mosquito spray from his day pack. "Is it okay if I attack it with this?"

There seems no other choice, because I do not fancy a long trip back to the Sandakan hospital with the leech in my ear down near the eardrum and feasting on my blood. He sprays inside the canal, and a moment later blood pours out of the ear. There is no pain, because nature has equipped the leech with an in-house anesthetic. It injects this into the victim as it feeds on the blood so that you can have several leeches hooked to your body, as I did once in Africa, and not notice their plunder.

"It's just thrown up much of the blood it sucked from you. We'll wait to see if the leech emerges or goes deeper into the canal."

It chooses the latter. I pour water down the canal and again it moves a little way up, so Mincho can again annoy it with the mosquito spray. More blood pours from the ear. Mincho shines the flashlight down my ear again, then takes a pair of small twigs and uses them like chopsticks to haul out the leech. It is swollen with my blood and Mincho tosses it into the forest.

Only now do I remember advice Wasse, my pygmy friend, had given me back in the Central African Republic when we had gone into the forest together. "When you rest, put leaves in your ears and up your nostrils," he said. "That will protect you from spiders, sweat bees, and leeches."

Mincho and I return to the field assistants' hut, where I lay down on the bare boards with the right ear resting against the floor. He gives me several wads of cotton and for more than two hours the blood drips from my ear, soaking the pads with gore. "The leeches have an antico-agulant which they inject into you to make the blood flow freely," he says. The blood will seep out from the tiny puncture mark inside the ear for another six days.

We eat a lunch of boiled rice and river prawn and then head back into the forest to link up with Jenny and her youngster. Mincho uses his cell phone texting to get the position from the field assistants who have re-mained with the orangutans. In search of fruiting trees, the orangutans have crossed to the other side of the stream, which is about 10 yards wide, using the forest canopy as a natural bridge. The orangutan project has constructed a rope bridge over the stream, and we edge across it.

Mincho is reminded of Jenny's older son, Stephen. "He was very naughty and loved playing tricks on us," he tells me when we reach the other side. "One time when he was about four years old, we were tracking him and Jenny when the rope bridge wasn't built. He crossed to the other side of the stream using the canopy branches with Jenny and then sat in a tree watching as we waded across chest deep in the muddy water. The moment we got across, he used the canopy to cross back to the other side of the stream. Then he sat in a tree by the water and watched us struggle across to his side. The moment we got there, once again he crossed back over the stream. He did this six times until, tired of teasing us, he rejoined Jenny."

Orangutans have no fear of water and often wade through streams where there is no canopy highway. Biruté Galdikas has even seen orang-utans up to their chins in water, but they cannot swim and will step on submerged logs as a natural bridge to cross a stream.

Despite the rest, I quickly become exhausted. Thankfully, Jenny and Mallotus have not moved far—they are up in the canopy, and I watch them through binoculars. We ramble together through the rainforest, me on the ground and mother and son moving from tree to tree high in the canopy. Their rhythm of life is calm, unhurried, focused.

As dusk nears, Jenny and Mallotus head toward a stand of tall trees nearby. "That's probably where she'll build her night nest," Mincho predicts. Then his eyebrows rise in surprise, and silently he retreats about 10 yards into the shadows. When I move to join him, he shakes his head and points to the low branches of a tree near me. Jenny and Mallotus are making their way along it toward me. When she is about two yards above my head, Jenny stops, sits on a branch, leans forward, and peers at me with thoughtful eyes.

Mallotus swings from a nearby branch and seems unconcerned about my closeness, but Jenny's eyes penetrate me. My eyes meet hers and the minutes slip by as we gaze at each other, me in great wonderment and she with emotions I cannot even guess at. In the final moments of my long quest to be among the great apes, Jenny has gifted me the privilege of this rare close encounter with orangutans in the wild.

Mallotus must be getting sleepy, because he moves to Jenny and clings to the side of her stomach, taking a firm grip on the fur. She turns away from me and uses the branch she is perched on to sway the tree to and fro until she reaches out and grabs a branch from the nearest tree, so close that I could reach up and touch her.

By now the sky has turned a moody gray and, with Mallotus in tow, Jenny climbs high into a tree and disappears into the leaves. Mincho has a stopwatch ready. "She's about to make her night nest," he says. She breaks the first branch to form the nest's foundation, and it must be sturdy because the sound, a sharp crack, echoes around the jungle. We hear her break more branches and then silence. Mincho looks at the stopwatch. "One minute and twenty-five seconds," he says. "The nests are very strong. One of our field assistants climbed up to one and jumped up and down in it. He weighs about 150 pounds and it took his weight easily."

We trek back to the stream, where the boatman is waiting for us. I keep silent on the way to the riverside lodge, moved by a mingled sense of achievement and melancholy now that my quest to be among the great apes has come to an end. With the great apes in the wild disappearing fast, will my daughter's children and their children have the same chance I did to see them born free and living in their jungle homes?

A Call to Arms

The great apes in the wilds of Africa and Southeast Asia are disappearing fast. Their numbers dropped from several million in 1900 to a few hundred thousand in 2010. One of nature's glories, one of evolution's masterpieces, will almost vanish from the wild because most of us have our backs turned on their plight, either in ignorance or in willful neglect.

Do the bonobos, gorillas, chimpanzees, and orangutans in the wild have any hope? How best can we save them? Dave Greer has worked to rescue the great apes in Africa for more than a decade, and he is now coordinator for the World Wildlife Fund's African Great Apes' Project. He gave me this analysis:

"We are allowing the great apes to become more threatened with each passing year. Some populations recently at risk are now all but eliminated. A recent study from the Ivory Coast, once considered a stronghold for the western subspecies of chimpanzee, indicated a 90 percent decline in chimpanzee numbers in less than two decades. Gorilla and human populations in eastern Democratic Republic of Congo are under constant threat, caught in the crossfire of a seemingly never-ending civil war and the demand for food, fuel, and shelter. Orangutans in Indonesia are losing out to palm oil plantations as national parks are clear-cut to make way for increased bio-fuel and cooking oil production.

"There are isolated cases of conservation success which elicit some hope. Mountain gorilla populations have increased slightly in recent years. Lowland gorilla numbers remain stable in some of the isolated, uninhabitable swamps of the Congo Basin. Unfortunately, logging companies are now bridging these same swamps in order to access the remaining valuable timber species. And these same gorillas are relentlessly threatened by habitat loss and disease transmission, exacerbated by human encroachment.

"Perhaps most importantly, the majority of great ape range-country governments fall woefully short in their attempt to ensure the strict application of national laws meant to safeguard internationally protected species such as gorillas, chimpanzees, and orangutans. Illegal bushmeat vendors and pet traffickers thus operate with impunity, often facilitated by the unchecked activities of logging and mining companies.

"Consequently, the international community must further commit itself to assisting and promoting responsible governments that harbor and actively work to protect great apes. Indeed, it must insist that all great ape range-governments strengthen their efforts in defense of great apes. International companies doing business in countries harboring great apes must look beyond merely their profit margins; establishing partnerships in developing nations should require meeting ethical standards that favor both human rights and biodiversity preservation.

"Perhaps most importantly, there must be a systematic mechanism of accountability developed which commits range-country governments to adhere to national wildlife laws and international treaties. If we continue to refuse to hold the various stakeholders accountable, then looking at great apes through steel bars and thick Plexiglas windows may be the only way our children will ever get to admire our closest living relatives."

Dave is not alone with this vision of an approaching Armageddon for the great apes. Rob Muir, the brave conservationist, told me that within half a century, "the only great apes left will be in zoos, sanctuaries like Bonobo Paradise in Kinshasa, and in heavily guarded pockets of habitat such as the Virunga slopes in Rwanda and the rainforests at Bai Hokou in the Central African Republic."

Harvard's Professor Richard Wrangham mostly agrees: "The orangutans, western chimpanzees, and maybe chimpanzees in East Africa (Uganda, Tanzania, Rwanda) will probably be largely restricted to rather sharply defined areas with crisp borders, perhaps as big as a few hundred square kilometers."

But Richard is cautiously hopeful, adding, "In the Congo Basin I would be more optimistic, particularly about a few remote and swampy areas, such as in northern Congo Republic. Maybe I am naïve, but I hold out hope for some areas that are not easily logged and farmed, where poor transport may continue to give reprieve until at least the second half of the twenty-first century."

The possibility of this reprieve for one of the embattled species came late in 2008. The Bronx Zoo's Wildlife Conservation Society grabbed world headlines when it announced that its researchers had discovered a hidden gorilla paradise, deep in the swamps of the Congo. It estimated that over 100,000 previously unknown western lowland gorillas are safely living there. Hope for the great apes surged. With this news, the estimated numbers of the subspecies doubled to almost a quarter of a million.

Dave Greer, who knows both the area where the gorillas were found and the WCS researchers, is wary of the startling numbers. But the discovery at least offers a glimpse of hope, though Dave believes we can expect few such discoveries of great ape paradises in the future. Fiona Maisels, the WCS central African coordinator, predicts, "Unless drastic action is taken, more than eighty percent of western lowland gorillas will have gone in just three gorilla generations."

Ultimately, it will be the African and Southeast Asian leaders who decide whether or not their great apes cling to survival in the wild or largely disappear, and the brutal reality is that their chances are slim. Whatever the good intentions of many of the crusaders against the bushmeat, timber, palm oil, and charcoal trades, these blights on the great apes will not easily be wiped out. It is more likely that the threats will grow dramatically in the coming decades as populations in the great apes' homelands continue their relentless growth with ever more people needing more meat for food and fuel for their fires.

So, how effective then is Rob's and Richard's solution—zoos, sanctuaries, and heavily guarded natural habitats in national parks?

The zoo great apes are safe, and they will continue to breed in controlled matings to avoid inbreeding. Several thousand great apes are kept behind bars in zoos around the world, but at best they are useful for conservation education and as repositories of their species' precious DNA even with the lack of diversity. They are largely zoo-born and know little of life beyond their barred cages. They lack the many complex cultures that primatologists have observed enriching the lives of their kin in the wild. Placed into the wild, most would perish.

Western zoos have come a long way since the nineteenth and early twentieth centuries, when the animals were kept in small concrete enclosures and offered little mental enrichment. Primatology discoveries in the latter half of the twentieth century have shown that the great apes in particular are complex, thinking creatures, and zookeepers have responded by attempting with grassy enclosures and many kinds of food for thought—puzzles, "termite" fishing, and rope assemblies—to keep them happy and their minds stimulated. But lack of space continues to be the biggest problem, and the only humane solution seems to be the moving of great apes in smaller city-bound zoos to safari-type zoos where the wide-open spaces give them room to romp. Just as the great apes in the wild must be protected, so must the great apes held captive in zoos be respected.

Then there are the great ape sanctuaries, over a dozen in Africa and several in Southeast Asia. They harbor about 2,000 great apes, mostly orphans. Like the zoo animals, few will be released back into the wild, but they are also repositories of great ape DNA. Many more sanctuaries are needed, though, because many are now overcrowded, forcing their orphans to suffer high stress.

The millions of dollars yearly spent on keeping the sanctuary orphans in conditions that range from woeful, as in Limbe, to very good, as in Mefou and Florida, could be far more effectively used employing armed guards to protect the wild great apes in their natural habitats. But it is not a matter of choice between the two. The challenge is how best to support great ape sanctuaries and also protect great apes living in the wild.

That leaves great apes still living in their natural habitats, and one hopes that the outside world will urgently agree to fund their protection. Africa has a network of dozens of national parks that extend from the Sahara Desert in Niger in the far north to South Africa at the bottom of the continent. The best are in South Africa with their emblematic lions, elephants, rhinos, cheetahs, and other popular animals running wild in the parks. Massive foreign "safari" tourism provides the funds to operate these parks.

At Hluhluwe Umfolozi National Park in Zululand, I observed a well-trained cadre of professional rangers smooth the way for safari tourists to have unforgettable experiences seeing the many animals, including the big five—lions, leopards, elephants, buffaloes, and rhinos, running free in 237,000 acres of scrubby forests and plains. At the same time, the rangers supervise the successful breeding within the park of endangered species such as the black rhino.

You have to travel more than 2,000 miles up the continent to equatorial Africa to find the great apes in their wilderness habitats, and almost all the countries that have them are ravaged by corrupt dictatorships, rebellions, and extreme poverty. There are precious few local funds to devote to preserving great apes in the wild, and safari tourists are frightened off by the threat of violence. Most of the impetus and money to protect the great apes there comes from foreign conservation bodies such as the World Wildlife Fund, the Wildlife Conservation Society, the Jane Goodall Institute, and the African Conservation Foundation.

They face immense problems in protecting the great apes because poachers run rampant in almost all the national parks in equatorial Africa. In the *African Journal of Ecology*, American researcher Tim Caro and others outlined their investigation into Africa's national parks. They found that in "Western-Central Africa," the national parks, "bastions of wildlife diversity," are being ravaged by bushmeat poachers. "This bushmeat used to cover local consumption only, [but] now it includes tables in far-off cities that extend to London and Paris." The bushmeat is sold by the ton to large African immigrant communities in the United States and Europe who hanker for a taste of the wild from their mother countries. The researchers see a restricted hope in "isolated pockets of large mammal diversity living at low population sizes."

But since Dave Greer left Bayanga in the Central African Republic, the armed ranger project he revitalized there has fallen fallow once more. At Kibale in Uganda, settlements are pushing against the national park's boundaries as the population increases and poachers do their deadly business in its forests almost without opposition. In Tanzania, the Gombe Stream National Park with its famous wild chimpanzees has been squeezed into a very narrow strip of forest facing Lake Tanganyika. Not far away, the Mahale Mountains National Park is also under threat from human settlement. In Senegal, the Ivory Coast, and both Congos, chimpanzee and gorilla populations are dwindling, under attack from poachers and new settlements. Remote swampy areas remain mostly untouched, but the loggers have these in their sights.

Gabon, on the west coast, shows the most promise. In 2002, President Omar Bongo quarantined 10 percent of the country's land, mostly swamp jungles and forest, and parceled it into 13 national parks. Bongo devoted the parks to protecting the nation's wild animals, including significant populations of chimpanzees and western lowland gorillas.

He was inspired by a 454-day mega-transect of Gabon's wilderness by Mike Fay, an American conservationist. Fay covered 2,000 miles on foot in one of the greatest exploration feats of our time, and discovered untouched forests and swampland teeming with wild creatures including great apes. He called it "the last wild place on Earth."

But Bongo died in mid–2009, and there is no guarantee that Gabon's future leaders will honor his pledge to keep the national parks untouched. Also, Gabon shares borders with Cameroon, Equatorial Guinea, and the Republic of Congo, and poachers cross the borders with little to challenge their plunder of the wild animals in Gabon's national parks. So, Gabon needs massive outside financial help to raise, train, arm, and fund a large cadre of local rangers based inside the parks.

In the Virungas, the mountain gorillas on the Congo side of the border are under threat from several rebellions and their fate is fragile, but their kin on the Rwandan side are flourishing for now. The same goes for the mountain gorillas in nearby Uganda. In both places, rangers accompanied by soldiers from the national army patrol the national parks, funded by some of the thousands of American dollars collected daily from foreign tourists as fees to visit the gorillas.

The United Nations has set up the Great Apes Survival Partnership (GRASP) with headquarters in Nairobi, but it has had little effect in its primary aim, saving the great apes. It prompts grandstanding African leaders to vow to protect their countries' great apes at conferences it organizes, but with little effective follow-up. Dave Greer told me that at one GRASP conference held recently in Paris, the lead speaker was the forestry minister from a Central African country where Dave had worked at a senior level in great ape conservation. He flashed a grim smile when claiming he knew that the minister profits immensely and corruptly from approving the destruction of the great apes' habitats by loggers.

While there is some chance for Africa's chimpanzees, gorillas, and bonobos, hope has almost run out for most of Southeast Asia's orangutans. In Sumatra, hardly any will live in the wild by 2060. Many of the wild orangutans in Indonesian Borneo and Sarawak will also have disappeared, victims of bureaucratic and commercial greed. In Sabah, the small Malaysian state in north Borneo, the orangutans have a chance because the government leaders have devoted funds and land to saving the state's last wild orangutans, numbering several thousand individuals.

The world's disinterest in the deteriorating fate of the great apes in the wild is hard for me to understand. Our closest kin, they have been mostly abandoned, left to their grim fate. We are privileged to have them among us, and yet the world community has placed their protection a long way down the list of global concerns.

I can never forget the deeply moving gesture of Humba, the huge Congo mountain gorilla, just weeks before his brother, Rugendo, and his family's adult females were shot dead. Humba could have charged and bashed me, or broken my neck with just one bite from his enormous canines. Instead, as I prepared to leave his jungle home, he approached to within a yard of me with leaves in his mouth, his sign of peace, and gave me a lingering meaningful stare.

The world should not turn its back on the plight of Humba and our other great ape cousins—the Bornean orangutan Jenny, the CAR western lowland gorilla Makumba, the unnamed Cameroonian Cross River silverback, the Congo bonobo Raphael, the Ugandan chimpanzee Outamba, and all their kin in the wilds. That would condemn them to virtual extinction and shame us forever.

Bibliography

Ankel-Simons, Friderun. *Primate Anatomy: An Introduction*. San Diego: Academic Press, 2000.

Ballantyne, R. M. *The Gorilla Hunters*. London: Dean and Son, 1935.

Baumgartel, Walter. *Up Among the Mountain Gorillas*. New York: Hawthorn Books, 1976.

Caldecott, Julian, and Lera Miles. *World Atlas of Great Apes and Their Conservation*. London: University of California Press, 2005.

Chomsky, Noam. *Language and Problems of Knowledge: The Managua Lectures*. Cambridge, MA: M.I.T. Press, 1988.

Coolidge, Harold J. "Pygmy Chimpanzee from South of the Congo River." *The American Journal of Physical Anthropology*, 17, no. 1 (1933).

Darwin, Charles. *The Descent of Man and Selection in Relation to Sex*. London: John Murray, 1871.

De Waal, Frans, and Frans Lanting. *Bonobo: The Forgotten Ape*. Berkeley: University of California Press, 1997.

Dixon, Alan. F. *Primate Sexuality*. New York: Oxford University Press, 1998.

Fields, William, et al. *Kanzi's Primal Language*. New York: Palgrave Macmillan, 2005.

Fossey, Dian. *Gorillas in the Mist*. Ringwood, Australia: Penguin Books, 1988.

Galdikas, Biruté Mary. *Reflections of Eden*. Boston: Little Brown, 1995.

———. *Great Ape Odyssey*. New York: Harry N. Abrams, 2005.

Garner, R. L. *Gorillas and Chimpanzees*. London: Osgood and McIlvaine, 1896.

———. *Apes and Monkeys*. London: Osgood and McIlvaine, 1898.

Ghiglieri, Michael P. *East of the Mountains of the Moon*. New York: The Free Press, 1988.

Gomez, Juan Carlos. *Apes, Monkeys, Children and the Growth of Mind*. Cambridge, MA: Harvard University Press, 2004.

Goodall, Jane. *My Friends the Wild Chimpanzees*. Washington, DC: The National Geographic Society, 1970.

———. *The Chimpanzees of Gombe*. Cambridge, MA: Harvard University Press, 1986.

Jane Goodall Institute. *40 Years at Gombe*. New York: Stewart, Tabori and Chang, 1999.

Kano, Takayoshi. *The Last Ape*. Stanford, CA: Stanford University Press, 1992.

Kaplan, Gisela, and Lesley J. Rogers. *The Orangutans*. Cambridge, MA: Perseus Publishing, 2000.

Payne, Sunaidi, and Cede Prudente. *Orangutans: Behavior, Ecology and Conservation*. Cambridge, MA: M.I.T. Press, 2008.

Peterson, Dale. *Jane Goodall: The Woman Who Redefined Man*. New York: Houghton Mifflin, 2006.

Robbins, Martha M., Pascale Sicotte, and Kelly J. Stewart. *Mountain Gorillas: Three Decades of Research at Karisoke*. Cambridge, England: Cambridge University Press, 2001.

Russon, Anne E. *Orangutans: Wizards of the Rain Forest*. London: Robert Hale, 1999.

Savage-Rumbaugh, Sue, and Roger Lewin. *Kanzi: The Ape at the Brink of the Human Mind*. Hoboken, NJ: John Wiley and Sons, 1994.

Savage-Rumbaugh, Sue, et al. *Apes, Language and the Human Mind*. New York: Oxford University Press, 1998.

Schuster, Gerd, et al. *Thinkers of the Jungle*. Konigswinter: Ullmann-publishing, 2007.

Shumaker, Robert W., and Benjamin B. Beck. *Primates in Question*. Washington, DC: Smithsonian Books, 2003.

Smuts, Barbara B., et al. *Primate Societies*. Chicago: University of Chicago Press, 1987.

Stanford, Craig B. *Chimpanzee and Red Colobus*. Cambridge, MA: Harvard University Press, 1998.

Taylor, Barbara. *Great Apes*. London: Lorenz Books, 2001.

Taylor Parker, Sue, et al. *Mentalities of Gorillas and Orangutans*. Cambridge, England: Cambridge University Press, 1999.

Wallace, Alfred Russel. *Darwinism*. London: Macmillan, 1889.

Weber, Bill, and Amy Vedder. *In the Kingdom of Gorillas*. New York: Simon and Schuster, 2001.

Wrangham, Richard, and Dale Peterson. *Demonic Males: Apes and the Origins of Human Violence*. Boston, MA: Houghton Mifflin, 1996.

Wrangham, Richard, et al. *Chimpanzee Culture*. Cambridge, MA: Harvard University Press, 1994.

Yerkes, Robert M. *Almost Human*. London: J. Cape, 1925.

Index